Some Physical Constants

Speed of light	c	3.00×10^8 m/s
Gravitational constant	G	6.67×10^{-11} N·m²/kg²
Coulomb's constant	$1/4\pi\varepsilon_0$	8.99×10^9 N·m²/C²
Permittivity constant	ε_0	8.85×10^{-12} C²/(N·m²)
Permeability constant	μ_0	$4\pi \times 10^{-7}$ N/A²
Planck's constant	h	6.63×10^{-34} J·s
Boltzmann's constant	k_B	1.38×10^{-23} J/K
Elementary charge	e	1.602×10^{-19} C
Electron mass	m_e	9.11×10^{-31} kg
Proton mass	m_p	1.673×10^{-27} kg
Neutron mass	m_n	1.675×10^{-27} kg
Avogadro's number	N_A	6.02×10^{23}

Standard Metric Prefixes
(for powers of 10)

Power	Prefix	Symbol
10^{18}	exa	E
10^{15}	peta	P
10^{12}	tera	T
10^{9}	giga	G
10^{6}	mega	M
10^{3}	kilo	k
10^{-2}	centi	c
10^{-3}	milli	m
10^{-6}	micro	μ
10^{-9}	nano	n
10^{-12}	pico	p
10^{-15}	femto	f
10^{-18}	atto	a

Commonly Used Physical Data

Gravitational field strength $g = |\vec{g}|$ (near the earth's surface)	9.80 N/kg = 9.80 m/s²
Mass of the earth M_e	5.98×10^{24} kg
Radius of the earth R_e	6380 km (equatorial)
Mass of the sun $M_\odot$	1.99×10^{30} kg
Radius of the sun $R_\odot$	696,000 km
Mass of the moon	7.36×10^{22} kg
Radius of the moon	1740 km
Distance to the moon	3.84×10^8 m
Distance to the sun	1.50×10^{11} m
Density of water†	1000 kg/m³ = 1 g/cm³
Density of air†	1.2 kg/m³
Absolute zero	0 K = −273.15°C = −459.67°F
Freezing point of water‡	273.15 K = 0°C = 32°F
Boiling point of water‡	373.15 K = 100°C = 212°F
Normal atmospheric pressure	101.3 kPa

†At normal atmospheric pressure and 20°C.
‡At normal atmospheric pressure.

Useful Conversion Factors

1 meter = 1 m = 100 cm = 39.4 in = 3.28 ft
1 mile = 1 mi = 1609 m = 1.609 km = 5280 ft
1 inch = 1 in = 2.54 cm
1 light-year = 1 ly = 9.46 Pm = 0.946×10^{16} m
1 minute = 1 min = 60 s
1 hour = 1 h = 60 min = 3600 s
1 day = 1 d = 24 h = 86.4 ks = 86,400 s
1 year = 1 y = 365.25 d = 31.6 Ms = 3.16×10^7 s
1 newton = 1 N = 1 kg·m/s² = 0.225 lb
1 joule = 1 J = 1 N·m = 1 kg·m²/s² = 0.239 cal
1 watt = 1 W = 1 J/s
1 pascal = 1 Pa = 1 N/m² = 1.45×10^{-4} psi
1 kelvin (temperature difference) = 1 K = 1°C = 1.8°F
1 radian = 1 rad = 57.3° = 0.1592 rev
1 revolution = 1 rev = 2π rad = 360°
1 cycle = 2π rad
1 hertz = 1 Hz = 1 cycle/s

1 m/s = 2.24 mi/h = 3.28 ft/s
1 mi/h = 1.61 km/h = 0.447 m/s = 1.47 ft/s
1 liter = 1 l = (10 cm)³ = 10^{-3} m³ = 0.0353 ft³
1 ft³ = 1728 in³ = 0.0283 m³
1 gallon = 1 gal = 0.00379 m³ = 3.79 l ≈ 3.8 kg H₂O
Weight of 1-kg object near the earth = 9.8 N = 2.2 lb

1 pound = 1 lb = 4.45 N
1 calorie = energy needed to raise the temperature of 1 g
        of H₂O by 1 K = 4.186 J
1 horsepower = 1 hp = 746 W
1 pound per square inch = 6895 Pa
1 food calorie = 1 Cal = 1 kcal = 1000 cal = 4186 J
1 electron volt = 1 eV = 1.602×10^{-19} J

$$T = \left(\frac{1\text{K}}{1°\text{C}}\right)(T_{[C]} + 273.15°\text{C}) \qquad T_{[C]} = \left(\frac{5°\text{C}}{9°\text{F}}\right)(T_{[F]} - 32°\text{F})$$

$$T = \left(\frac{5\text{K}}{9°\text{F}}\right)(T_{[F]} + 459.67°\text{F}) \qquad T_{[F]} = 32°\text{F} + \left(\frac{9°\text{F}}{5°\text{C}}\right)T_{[C]}$$

Useful Astronomical Data

(1 AU $\equiv$ mean distance from earth to sun $\equiv 1.50 \times 10^{11}$ m)

Object	Mass	Mean Radius	Semimajor Axis*	Orbital Period	Orbital Eccentricity
Sun	1.99×10^{30} kg	696,000 km	—	—	—
Moon	7.36×10^{22} kg	1740 km	384,000 km	27.3 d	0.055
Mercury	$0.0553 M_E$	2440 km	0.387 AU	0.241 y	0.206
Venus	$0.815 M_E$	6052 km	0.723 AU	0.615 y	0.007
Earth	5.98×10^{24} kg $\equiv M_E$	6371 km	1.000 AU	1.000 y	0.017
Mars	$0.107 M_E$	3390 km	1.524 AU	1.88 y	0.093
Jupiter	$318 M_E$	69,900 km	5.204 AU	11.9 y	0.049
Saturn	$95.2 M_E$	58,200 km	9.582 AU	29.5 y	0.056
Uranus	$14.5 M_E$	25,400 km	19.201 AU	84.0 y	0.046
Neptune	$17.2 M_E$	24,600 km	30.047 AU	165 y	0.009
Pluto/Charon	$0.00245 M_E$	1187/606 km	39.482 AU	248 y	0.254
Ceres (asteroid)	9.5×10^{20} kg	470 km	2.767 AU	4.60 y	0.079
Halley's comet	2.2×10^{14} kg	≈ 7 km	17.94 AU	76.1 y	0.967

Based mostly on NASA data from <nssdc.gsfc.nasa.gov/planetary/>. Pluto/Charon data is from the *New Horizons* mission (as of October 2015).

*An orbit's semimajor axis is approximately equal to its mean orbital radius.

Six Ideas That Shaped Physics

Six Ideas That Shaped Physics

Unit N: The Laws of Physics Are Universal

Fourth Edition

Thomas A. Moore

SIX IDEAS THAT SHAPED PHYSICS, UNIT N:
THE LAWS OF PHYSICS ARE UNIVERSAL, FOURTH EDITION

Published by McGraw Hill LLC, 1325 Avenue of the Americas, New York, NY 10019. Copyright © 2023 by McGraw Hill LLC. All rights reserved. Printed in the United States of America. Previous editions © 2017, 2003, and 1998. No part of this publication may be reproduced or distributed in any form or by any means, or stored in a database or retrieval system, without the prior written consent of McGraw Hill LLC, including, but not limited to, in any network or other electronic storage or transmission, or broadcast for distance learning.

Some ancillaries, including electronic and print components, may not be available to customers outside the United States.

This book is printed on acid-free paper.

1 2 3 4 5 6 7 8 9 LWI 27 26 25 24 23 22

ISBN 978-1-264-87682-2 (bound edition)
MHID 1-264-87682-3 (bound edition)
ISBN 978-1-264-86660-1 (loose-leaf edition)
MHID 1-264-86660-7 (loose-leaf edition)

Portfolio Manager: *Beth Bettcher*
Product Developer: *Theresa Collins*
Marketing Manager: *Lisa Granger*
Content Project Managers: *Jeni McAtee, Samanthi Donisi*
Buyer: *Sandy Ludovissy*
Designer: *David Hash*
Content Licensing Specialist: *Beth Cray*
Cover Image: *NASA/Johns Hopkins University Applied Physics Laboratory/Southwest Research Institute/Goddard Space Flight Center*
Compositor: *Aptara®, Inc.*

Dedication

To Brittany
whose intuitive understanding of Newtonian
physics is part of what makes her awesome

Library of Congress Cataloging-in-Publication Data

Names: Moore, Thomas A. (Thomas Andrew), author.
Title: Six ideas that shaped physics. Unit N, The laws of physics are
    universal / Thomas A. Moore.
Other titles: Laws of physics are universal
Description: Fourth edition. | New York, NY : McGraw Hill LLC, [2023] |
    Includes index.
Identifiers: LCCN 2021029359 (print) | LCCN 2021029360 (ebook) | ISBN
    9781264876822 (hardcover) | ISBN 9781264866601 (spiral bound) | ISBN
    9781264873357 (ebook) | ISBN 9781264864584 (ebook other)
Subjects: LCSH: Mechanics—Textbooks.
Classification: LCC QC125.2 .M66 2023 (print) | LCC QC125.2 (ebook) | DDC
    531—dc23
LC record available at https://lccn.loc.gov/2021029359
LC ebook record available at https://lccn.loc.gov/2021029360

mheducation.com/highered

Contents: Unit N
The Laws of Physics Are Universal

About the Author

Thomas A. Moore graduated from Carleton College (magna cum laude with Distinction in Physics) in 1976. He won a Danforth Fellowship that year that supported his graduate education at Yale University, where he earned a Ph.D. in 1981. He taught at Carleton College and Luther College before taking his current position at Pomona College in 1987, where he won a Wig Award for Distinguished Teaching in 1991. He served as an active member of the steering committee for the national Introductory University Physics Project (IUPP) from 1987 through 1995. This textbook grew out of a model curriculum that he developed for that project in 1989, which was one of only four selected for further development and testing by IUPP.

He has published a number of articles about astrophysical sources of gravitational waves, detection of gravitational waves, and new approaches to teaching physics, as well as a book on general relativity entitled *A General Relativity Workbook* (University Science Books, 2013). He has also served as a reviewer and as an associate editor for *American Journal of Physics*. He currently lives in Claremont, California, with his wife Joyce, a retired pastor. When he is not teaching, doing research, or writing, he enjoys reading, hiking, calling contradances, and playing Irish traditional music.

(Credit: Courtesy of Thomas Moore)

Preface

Introduction

This volume is one of six that together comprise the text materials for *Six Ideas That Shaped Physics*, a unique approach to the two- or three-semester calculus-based introductory physics course. I have designed this curriculum (for which these volumes only serve as the text component) to support an introductory course that combines three elements:

- Inclusion of 20th-century physics topics,
- A thoroughly 21st-century perspective on even classical topics, and
- Support for a student-centered and active-learning-based classroom.

This course is based on the premises that innovative metaphors for teaching basic concepts, explicitly instructing students in the processes of constructing physical models, and active learning can help students learn the subject much more effectively. Physics education research has guided the presentation of all topics. Moreover, because such research has consistently underlined the importance of active learning, I have sought to provide tools for professors (both in the text and online) to make creating a coherent and self-consistent course structure based on a student-centered classroom as easy and practical as possible. All of the materials have been tested, evaluated, and rewritten multiple times: the result is the culmination of more than 30 years of continual testing and revision.

Rather than oversimplifying the material, I have sought to make physics more accessible by providing students with the tools and guidance to become more sophisticated in their thinking. This book helps students to step beyond rote thinking patterns to develop flexible and powerful, conceptual reasoning and model-building skills. My experience and that of other users is that normal students in a wide range of institutional settings can (with appropriate support and practice) successfully learn these skills.

Each of six volumes in the text portion of this course is focused on a single core concept that has been crucial in making physics what it is today. The six volumes and their corresponding ideas are as follows:

Unit C: **C**onservation laws constrain interactions
Unit N: The laws of physics are universal (**N**ewtonian mechanics)
Unit R: The laws of physics are frame-independent (**R**elativity)
Unit E: **E**lectricity and magnetism are unified
Unit Q: Matter behaves like waves (**Q**uantum physics)
Unit T: Some processes are irreversible (**T**hermal physics)

I have listed the units in the order that I *recommend* that they be taught, but I have also constructed units R, E, Q, and T to be sufficiently independent so that they can be taught in any order after units C and N. (This is why the units are lettered as opposed to numbered.) There are *six* units (as opposed to five or seven) to make it possible to easily divide the course into two semesters, three quarters, or three semesters. This unit organization therefore not only makes it possible to dole out the text in small, easily-handled pieces and provide a great deal of flexibility in fitting the course to a given schedule, but also carries its own important pedagogical message: *physics is organized hierarchically*, structured around only a handful of core ideas and metaphors.

An important feature of all of the volumes is that each chapter represents a logical unit that one might hope to handle in a single 50-minute class session, providing guidance about pacing based on decades of experience. This organization also gives instructors increased flexibility in designing a well-paced course in any particular institutional setting, since a number of chapters have been designed so that they can be omitted without loss of continuity. The preface to each unit, the chapter headers, and the instructor's manual all provide guidance about chapter dependencies.

Finally, let me emphasize again that the text materials are just one part of the comprehensive *Six Ideas* curriculum. On the *Six Ideas* website, at

www.physics.pomona.edu/sixideas/

you will find a wealth of supporting resources. The most important of these is a detailed instructor's manual that provides guidance (based on *Six Ideas* users' experiences over more than 30 years) about how to construct an effective course. This manual exposes the important issues and raises the questions that an instructor should consider in creating an effective *Six Ideas* course at a particular institution. The site also provides software that allows instructors to post selected problem solutions online where only their students can access them and assign each solution a time window for viewing. Web-based computer applets on the site provide experiences that support student learning in important ways. The site also provides a (steadily increasing) number of other resources that instructors and students may find valuable.

There is a preface for students appearing just before the first chapter of each unit that explains some important features of the text and assumptions behind the course. I recommend that *everyone* read it.

Comments About the Fourth Edition

Our main goals in this edition have been to

- simplify certain difficult sections in units T, Q, E, and C,
- make it easier for instructors to drop certain chapters,
- reduce the pace in unit E,
- add some new problems (and cull a few less-effective ones), and
- add more end-of-chapter problems to Connect for each unit.

In addition (under the chapter number on each chapter's first page), I have noted how crucial that chapter is for what follows. The categories are

- **Core** (essential for understanding the unit)
- **Extension** (extends core material in valuable ways, and may be useful for future chapters in this category, but not essential)
- **Optional** (interesting but not needed for any future chapter).

Information clarifying the chapter's designation and chapter dependencies typically appears for chapters in the non-Core categories.

Specifically About Unit N

This unit is structured on the premise that students have already studied unit C and draws on ideas from almost all the chapters of that unit. Chapter C7 on angular momentum is technically optional (and is not needed for any other unit) but the expression for torque as $\vec{\tau} = \vec{r} \times \vec{F}$ presented in that chapter is valuable for chapters N4, N11, and N12 in this unit. I have tried to write these chapters so that chapter C7 is not strictly necessary, but if one has omitted chapter C7, one may need to spend some time building students' understanding of this equation.

Understanding the core concepts of Newtonian mechanics and the central idea of the unit requires discussing chapters N1–N11 at least. Chapter N12 has been constructed to be optional, but it is still highly recommended.

The Newton web-based computer application, which one can find on the Resources page of the Six Ideas website (www.physics.pomona.edu/sixideas/), makes it possible for students to explore more realistic problems than would otherwise be possible at this level. It implements the "trajectory diagram" algorithm discussed in chapter N3. I strongly recommend that one emphasize this algorithm when teaching chapter N3 to prepare students for later uses of the Newton app that are built into the text.

The fourth edition of this unit has not been changed very much from the third edition, though I have rewritten a handful of problems to improve their clarity.

Appreciation

Thanking everyone who has offered important and greatly appreciated help with this project over the past three decades would be much too long to provide here. So, as in previous editions, I will focus on thanking those who have helped with this particular edition.

Thanks to my colleagues David Tanenbaum and Dwight Whitaker who offered good ideas and thoughtful advice for this edition. I'd like to thank Marisa Dobbeleare and especially Megan Platt and Beth Bettcher at McGraw-Hill for having faith in the *Six Ideas* project and starting the push for this edition. Theresa Collins has been superb at guiding the project at the detail level. Many others at McGraw-Hill and its contractors, including Jeni McAtee, Sarita Yadav, Ashish Vyas, and Anand Singh, were instrumental in producing this particular edition. Finally, very special thanks to my wife Joyce, who (as always) has sacrificed, supported me, and loved me during my work on this edition. I am very grateful to you all!

Thomas A. Moore
Claremont, California

Digital Learning Tools

Proctorio
Remote Proctoring & Browser-Locking Capabilities

Remote proctoring and browser-locking capabilities, hosted by Proctorio within Connect, provide control of the assessment environment by enabling security options and verifying the identity of the student.

Seamlessly integrated within Connect, these services allow instructors to control students' assessment experience by restricting browser activity, recording students' activity, and verifying students are doing their own work.

Instant and detailed reporting gives instructors an at-a-glance view of potential academic integrity concerns, thereby avoiding personal bias and supporting evidence-based claims.

 ## ReadAnywhere

Read or study when it's convenient for you with McGraw Hill's free ReadAnywhere app. Available for iOS or Android smartphones or tablets, ReadAnywhere gives users access to McGraw Hill tools including the eBook and SmartBook 2.0 or Adaptive Learning Assignments in Connect. Take notes, highlight, and complete assignments offline – all of your work will sync when you open the app with WiFi access. Log in with your McGraw Hill Connect username and password to start learning – anytime, anywhere!

Tegrity: Lectures 24/7

Tegrity in Connect is a tool that makes class time available 24/7 by automatically capturing every lecture. With a simple one-click start-and-stop process, you capture all computer screens and corresponding audio in a format that is easy to search, frame by frame. Students can replay any part of any class with easy-to-use, browser-based viewing on a PC, Mac, iPod, or other mobile device.

Educators know that the more students can see, hear, and experience class resources, the better they learn. In fact, studies prove it. Tegrity's unique search feature helps students efficiently find what they need, when they need it, across an entire semester of class recordings. Help turn your students' study time into learning moments immediately supported by your lecture. With Tegrity, you also increase intent listening and class participation by easing students' concerns about note-taking. Using Tegrity in Connect will make it more likely you will see students' faces, not the tops of their heads.

Writing Assignment

Available within Connect and Connect Master, the Writing Assignment tool delivers a learning experience to help students improve their written communication skills and conceptual understanding. As an instructor you can assign, monitor, grade, and provide feedback on writing more efficiently and effectively.

Create
Your Book, Your Way

McGraw Hill's Content Collections Powered by Create® is a self-service website that enables instructors to create custom course materials—print and eBooks—by drawing upon McGraw Hill's comprehensive, cross-disciplinary content. Choose what you want from our high-quality textbooks, articles, and cases. Combine it with your own content quickly and easily, and tap into other rights-secured, third-party content such as readings, cases, and articles. Content can be arranged in a way that makes the most sense for your course and you can include the course name and information as well. Choose the best format for your course: color print, black-and-white print, or eBook. The eBook can be included in your Connect course and is available on the free ReadAnywhere app for smartphone or tablet access as well. When you are finished customizing, you will receive a free digital copy to review in just minutes! Visit McGraw Hill Create®—www.mcgrawhillcreate.com—today and begin building!

Instructors: Student Success Starts with You

Tools to enhance your unique voice

Want to build your own course? No problem. Prefer to use an OLC-aligned, prebuilt course? Easy. Want to make changes throughout the semester? Sure. And you'll save time with Connect's auto-grading too.

65%
Less Time Grading

Laptop: McGraw Hill; Woman/dog: George Doyle/Getty Images

Study made personal

Incorporate adaptive study resources like SmartBook® 2.0 into your course and help your students be better prepared in less time. Learn more about the powerful personalized learning experience available in SmartBook 2.0 at **www.mheducation.com/highered/connect/smartbook**

Affordable solutions, added value

Make technology work for you with LMS integration for single sign-on access, mobile access to the digital textbook, and reports to quickly show you how each of your students is doing. And with our Inclusive Access program you can provide all these tools at a discount to your students. Ask your McGraw Hill representative for more information.

Padlock: Jobalou/Getty Images

Solutions for your challenges

A product isn't a solution. Real solutions are affordable, reliable, and come with training and ongoing support when you need it and how you want it. Visit **www. supportateverystep.com** for videos and resources both you and your students can use throughout the semester.

Checkmark: Jobalou/Getty Images

Students: Get Learning That Fits You

Effective tools for efficient studying

Connect is designed to help you be more productive with simple, flexible, and intuitive tools that maximize your study time and meet your individual learning needs. Get learning that works for you with Connect.

Study anytime, anywhere

Download the free ReadAnywhere app and access your online eBook, SmartBook 2.0, or Adaptive Learning Assignments when it's convenient, even if you're offline. And since the app automatically syncs with your Connect account, all of your work is available every time you open it. Find out more at **www.mheducation.com/readanywhere**

"I really liked this app—it made it easy to study when you don't have your text-book in front of you."

- Jordan Cunningham,
  Eastern Washington University

Everything you need in one place

Your Connect course has everything you need—whether reading on your digital eBook or completing assignments for class, Connect makes it easy to get your work done.

Learning for everyone

McGraw Hill works directly with Accessibility Services Departments and faculty to meet the learning needs of all students. Please contact your Accessibility Services Office and ask them to email accessibility@mheducation.com, or visit **www.mheducation.com/about/accessibility** for more information.

Introduction for Students

Introduction

Welcome to *Six Ideas That Shaped Physics!* This text has a number of features that may be different from science texts you may have encountered previously. This section describes those features and how to use them effectively.

Why Is This Text Different?

Why *active learning* is crucial

Research into physics education consistently shows that people learn physics most effectively through *activities* where they practice applying physical reasoning and model-building skills in realistic situations. This is because physics is not a body of facts to absorb, but rather a set of thinking skills acquired through practice. You cannot learn such skills by listening to factual lectures any more than you can learn to play the piano by listening to concerts!

This text, therefore, has been designed to support *active learning* both inside and outside the classroom. It does this by providing (1) resources for various kinds of learning activities, (2) features that encourage active reading, and (3) features that make it as easy as possible to use the text (as opposed to lectures) as the primary source of information, so that you can spend class time doing activities that will actually help you learn.

The Text as Primary Source

Features that help you use the text as the primary source of information

To serve the last goal, I have adopted a conversational style that I hope you will find easy to read, and have tried to be concise without being too terse.

Certain text features help you keep track of the big picture. One of the key aspects of physics is that the concepts are organized *hierarchically*: some are more fundamental than others. This text is organized into six units, each of which explores the implications of a single deep idea that has shaped physics. Each unit's front cover states this **core idea** as part of the unit's title.

A two-page **chapter overview** provides a compact summary of that chapter's contents to give you the big picture before you get into the details and later when you review. **Sidebars** in the margins help clarify the purpose of sections of the main text at the subpage level and can help you quickly locate items later. I have highlighted technical terms in bold type (like **this**) when they first appear: their definitions usually appear nearby.

A physics **formula** consists of both a mathematical equation and a *conceptual frame* that gives the equation physical meaning. The most important formulas in this book (typically, those that might be relevant outside the current chapter) appear in **formula boxes**, which state the equation, its *purpose* (which describes the formula's meaning), a description of any *limitations* on the formula's applicability, and (optionally) some other useful *notes*. Treat everything in a box as a unit to be remembered and used together.

Active Reading

What is *active reading*?

Just as passively listening to a lecture does not help you really learn what you need to know about physics, you will not learn what you need by simply scanning your eyes over the page. **Active reading** is a crucial study skill for all kinds of technical literature. An active reader stops to pose internal questions such as these: Does this make sense? Is this consistent with my experience? Do I see how I might be able to use this idea? This text provides two important tools to make this process easier.

Use the **wide margins** to (1) record *questions* that arise as you read (so you can be sure to get them answered) and the *answers* you eventually receive, (2) flag important passages, (3) fill in missing mathematical steps, and (4) record insights. Writing in the margins will help keep you actively engaged as you read and supplement the sidebars when you review.

Each chapter contains three or four **in-text exercises**, which prompt you to develop the habit of *thinking* as you read (and also give you a break!). These exercises sometimes prompt you to fill in a crucial mathematical detail but often test whether you can *apply* what you are reading to realistic situations. When you encounter such an exercise, stop and try to work it out. When you are done (or after about 5 minutes or so), look at the answers at the end of the chapter for some immediate feedback. Doing these exercises is one of the more important things you can do to become an active reader.

SmartBook (TM) further supports active reading by continuously measuring what a student knows and presenting questions to help keep students engaged while acquiring new knowledge and reinforcing prior learning.

Features that support developing the habit of active reading

Class Activities and Homework

This book's *entire purpose* is to give you the background you need to do the kinds of *practice* activities (both in class and as homework) that you need to genuinely learn the material. *It is therefore ESSENTIAL that you read every assignment BEFORE you come to class*. This is *crucial* in a course based on this text (and probably more so than in previous science classes you have taken).

Read the text BEFORE class!

The homework problems at the end of each chapter provide for different kinds of practice experiences. **Two-minute problems** are short conceptual problems that provide practice in extracting the implications of what you have read. **Basic Skills** problems offer practice in straightforward applications of important formulas. Both can serve as the basis for classroom activities: the letters on the book's back cover help you communicate the answer to a two-minute problem to your professor (simply point to the letter!). **Modeling** problems give you practice in constructing coherent mental models of physical situations, and usually require combining several formulas to get an answer. **Derivation** problems give you practice in mathematically extracting useful consequences of formulas. **Rich-context** problems are like modeling problems, but with elements that make them more like realistic questions that you might actually encounter in life or work. They are especially suitable for collaborative work. **Advanced** problems challenge advanced students with questions that involve more subtle reasoning and/or difficult math.

Types of practice activities provided in the text

Note that this text contains perhaps fewer examples than you would like. This is because the goal is to teach you to *flexibly reason from basic principles*, not slavishly copy examples. You may find this hard at first, but real life does not present its puzzles neatly wrapped up as textbook examples. With practice, you will find your power to deal successfully with realistic, practical problems will grow until you yourself are astonished at how what had seemed impossible is now easy. *But it does take practice*, so work hard and be hopeful!

Six Ideas That Shaped Physics

Newton's Laws

Chapter Overview

Section N1.1: The Newtonian Synthesis

This section reviews the history of physics before Newton to underline the astonishing nature of Newton's achievement. Aristotle, whose work provided the background for Western thinking about motion until the 1500s, thought that when undisturbed, objects moved in ways consistent with their *natures*. In this view, terrestrial and celestial objects had different natures and thus obeyed different laws. But Newton, building on theoretical work by Copernicus, Galileo, and Descartes, and observational work by Kepler, was able to show that a theoretical model based on three simple laws of motion and a law of gravitation could explain *both* terrestrial motion and celestial motion more simply and accurately than any model previously proposed. This **Newtonian synthesis**, demonstrating that *the laws of physics are universal*, had an enormous impact on Western thought and led to the birth of physics as a science.

Section N1.2: Newton's Laws of Motion

Newton's three laws of motion state that

1. In the absence of external interactions, an object's (or system's) center of mass moves at a *constant velocity*.
2. The net external force on a system causes that system's center of mass to accelerate at a rate inversely proportional to the system's total mass.
3. The force that a given interaction A between objects 1 and 2 exerts on object 1 is equal in magnitude and opposite in direction to the force it exerts on object 2.

The section shows how these follow from the momentum transfer model of unit C.

Currently, physicists consider **Newton's first law** to be a law that distinguishes inertial from noninertial reference frames (see sections C4.4 and C4.5), but Newton wanted to make it clear that in his model the *natural* state of motion of objects (both celestial and terrestrial) is motion in a straight line at a constant velocity.

In mathematical form, **Newton's second law** says that

$$\sum \vec{F}_{\text{ext}} = M \vec{a}_{\text{CM}} \qquad \text{(N1.4)}$$

- **Purpose:** This equation links the acceleration $\vec{a}_{\text{CM}}$ of a system's center of mass to the system's mass M and the vector sum $\sum \vec{F}_{\text{ext}}$ of external forces acting on it.
- **Limitations:** The system must have a fixed mass and a speed that is small compared to that of light (as we will see in unit R).
- **Note:** Read this equation verbally as follows: "the net external force on an object *causes* its mass to accelerate."

We will spend the rest of the unit exploring applications of this law.

Newton's third law follows directly from the momentum transfer model, but sometimes leads to counterintuitive results. However, note that the effects of even equal forces exerted on different objects may not be the same!

Section N1.3: Vector Calculus

As Newton himself found, we can't apply Newton's model without **vector calculus**. In analogy to the definition of the time derivative of a function, we define the time derivative of a vector quantity $\vec{q}$ to be $d\vec{q}/dt \equiv \lim_{\Delta t \to 0} \{ [\vec{q}(t + \Delta t) - \vec{q}(t)]/\Delta t \}$.

Section N1.4: The Formal Definition of Velocity

$$\vec{v} \equiv \frac{d\vec{r}}{dt} \equiv \begin{bmatrix} \lim_{\Delta t \to 0} \dfrac{x(t+\Delta t) - x(t)}{\Delta t} \\[2mm] \lim_{\Delta t \to 0} \dfrac{y(t+\Delta t) - y(t)}{\Delta t} \\[2mm] \lim_{\Delta t \to 0} \dfrac{z(t+\Delta t) - z(t)}{\Delta t} \end{bmatrix} \equiv \begin{bmatrix} \dfrac{dx}{dt} \\[2mm] \dfrac{dy}{dt} \\[2mm] \dfrac{dz}{dt} \end{bmatrix} \tag{N1.11}$$

- **Purpose:** This equation formally defines an object's velocity $\vec{v}(t)$ at an instant t, where $\vec{r}(t) = [x(t), y(t), z(t)]$ is the position of the object's center of mass at time t, and $\vec{r}(t+\Delta t) = [x(t+\Delta t), y(t+\Delta t), z(t+\Delta t)]$ is the same evaluated at time $t+\Delta t$, where Δt is some finite interval of time. The symbols $d\vec{r}/dt$, dx/dt, dy/dt, and dz/dt now officially represent time derivatives.
- **Limitations:** This expression is a definition and so has no limitations.

An object's **average velocity** $\vec{\overline{v}} \equiv \Delta\vec{r}/\Delta t$ during a *finite* time interval Δt is a good approximation to the object's instantaneous velocity $\vec{v}$ halfway through the interval.

Section N1.5: The Formal Definition of Acceleration

$$\vec{a}(t) \equiv \frac{d\vec{v}}{dt} \equiv \begin{bmatrix} \lim_{\Delta t \to 0} \dfrac{v_x(t+\Delta t) - v_x(t)}{\Delta t} \\[2mm] \lim_{\Delta t \to 0} \dfrac{v_y(t+\Delta t) - v_y(t)}{\Delta t} \\[2mm] \lim_{\Delta t \to 0} \dfrac{v_z(t+\Delta t) - v_z(t)}{\Delta t} \end{bmatrix} \equiv \begin{bmatrix} \dfrac{dv_x}{dt} \\[2mm] \dfrac{dv_y}{dt} \\[2mm] \dfrac{dv_z}{dt} \end{bmatrix} \tag{N1.13}$$

- **Purpose:** This equation defines the acceleration $\vec{a}(t)$ of an object's center of mass at an instant of time t, where $\vec{v}(t)$, with components $v_x(t)$, $v_y(t)$, and $v_z(t)$, is the velocity of the object's center of mass at time t; $\vec{v}(t+\Delta t)$ (and analogously for its components) is the same evaluated at time $t+\Delta t$, where Δt is some finite interval of time. The quantity $d\vec{v}/dt$ and components dv_x/dt, dv_y/dt, and dv_z/dt represent formal time derivatives of the corresponding functions.
- **Limitations:** This is a definition, so it has no limitations.

Again, an object's **average acceleration** $\vec{\overline{a}} \equiv \Delta\vec{v}/\Delta t$ during a *finite* time interval Δt is a good approximation of its instantaneous acceleration $\vec{a}$ halfway through the interval.

Section N1.6: Uniform Circular Motion

In this section, we use the definition of acceleration to argue that

$$\vec{a} = |\vec{v}|^2/r \text{ toward the circle's center} \tag{N1.17}$$

- **Purpose:** This equation specifies the instantaneous acceleration $\vec{a}$ of an object in **uniform circular motion** at speed $|\vec{v}|$ in a circle of radius r.
- **Limitations:** To be *uniform* circular motion, $|\vec{v}|$ must be constant. This expression does not apply to *nonuniform* circular motion.

N1.1   The Newtonian Synthesis

In this unit, our focus shifts to the behavior of individual objects

In unit C, we studied how three great conservation laws constrain the behavior of isolated *systems* of objects in *spite* of the detailed characteristics of their interactions. In this unit, we will narrow our focus from systems to single objects: *how do interactions affect the motion of an individual object?*

This chapter describes the basic tools (Newton's three laws of motion and the definitions of velocity and acceleration) that we need to answer this question, connecting them to what we have already learned in unit C. The next two chapters will provide an overview of how we can apply these tools.

But before we dig into the details, I'd like to briefly review the history of humanity's work on this topic to underline Newton's intellectual impact.

Aristotelian physics

As mentioned in unit C, the Greek philosopher Aristotle (384–322 B.C.E.) was one of the first people to think about the laws of motion systematically. Aristotle's primary goal in his treatises *Physica* ("Physics") and *De Caelo* ("On the Heavens") was to understand motion in the context of a general philosophy of *change* and its relation to an object's *nature*. His model of change was actually more biological than what we would consider mechanical. Just as it is in an acorn's intrinsic nature to grow into an oak, Aristotle argued that objects move in ways that reflect their intrinsic natures. Heavy objects have a natural tendency to move toward the universe's center (the earth) and remain at rest there. Fire, by its nature, yearns to move away from the earth toward the heavens, where fiery objects naturally reside. Celestial bodies, because of their natures, move endlessly in the heavens in perfect circles.

Since a heavy object's nature is to be at rest, any *motion* requires a special cause, and the object comes back to rest after the cause ceases. Celestial objects, in contrast, naturally move in endless perfect circles. Note that *different laws of motion apply to terrestrial and celestial objects* in this model, because the natures of celestial and terrestrial objects are fundamentally different.

Though Aristotle's ingenious and careful work provided the *starting point* for Western thinking about motion until the 1500s, scholars did not unquestioningly accept it. People quickly saw that his simple picture of celestial objects moving in simple circles was inadequate. In 140 C.E., the Alexandrian astronomer Ptolemy published a much more sophisticated geocentric model of celestial motion that involved nested circular motions. Ptolemy's book was so comprehensive and well-written that it became *the* accepted text on astronomy until the 1600s (9th-century Arab astronomers referred to it as the *Almagest*, "The Greatest"). Medieval Western thinkers also noted problems with Aristotle's models of motion for terrestrial objects and proposed improvements. But these were incremental adjustments to a model that still treated celestial and terrestrial physics as being fundamentally different.

The Copernican model of the solar system

The Aristotelian consensus began to unravel in the mid-1500s, when the Polish scholar Nicolaus Copernicus proposed a model that put the sun, not the earth, at the universe's center. Copernicus hoped that this model could explain the observed behavior of the planets in terms of simple circular orbits instead of requiring the complex nesting of circular motions of Ptolemy's model. While his hope was ultimately in vain, astronomers found his model so handy for doing calculations that his book was widely circulated even as opposition mounted. Opposing scholars correctly recognized that Copernicus's model contradicted not only Ptolemy's great work but also the foundation of Aristotle's ideas of motion, since his idea of "natural" motion toward the earth or toward the heavens rested firmly on the assumption that the earth was the universe's fixed center. They were unwilling to overturn a 2000-year-old consensus just to make a few astronomical calculations simpler, particularly as no replacement for Aristotle's scheme was available.

However, the Italian physicist Galileo Galilei was convinced that Copernicus was right. In the late 1500s and early 1600s, he took the first steps toward an understanding of motion that would be consistent with this model, developing nascent forms of Newton's first law and the theory of relativity to ingeniously rebut Ptolemy's arguments that the earth must be at rest. Galileo was also on the leading edge of two hot trends in the "natural philosophy" of his time: (1) the desire to make science *quantitatively predictive* as well as merely descriptive and (2) using experiments to *test* hypotheses instead of relying on pure reason. These values led Galileo to do experiments that provided the first *quantitative* description of the motion of falling objects.

Several other crucial developments laid the foundation for Newton's work. In 1609, Johannes Kepler showed (to his surprise and some shock) that Copernicus's model was quantitatively consistent with observations only if the planets move in *ellipses* instead of perfect circles. By 1618, Kepler published three empirical laws that provided a complete description of planetary motion in Copernicus's model, finally making that model both simpler and more accurate than Ptolemy's. About the same time, observation of variation in the size and phases of Venus using the newly invented telescope enabled Galileo and others to argue convincingly that Venus (at least) must orbit the sun. Finally, the philosopher René Descartes began in the mid-1600s to express the radical idea that it should be possible to explain *all* phenomena in terms of moving and colliding particles. The stage was set for revolution.

In 1661, the 19-year-old Isaac Newton arrived in Cambridge to begin his college education. Newton learned Aristotle in class, but the ideas of Galileo and Descartes circulated unofficially. In 1665, the university closed because of the Black Plague, and Newton spent the next two years at his rural home. During this time, he invented vector calculus and applied it to the motion of the moon and planets. Starting with the idea that *all* objects (including planets) naturally move with constant velocities, Newton showed that the planets behaved as if they were attracted to the sun by a force that varied as the inverse square of their distance from the sun, a force that Newton saw as being the same as the gravitational force that causes an apple to drop from a tree. In spite of these successes, Newton did not publish anything at the time.

In 1679, after he had returned to Cambridge as a professor, disputes with his rival Robert Hooke goaded Newton to publish his work. In 1687, Newton published *Philosophiae Naturalis Principia Mathematica* ("Mathematical Principles of Natural Philosophy": see figure N1.1), which used his techniques of calculus, three simple laws of motion, and his law of inverse-square gravitation to explain all of Kepler's empirical laws of planetary motion as well as a variety of terrestrial phenomena, including the moon's influence on the tides and Galileo's empirical work on falling bodies. This was one of the most important books in the history of science. Newton's theory was the first to explain both terrestrial *and* celestial motion using the same theoretical framework. This unification, which we call the **Newtonian synthesis**, stands as one of the greatest intellectual achievements of all time. Indeed, the *Principia* was so impressive that most physicists at the time dropped what they were doing and began to explore the rich implications of his new theory. The *Principia* forged from a diversity of competing schools and perspectives a *single* community of scholars dedicated to *one* overarching theory, transforming physics from a branch of natural philosophy into a quantitative science.

This unit is devoted to exploring the application of Newton's laws (particularly his second law) to a variety of terrestrial and celestial phenomena. Our journey will culminate in chapters N11 and N12 with a demonstration that these laws indeed explain Kepler's empirical laws of planetary motion, allowing us to experience Newton's crowning achievement for ourselves!

Galileo and Kepler lay the foundations for Newton

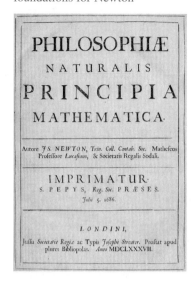

Figure N1.1
Title page of the first edition of Newton's *Principia*. (Credit: H.S. Photos/Alamy Stock Photo)

Newton invents a model that explains *both* terrestrial and celestial physics

N1.2   Newton's Laws of Motion

Newton's three laws of motion

In modern language, we can state Newton's laws of motion as follows:

1. In the absence of external interactions, an object's (or system's) center of mass moves at a *constant velocity*.
2. The net external force on a system causes that system's center of mass to accelerate at a rate inversely proportional to the system's total mass.
3. The force that a given interaction *A* between objects 1 and 2 exerts on 1 is equal in magnitude and opposite in direction to the force it exerts on 2.

We currently understand these laws to express implications of the principle of *momentum transfer* (or equivalently, the law of conservation of momentum). Let's see how this works for each law in turn.

The derivation of
Newton's first law

In chapter C4, we saw that the total momentum of any system of particles is equal to the system's total mass times the velocity of its center of mass:

$$M\vec{v}_{\mathrm{CM}} = \vec{p}_1 + \cdots + \vec{p}_N \equiv \vec{p}_{\mathrm{tot}} \qquad (\mathrm{N1.1})$$

This equation tells us that we can calculate the total momentum of an extended object (or even a system of objects!) by treating it as if it were a point particle of equal total mass located at the center of mass. Now, conservation of momentum implies that *the total momentum of an isolated system is conserved*. Since equation N1.1 implies that $\vec{v}_{\mathrm{CM}} = \vec{p}_{\mathrm{tot}}/M$, and M and $\vec{p}_{\mathrm{tot}}$ have fixed values for an isolated object, we see that an isolated object's (center-of-mass) velocity must be constant. This is **Newton's first law** for extended objects.

Though this law is in some sense a corollary of his second law (as we will see shortly), Newton had an important *rhetorical* reason for stating his first law as he did: it clearly and vividly contradicts the previously accepted Aristotelian assumptions about both terrestrial and celestial motion. Newton wanted to make it clear that the *natural* state of a terrestrial object or a celestial object is *not* to be at rest as close to the center of the earth as possible or unending motion in a *circle*, respectively, but rather unending motion in a *straight line*. Given his context, making this point clearly was essential.

From a contemporary perspective, Newton's first law not only states a useful consequence of the law of conservation of momentum but also plays an essential role in the definition of an *inertial reference frame* (see chapter N8).

The derivation of
Newton's second law

Let's turn now to Newton's second law. Recall that when external interactions *A*, *B*, ... act on a system, they transfer impulses $[d\vec{p}]_A$, $[d\vec{p}]_B$, ... to that system during a given time interval dt. If *A*, *B*, ... comprise the *complete* set of external interactions acting on the system, conservation of momentum then implies that the change in the system's total momentum must be

$$[d\vec{p}]_A + [d\vec{p}]_B + \cdots = d\vec{p}_{\mathrm{tot}} \qquad (\mathrm{N1.2})$$

But, assuming that the system's mass does not change during this time, then equation N1.1 implies that $d\vec{p}_{\mathrm{tot}} = M\,d\vec{v}_{\mathrm{CM}}$, where $d\vec{v}_{\mathrm{CM}}$ is the change in the velocity of the system's center of mass. If we substitute this into the above, divide both sides by dt, and use the definition of force, we have

$$\frac{[d\vec{p}]_A}{dt} + \frac{[d\vec{p}]_B}{dt} + \cdots \equiv \vec{F}_A + \vec{F}_B + \cdots = \frac{d\vec{p}_{\mathrm{tot}}}{dt} = M\frac{d\vec{v}_{\mathrm{CM}}}{dt} \qquad (\mathrm{N1.3})$$

Let's compactly write $\vec{F}_A + \vec{F}_B + \cdots \equiv \sum \vec{F}_{\mathrm{ext}}$. We also define the rate at which an object's velocity changes to be its **acceleration**: $d\vec{v}_{\mathrm{CM}}/dt \equiv \vec{a}_{\mathrm{CM}}$. Substituting these quantities into equation N1.3 above yields the mathematical statement of **Newton's second law**:

$$\sum \vec{F}_{\text{ext}} = M\,\vec{a}_{\text{CM}} \qquad\qquad (N1.4)$$

- **Purpose:** This equation links the acceleration $\vec{a}_{\text{CM}}$ of a system's center of mass to the system's total mass M and the vector sum $\sum \vec{F}_{\text{ext}}$ of all the external forces acting on it.
- **Limitations:** The system must have a fixed mass and a speed that is small compared to that of light (as we will see in unit R).
- **Note:** Read this equation verbally as follows: "the net external force on an object *causes* its mass to accelerate."

(Note that if a system is isolated, $\sum \vec{F}_{\text{ext}} = 0$, and equation N1.4 tells us that the system's velocity is constant, which is Newton's *first* law.)

Newton's second law contradicts common subconscious notions about force and motion in important ways that are worth making explicit. First, it asserts that forces *cause* acceleration. At the literal level, equation N1.4 expresses a quantitative equality between two sides of a mathematical equation. At the more important conceptual level, though, it expresses a *causal* relationship. The net external force is not in any sense "equivalent" to a mass accelerating (these are completely distinct ideas); rather, the net external force on an object *causes* its center of mass to accelerate (see the final note).

Moreover (contrary to the Aristotelian model), forces do not cause *motion*, they cause *acceleration*. Newton's first and second laws address this issue from two directions, the first law declaring that rectilinear motion is *natural* and needs no explanatory cause and the second proclaiming that forces cause an object's motion to *change* away from its natural motion.

Finally, Newton's second law tells us that it is the *net* force (the *vector sum* of *all* external forces) on the object that causes this acceleration, not the strongest force or the most recently applied force. A force does not *overcome* another force to cause an object to move in a certain way: rather, all forces acting on the object act *in concert* to direct the object's acceleration.

Let's turn now to Newton's third law. Conservation of momentum implies that if an interaction A acts between two objects 1 and 2, then any impulse that the interaction transfers *to* object 1 must come *out* of object 2: $[d\vec{p}_1]_A = -[d\vec{p}_2]_A$. If this is true, then the *rates* at which momentum flows into each object due to that interaction must also be equal in magnitude and opposite in direction. Applying the definition of force, we see that

$$\vec{F}_{1,A} \equiv \frac{[d\vec{p}_1]_A}{dt} = -\left(\frac{-[d\vec{p}_2]_A}{dt}\right) \equiv -\vec{F}_{2,A} \qquad\qquad (N1.5)$$

This is the mathematical statement of **Newton's third law**.

While this seems logical and straightforward in the context of the momentum transfer model, applying this law to realistic situations sometimes yields counterintuitive results. For example, suppose a 10,000-kg truck traveling at 65 mi/h hits a parked 500-kg Volkswagen Beetle. Which vehicle exerts the stronger force on the other during the collision? The third law says that the forces that their contact interaction exerts on each vehicle *must have the same magnitude:* any momentum that the interaction transfers out of the truck must go into the Beetle at the same rate!

But this seems absurd! We know that this interaction will smash the Beetle flat and fling it some distance, while hardly fazing the truck. How can the forces possibly be the same? The answer is that just because the *forces* are the same doesn't mean that the objects' *responses* must be the same. As the Beetle is

Anti-Aristotelian aspects of Newton's second law

The derivation of Newton's third law

This law can seem counterintuitive

20 times less massive than the truck, a given amount of momentum flowing into the Beetle causes a change in its velocity that is 20 times larger than the same flow changes the truck's velocity. Forces of the same magnitude can therefore have entirely different effects!

As we have seen, we can derive all three laws from conservation of momentum, which we now consider to be the more fundamental principle. Newton, however, took these laws as the basic assumptions of his model, and derived conservation of momentum from them! Newton's approach is not wrong. *Every physical model is based on assumptions, and one can build the same model from different sets of assumptions*. The choice between such logically equivalent approaches is really about what one wants to emphasize. (Physicists currently prefer to emphasize momentum.)

However we get to these laws, though, our goal in this unit is to see how we can *use* them (particularly the second law) to solve problems that would be harder to solve using an approach directly based on conservation laws.

N1.3   Vector Calculus

As Newton himself found, we cannot go very far toward this goal without **vector calculus**. Newton's second law intrinsically involves the rate at which an object's velocity changes, and calculus provides the appropriate mathematics for handling such rates. Moreover, since velocity is a vector, we need to know how to apply calculus to *vector* quantities. (Happily, we don't have to invent vector calculus from scratch the way he did!)

The time derivative of an ordinary mathematical function

Fortunately, the vector calculus we need is not very different from what you learn in an ordinary calculus class. Let's begin with a brief review. Consider an ordinary quantity $f(t)$ that varies with time t. We define the **time derivative** df/dt of such a quantity at a given instant of time t as follows:

$$\frac{df}{dt} \equiv \lim_{\Delta t \to 0} \frac{f(t + \Delta t) - f(t)}{\Delta t} \tag{N1.6}$$

In words, this says that we first compute the change $\Delta f = f(t + \Delta t) - f(t)$ in the quantity's value during a time interval Δt starting at time t, then calculate the value of the ratio $\Delta f/\Delta t$ and take the limiting value of this ratio as Δt approaches zero. The derivatives of ordinary functions of a single variable are typically described in an introductory calculus course. Appendix NA at the end of this volume provides a more detailed review if you need it.

The time derivative of a vector

We can easily extend this to vector quantities. Let $\vec{q}$ be a vector quantity that varies with time t. In analogy to equation N1.6, we define the time derivative $d\vec{q}/dt$ of $\vec{q}$ to be

$$\frac{d\vec{q}}{dt} \equiv \lim_{\Delta t \to 0} \frac{\vec{q}(t + \Delta t) - \vec{q}(t)}{\Delta t} \tag{N1.7}$$

In words, we first find the *change* $\Delta\vec{q} = \vec{q}(t + \Delta t) - \vec{q}(t)$ in the vector $\vec{q}$ during the time interval Δt starting at time t, calculate the ratio $\Delta\vec{q}/\Delta t$ (which is a vector), and finally determine this ratio's limiting value as Δt goes to zero.

The component definition of the vector difference implies that

$$\frac{d\vec{q}}{dt} \equiv \lim_{\Delta t \to 0} \frac{1}{\Delta t} \begin{bmatrix} q_x(t + \Delta t) - q_x(t) \\ q_y(t + \Delta t) - q_y(t) \\ q_z(t + \Delta t) - q_z(t) \end{bmatrix} = \begin{bmatrix} \lim_{\Delta t \to 0} \dfrac{q_x(t + \Delta t) - q_x(t)}{\Delta t} \\ \lim_{\Delta t \to 0} \dfrac{q_y(t + \Delta t) - q_y(t)}{\Delta t} \\ \lim_{\Delta t \to 0} \dfrac{q_z(t + \Delta t) - q_z(t)}{\Delta t} \end{bmatrix} \equiv \begin{bmatrix} \dfrac{dq_x}{dt} \\ \dfrac{dq_y}{dt} \\ \dfrac{dq_z}{dt} \end{bmatrix} \tag{N1.8}$$

In words, this says that the *components* of the time derivative of the vector $d\vec{q}/dt$ are just the ordinary time derivatives of the components of $\vec{q}$ treated as ordinary functions. This means that we can easily apply what we know about ordinary calculus to vector calculus.

N1.4   The Formal Definition of Velocity

In unit C, we defined an object's velocity vector $\vec{v}$ at a time t to be

$$\vec{v} \equiv \frac{d\vec{r}}{dt} \equiv \frac{\text{small displacement}}{\text{short time interval}} \qquad (\text{N1.9})$$

The definition of velocity we used in unit C

where $d\vec{r}$ is the object's displacement during the time interval dt, which in turn (1) encloses the instant t in question and (2) is "sufficiently short" that the velocity doesn't change significantly during the interval.

This definition works intuitively and was adequate for our purposes in unit C, but is really a bit fuzzy. What constitutes a "sufficiently short" dt is poorly defined, partly because the criterion refers to the quantity (velocity) we are trying to define! While we may *intuitively* know what this means, it is not good enough to be used as a basis for a mathematically precise definition. Indeed, the lack of a precise definition of velocity *at an instant* was a significant stumbling block for Western physicists until Newton.

The concept of the *time derivative of position* provides a natural way to put the concept of velocity on a mathematically firm foundation. We define an object's **instantaneous velocity** vector at a given instant of time t to be

$$\vec{v} \equiv \frac{d\vec{r}}{dt} \equiv \lim_{\Delta t \to 0} \frac{\Delta \vec{r}}{\Delta t} \equiv \lim_{\Delta t \to 0} \frac{r(t+\Delta t) - \vec{r}(t)}{\Delta t} \qquad (\text{N1.10})$$

The formal definition of instantaneous velocity

where $\vec{r}(t)$ is the object's position vector as a function of time. This definition is essentially the same as that given by equation N1.9 except that *no* finite interval Δt is considered "sufficiently short": we define $\vec{v}$ by the *limiting value* that $\Delta\vec{r}/\Delta t$ approaches as Δt goes to *zero*. This definition thus entirely avoids the problem of defining when an interval is "sufficiently short." It also makes it mathematically clear what we can possibly mean by an object's velocity *at an instant*, an idea that superficially seems to be incompatible with the definition of velocity as the object's *displacement* during a (nonzero!) *interval* of time divided by the duration of that interval.

According to equation N1.10, this definition implies that

$$\vec{v} \equiv \frac{d\vec{r}}{dt} \equiv \begin{bmatrix} \lim_{\Delta t \to 0} \dfrac{x(t+\Delta t) - x(t)}{\Delta t} \\[2mm] \lim_{\Delta t \to 0} \dfrac{y(t+\Delta t) - y(t)}{\Delta t} \\[2mm] \lim_{\Delta t \to 0} \dfrac{z(t+\Delta t) - z(t)}{\Delta t} \end{bmatrix} \equiv \begin{bmatrix} \dfrac{dx}{dt} \\[2mm] \dfrac{dy}{dt} \\[2mm] \dfrac{dz}{dt} \end{bmatrix} \qquad (\text{N1.11})$$

- **Purpose:** This equation formally defines an object's velocity $\vec{v}(t)$ at an instant t, where $\vec{r}(t) = [x(t), y(t), z(t)]$ is the position of the object's center of mass at time t, and $\vec{r}(t+\Delta t) = [x(t+\Delta t), y(t+\Delta t), z(t+\Delta t)]$ is the same evaluated at time $t+\Delta t$, where Δt is some finite interval of time. The symbols $d\vec{r}/dt$, dx/dt, dy/dt, and dz/dt now officially represent time derivatives.
- **Limitations:** This expression is a definition and so has no limitations.

We will call the components $v_x \equiv dx/dt$, $v_y \equiv dy/dt$, and $v_z \equiv dz/dt$ the object's **x-velocity**, **y-velocity**, and **z-velocity** at time t, respectively. The object's instantaneous speed at time t is

$$|\vec{v}| \equiv \sqrt{v_x^2 + v_y^2 + v_z^2} = \sqrt{\left(\frac{dx}{dt}\right)^2 + \left(\frac{dy}{dt}\right)^2 + \left(\frac{dz}{dt}\right)^2} \qquad \text{(N1.12)}$$

Example N1.1

Problem: Suppose the components of an object's position as a function of time are given by $x(t) = at^2 + b$, $y(t) = ct$, and $z(t) = 0$, where $a = 2.0$ m/s^2, $b = 5.2$ m, and $c = 1.6$ m/s. What is the object's speed at time $t = 0$?

Solution Taking the time derivative of each of these position components (using the methods discussed in appendix NA), we find that the components of the object's velocity (as functions of time) are $v_x(t) = dx/dt = 2at$, $v_y(t) = dy/dt = c$, and $v_z(t) = dz/dt = 0$. Evaluating these at time $t = 0$, we find that $v_x(0) = 2a \cdot 0 = 0$, $v_y(0) = c$, and $v_z(0) = 0$. Therefore, the object's speed at time $t = 0$ is $|\vec{v}(0)| = [v_x^2 + v_y^2 + v_z^2]^{1/2} = [0 + c^2 + 0]^{1/2} = c = 1.6$ m/s.

Exercise N1X.1

Now *you* try. Suppose that the components of an object's position vector as functions of time are given by $x(t) = at^2 + b$, $y(t) = -ct$, and $z(t) = -at^2 + ct$, where $a = 1.5$ m/s^2, $b = 3.0$ m, and $c = 4.0$ m/s. Find the components of this object's velocity and its speed at $t = 0$ and $t = 2.0$ s.

As figure N1.2 illustrates, the ratio $\Delta\vec{r}/\Delta t$ computed for any finite time interval Δt starting at time t is only *approximately* equal to the object's instantaneous velocity $\vec{v}$ at time t (though the approximation gets better as Δt gets smaller). We will call $\mathbf{\Psi} \equiv \Delta\vec{r}/\Delta t$ the object's **average velocity** during the time interval Δt, where the slash (symbolically evoking a "drawn out" interval) distinguishes average velocity from the symbol $\vec{v}$, which from now on we will use exclusively for *instantaneous* velocity. (A more common symbol for average velocity is $\vec{v}_{avg}$, but this symbol would be awkward in what follows.)

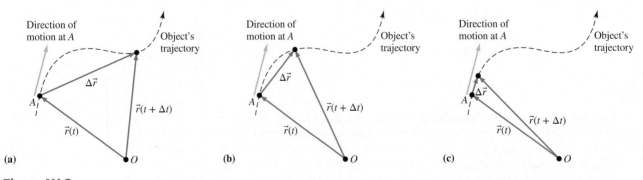

Figure N1.2
The shorter the time interval Δt, the closer that the direction of $\Delta\vec{r}/\Delta t$ (which is the same as that of r) becomes to the direction of the object's velocity $\vec{v}$ at time t (which is the direction of the object's motion as it passes point A).

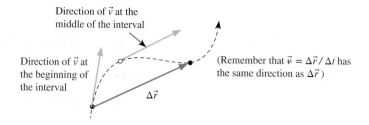

Direction of $\vec{v}$ at the middle of the interval

Direction of $\vec{v}$ at the beginning of the interval

(Remember that $\vec{v} = \Delta\vec{r}/\Delta t$ has the same direction as $\Delta\vec{r}$)

$\Delta\vec{r}$

Figure N1.3
The direction of an object's average velocity $\vec{v}$ during an interval Δt is generally closer to the direction of its instantaneous velocity $\vec{v}$ at the *middle* rather than the beginning of the interval.

As figure N1.3 illustrates, $\vec{v} \equiv \Delta\vec{r}/\Delta t$ for a nonzero Δt generally most closely approximates the object's instantaneous velocity at an instant t that is *halfway* through the interval Δt (in both magnitude and direction, as it turns out). This is important to know, because in physics we often have to estimate instantaneous velocities by computing average velocities over small but finite time intervals. This rule tells us how to do this most accurately. We will find this quite useful in subsequent chapters.

Problem: The diagram below shows a top view of a ball rolling toward the right along an inclined track, showing the ball's position every 0.1 s. Is the ball's speed increasing, decreasing, or staying constant? Estimate the ball's speed at $t = 0.3$ s.

Example N1.2

| $t = 0$ | 0.1 s | 0.2 s | 0.3 s | 0.4 s | 0.5 s | 0.7 s |

| 0 | 20 | 40 | 60 | 80 | 100 | 120 | 140 cm |

Solution Since the ball's displacement during successive intervals gets smaller, the ball is slowing down. Its speed at $t = 0.3$ s will be most closely approximated by the magnitude of $\Delta\vec{r}/\Delta t$ for the shortest interval having that instant as its midpoint. Between $t = 0.2$ s and $t = 0.4$ s, the ball travels 45 cm, so $|\vec{v}(0.3 \text{ s})| \approx |\Delta\vec{r}|/\Delta t$ during this interval $\approx (45 \text{ cm})/(0.2 \text{ s}) = 225$ cm/s.

The picture in example N1.2, which shows an object's position at a sequence of equally spaced intervals of time, is an example of what we call a **motion diagram**. We'll talk about these diagrams more in the next chapter.

N1.5 The Formal Definition of Acceleration

The word *acceleration* in everyday English carries a connotation of "speeding up." Its meaning in *physics* is both more general and more precise:

> An object's **acceleration** at an instant is the *rate of change of its velocity*, a vector that expresses *how rapidly* and *in what direction* the object's velocity vector is changing at that instant.

The physics definition of acceleration in words

In physics, then, we use acceleration to describe *any* change in the magnitude and/or direction of an object's velocity. A car that is speeding up is indeed accelerating (because the magnitude of its velocity is increasing), but according to the physics definition of the word, a car that is slowing down is *also* accelerating, because the magnitude of its velocity vector is changing (in this case decreasing). Even a car moving at a constant speed can be accelerating if it is going around a bend in the road (in this case, because the *direction* of its velocity is changing). If an object's velocity vector changes in *any* way, the object is accelerating.

The mathematical definition of the acceleration vector

If an object's **instantaneous acceleration** vector $\vec{a}(t)$ at an instant of time t is the *rate of change of its velocity* $d\vec{v}/dt \equiv \lim_{\Delta t \to 0}\{[\vec{v}(t+\Delta t) - \vec{v}(t)]/\Delta t\}$, then

$$\vec{a}(t) \equiv \frac{d\vec{v}}{dt} \equiv \begin{bmatrix} \lim_{\Delta t \to 0} \dfrac{v_x(t+\Delta t) - v_x(t)}{\Delta t} \\ \lim_{\Delta t \to 0} \dfrac{v_y(t+\Delta t) - v_y(t)}{\Delta t} \\ \lim_{\Delta t \to 0} \dfrac{v_z(t+\Delta t) - v_z(t)}{\Delta t} \end{bmatrix} \equiv \begin{bmatrix} \dfrac{dv_x}{dt} \\ \dfrac{dv_y}{dt} \\ \dfrac{dv_z}{dt} \end{bmatrix} \qquad (N1.13)$$

- **Purpose:** This equation defines the acceleration $\vec{a}(t)$ of an object's center of mass at an instant of time t, where $\vec{v}(t)$, with components $v_x(t)$, $v_y(t)$, and $v_z(t)$, is the velocity of the object's center of mass at time t; $\vec{v}(t+\Delta t)$ (and analogously for its components) is the same evaluated at time $t+\Delta t$, where Δt is some finite interval of time. The quantity $d\vec{v}/dt$ and components dv_x/dt, dv_y/dt, and dv_z/dt represent formal time derivatives of the corresponding functions.
- **Limitations:** This is a definition, so it has no limitations.

In words, this equation says that an object's acceleration at time t is the limiting value as Δt goes to 0 of the *change* $\Delta\vec{v}$ in the object's instantaneous velocity between times t and $t+\Delta t$, divided by the duration Δt of that interval. Since velocity is measured in meters per second (m/s), the SI units of acceleration are meters per second squared (m/s^2).

We also see that if we know the components of an object's velocity vector as a function of time, we can compute the components of its acceleration simply by taking the time derivatives of the corresponding velocity components, treating them as ordinary simple functions of t.

The definition of the average acceleration

Just as an object's average velocity $\vec{v} \equiv \Delta\vec{r}/\Delta t$ for a given nonzero (but fairly short) time interval Δt best approximates that object's instantaneous velocity midway through the interval, so an object's **average acceleration** $\vec{a} \equiv \Delta\vec{v}/dt$ during a given nonzero (but reasonably short) interval Δt best approximates its instantaneous acceleration midway through the interval:

$$\vec{a}(t + \tfrac{1}{2}\Delta t) \approx \vec{a} \equiv \frac{\Delta\vec{v}}{\Delta t} \equiv \frac{\vec{v}(t+\Delta t) - \vec{v}(t)}{\Delta t} \qquad (N1.14)$$

So if we know an object's velocity at two instants of time, we can *estimate* that object's acceleration halfway through the interval by computing the difference $\Delta\vec{v}$ between those velocities and dividing by Δt. This approximation improves as Δt becomes smaller compared to the time required for the acceleration to change significantly. If the object's acceleration is *constant* during Δt, then the instantaneous acceleration throughout the interval is the *same* as this average acceleration.

In situations where we need only to *estimate* an object's acceleration, equation N1.14 provides a quick way to connect an object's acceleration to the change in its velocity *without* doing any calculus.

The following examples illustrate the application of this idea in several different contexts.

Problem: A person driving eastward on a straight road at 20 m/s (44 mi/h) sees the brake lights of the car in front go on and so applies the brakes for 1.5 s. This slows the 1000-kg car to 14 m/s. What was the magnitude and direction of the average net external force acting on the car during this period?

Solution The car's initial velocity vector is $\vec{v}_i = 20$ m/s eastward; its final velocity vector is $\vec{v}_f = 14$ m/s eastward. As the vector construction in figure N1.4 illustrates, this means that the car's change in velocity is $\Delta\vec{v} = 6.0$ m/s *westward*. (Remember that difference $\Delta\vec{v}$ between the final and initial velocities is the vector we would have to add to the initial velocity to get the final velocity.) Therefore, $\vec{a} = (6.0$ m/s west$)/(1.5$ s$) = 4.0$ m/s^2 westward. By Newton's second law and since the car's mass is $m = 1000$ kg, the average net external force acting on the car must be $m\vec{a} = (1000$ kg$)/(4.0$ m/s$^2) = 4000$ kg·m/s$^2 = 4000$ N westward in this case.

Example N1.3

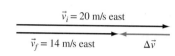

Figure N1.4
The car's *change* in velocity in this example points westward, according to the definition of the vector difference.

Problem: A bicyclist travels at a constant speed of 12 m/s around a bend in the road during a 34-s interval of time. If the cyclist was traveling northward at the beginning of the bend and westward at the end, what are the magnitude and direction of the cyclist's average acceleration during this interval?

Solution Let $\vec{v}_i \equiv \vec{v}(t)$ and $\vec{v}_f = \vec{v}(t+\Delta t)$ be the the cyclist's initial and final velocities, respectively. Note that $\Delta\vec{v} \equiv \vec{v}_f - \vec{v}_i \Rightarrow \vec{v}_f = \vec{v}_i + \Delta\vec{v}$. Therefore, we can construct $\Delta\vec{v}$ by putting $\vec{v}_i$ and $\vec{v}_f$ tail-to-tail and constructing $\Delta\vec{v}$ from the head of $\vec{v}_i$ to the head of $\vec{v}_f$ (this is a valuable technique to remember for the future). Figure N1.5 shows that $\Delta\vec{v}$ in this case is a vector pointing *southwest*. Using the Pythagorean theorem, that vector's magnitude is

$$|\Delta\vec{v}| = \sqrt{(12 \text{ m/s})^2 + (12 \text{ m/s})^2} = 17 \text{ m/s} \qquad (N1.15)$$

The magnitude of the bicyclist's average acceleration is therefore $|\Delta\vec{v}|/\Delta t = (17$ m/s$)/(34$ s$) = 0.50$ m/s^2. So the cyclist's average acceleration during this interval is 0.5 m/s^2 southwest (even though the cyclist's speed is constant!).

Example N1.4

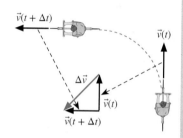

Figure N1.5
The cyclist's change in velocity.

Exercise N1X.2

Suppose you throw a ball vertically into the air. The ball leaves your hand traveling upward at a speed of 15 m/s. Three seconds later, the ball passes you going downward with the same speed. What is the magnitude and direction of the ball's average acceleration during this time interval?

N1.6   Uniform Circular Motion

Circular motion (or approximately circular motion) appears in a broad variety of physical situations, such as atomic orbitals, bicycle wheels, hurricanes, satellite orbits, and galactic rotation. Historically, one of the ideas that helped Newton build his theory was growing awareness in the mid-1600s that even an object that is moving along a circular path (even if it has a constant speed) is accelerating, because the *direction* of its velocity is changing.

The definition of
uniform circular motion

In this section, we will use the definition of acceleration to determine how an object's acceleration in **uniform circular motion** (circular motion at a constant speed) depends on the object's speed and the radius of its circular path. Consider an object moving along the circular path shown in figure N1.6a, and assume that it passes point P at time t. Let points 1 and 2 correspond to the object's (center-of-mass) position at times $t - \frac{1}{2}\Delta t$ and $t + \frac{1}{2}\Delta t$, respectively, and let $\vec{v}_1$ and $\vec{v}_2$ be its instantaneous velocities at those times. If the object's speed is constant, then $|\vec{v}_1| = |\vec{v}_2|$. Note also that if θ is the angle through which the object moves during time Δt, then because $\vec{v}_1$ and $\vec{v}_2$ are both tangent to the circle, the angle between these vectors is also θ.

The direction of the
object's acceleration

Now, the object's acceleration $\vec{a}(t)$ at time t will be best approximated by its average acceleration between times $t - \frac{1}{2}\Delta t$ and $t + \frac{1}{2}\Delta t$, meaning that

$$\vec{a}(t) \approx \vec{a}_{12} = \frac{\vec{v}(t + \frac{1}{2}\Delta t) - \vec{v}(t - \frac{1}{2}\Delta t)}{\Delta t} = \frac{\vec{v}_2 - \vec{v}_1}{\Delta t} \tag{N1.16}$$

I have drawn the vector $\vec{a}(t)\,\Delta t = \vec{v}_2 - \vec{v}_1$ on the diagram so that the tail of $\vec{a}(t)$ is located right at the top of the circle. We see that because $|\vec{v}_1| = |\vec{v}_2|$, the vector $\vec{a}(t)$ *points directly toward the circle's center*. The direction of $\vec{a}$ will not change if we take the limit that $\Delta t \to 0$, so this is an exact result. Moreover, since all positions on the circle are equivalent, this must be true of the object's acceleration at all points on the circle: the acceleration of an object in uniform circular motion always points toward the circle's center.

How the acceleration
depends on radius

Now, figure N1.6b shows the situation if we decrease the circle's radius by a factor of 2 but keep the speed the same. The lengths of the arrows representing $\vec{v}_1$ and $\vec{v}_2$ are the same as before, but in the same time interval Δt, the object goes an angle 2θ around the smaller circle instead of θ. This makes the acceleration arrow $\vec{a}\,\Delta t$ approximately *twice* as large as before (and the approximation becomes more accurate as $\theta \to 0$). So we see that $|\vec{a}| \propto 1/r$.

How the acceleration
depends on speed

On the other hand, figure N1.6c shows the situation if we keep the circle's radius the same as it was in figure N1.6a but double the object's speed. Now the arrow lengths are double what they were before *and* the angle between them has also doubled (because the object goes twice as far around the circle's circumference in the same interval Δt). This makes the acceleration arrow $\vec{a}\,\Delta t$ approximately *four times* as large as before (and again the approximation becomes more accurate as $\theta \to 0$). So we see that $|\vec{a}| \propto |\vec{v}|^2$.

The simplest formula that embraces all these results is

The general formula

$$\vec{a} = \frac{|\vec{v}|^2}{r} \text{ toward the circle's center} \tag{N1.17}$$

- **Purpose:** This equation specifies the instantaneous acceleration $\vec{a}$ of an object in *uniform circular motion* at speed $|\vec{v}|$ in a circle of radius r.
- **Limitations:** To be *uniform* circular motion, $|\vec{v}|$ must be constant. This expression does not apply to *nonuniform* circular motion.

Note that the SI units of this expression are $(m/s)^2/m = m/s^2$, which are the correct units for acceleration. We could multiply the right side of this expression by any unitless constant and still get a result consistent with the requirements that $|\vec{a}| \propto 1/r$ and $|\vec{a}| \propto |\vec{v}|^2$, but this equation is correct as it stands (as you can show more rigorously by doing problem N1D.1).

Note that this result and Newton's second law means that *an object moving in a circle must experience a net inward force*. We will explore this idea in more depth in both the homework problems and in future chapters.

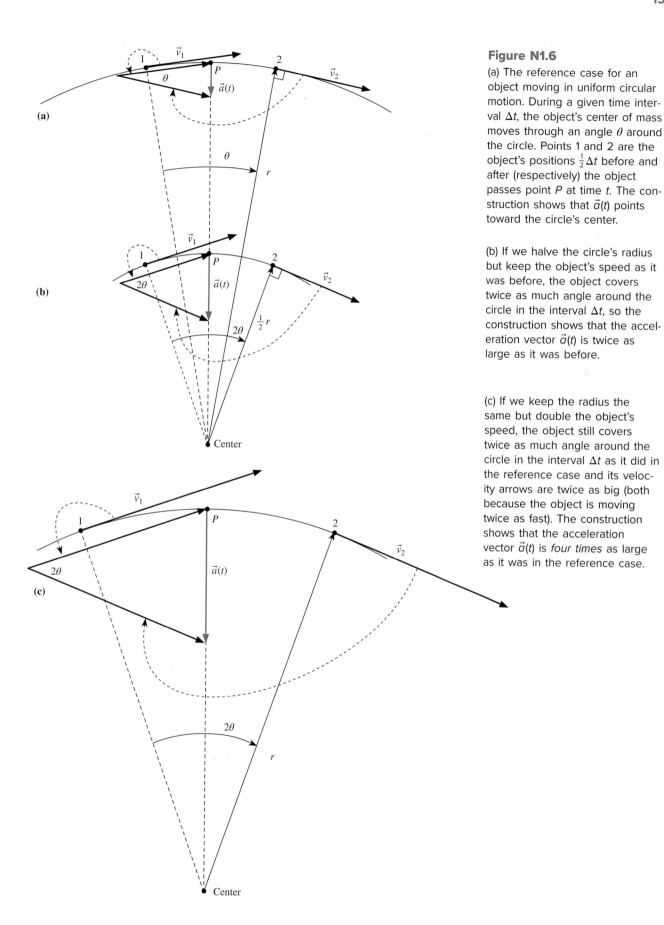

Figure N1.6

(a) The reference case for an object moving in uniform circular motion. During a given time interval Δt, the object's center of mass moves through an angle θ around the circle. Points 1 and 2 are the object's positions $\frac{1}{2}\Delta t$ before and after (respectively) the object passes point P at time t. The construction shows that $\vec{a}(t)$ points toward the circle's center.

(b) If we halve the circle's radius but keep the object's speed as it was before, the object covers twice as much angle around the circle in the interval Δt, so the construction shows that the acceleration vector $\vec{a}(t)$ is twice as large as it was before.

(c) If we keep the radius the same but double the object's speed, the object still covers twice as much angle around the circle in the interval Δt as it did in the reference case and its velocity arrows are twice as big (both because the object is moving twice as fast). The construction shows that the acceleration vector $\vec{a}(t)$ is *four times* as large as it was in the reference case.

TWO-MINUTE PROBLEMS

N1T.1 Do the following statements about the physical world express a primarily Aristotelian viewpoint (A), do they express a primarily Newtonian viewpoint (B), or are they consistent with both viewpoints (C)?
(a) If you push on something, it moves. If you push twice as hard, it moves twice as fast.
(b) A heavy object falls faster than a light object.
(c) The moon is held in its orbit around the earth by the force of the earth's gravity.
(d) If you push on an object, it moves. After you release it, it gradually comes to rest because friction drains away the force of the initial push.
(e) The speed of a falling object increases as it falls.

N1T.2 A spaceship with a mass of 20,000 kg is traveling in a straight line at a constant speed of 300 km/s in deep space. Its rockets engines must be: (A) off or (B) firing?

N1T.3 A 6-kg bowling ball moving at 3 m/s collides with a 1.4-kg bowling pin at rest. How do the magnitudes of the forces exerted by the collision on each object compare?
A. The force on the pin is larger than the force on the ball.
B. The force on the ball is larger than the force on the pin.
C. Both objects experience the same magnitude of force.
D. The collision exerts no force on the ball.
E. There is no way to tell.

N1T.4 A large car drags a small trailer in such a way that their common speed increases rapidly. Which tugs harder on the other?
A. The car tugs harder on the trailer than vice versa.
B. The trailer tugs harder on the car than vice versa.
C. Both tug equally on each other.
D. The trailer exerts no force on the car at all.
E. There is no way to tell.

N1T.5 A parent pushes a small child on a swing so that the child moves rapidly away while the parent remains at rest. How does the magnitude of the force that the child exerts on the parent compare to the magnitude of the force that the parent exerts on the child?
A. The force on the child is larger in magnitude.
B. The force on the parent is larger in magnitude.
C. These forces have equal magnitudes.
D. The child exerts zero force on the parent.
E. There is no way to tell.

N1T.6 Two marbles are rolling along parallel tracks (which may or may not be inclined). A stroboscopic photograph showing a top view of the positions of the marbles at equally spaced instants of time looks like this:

At what instant(s) of time do the marbles have the same instantaneous velocity?
A. At time t_B
B. At time t_G
C. At both t_B and t_G
D. Some instant between t_C and t_D
E. Some instant between t_D and t_E
F. Roughly time t_D

N1T.7 Two marbles are rolling along parallel tracks (which may or may not be inclined). A stroboscopic photograph showing a top view of the positions of the marbles at equally spaced instants of time looks like this:

At what instant(s) of time do the marbles have the same instantaneous velocity?
A. At time t_B
B. At time t_F
C. At both t_B and t_F
D. Some instant between t_C and t_D
E. Some instant between t_D and t_E
F. Roughly time t_D

N1T.8 An object travels halfway around a circle at a constant instantaneous speed $|\vec{v}|$. What is the magnitude $|\vec{v}|$ of its average velocity during this time interval?
A. $|\vec{v}|$
B. $1.41\,|\vec{v}|$
C. $1.57\,|\vec{v}|$
D. $|\vec{v}|/1.41$
E. $|\vec{v}|/1.57$
F. Some other multiple of $|\vec{v}|$ (specify)
T. We do not have enough information to answer.

N1T.9 An object can have a constant speed and still be accelerating. True (T) or false (F)?

N1T.10 An object's acceleration vector always points in the direction that it is moving. T or F?

N1T.11 An object can have zero velocity (at an instant) and still be accelerating. T or F?

N1T.12 An object's x-velocity can be positive at the same time that its x-acceleration is negative. T or F?

N1T.13 An object falling vertically at a speed of 20 m/s lands in a snowbank and comes to rest 0.5 s later. The object's average acceleration during this interval is
A. Upward
B. Downward
C. Zero

N1T.14 Suppose you throw a ball vertically into the air. At the exact instant that the ball reaches its highest position, select from the choices below the most accurate description of
(a) The ball's instantaneous velocity.
(b) The ball's instantaneous acceleration.
(c) The gravitational force on the ball.
(d) The net external force on the ball.
A. Upward
B. Downward
C. Zero
D. Other (specify)

N1T.15 The diagram below shows a top view of the moon's nearly circular orbit around the earth. The moon orbits the earth at a nearly constant speed.

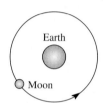
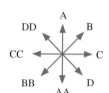

E. Up out of the page
F. Down into the page
T. Zero

At the point shown, use the arrow choices shown to the right to indicate the direction of
(a) The moon's velocity.
(b) The moon's acceleration.
(c) The gravitational force on the moon due to the earth.
(d) The net external force on the moon.
(e) The gravitational force on the earth due to the moon.
(To choose a double-letter answer using the letters on the back of the book, point with two fingers to the letter.)

N1T.16 Figure N1.6 shows the acceleration for an object moving *clockwise* around the circle. If the object moves *counterclockwise* around the circle, its acceleration points outward instead of inward. T or F?

N1T.17 An object travels at a constant speed $|\vec{v}|$ exactly once around a circle of radius r. The magnitude $|\vec{a}|$ of the object's average acceleration is:
A. $|\vec{v}|^2/r$
B. Zero
C. Something else (specify)

HOMEWORK PROBLEMS

Basic Skills

N1B.1 Imagine a cannon that is free to roll on wheels. According to Newton's third law, the interaction between the cannon and a cannonball as the latter is fired should exert equal forces on both.
(a) If this is so, why does the cannonball fly away at a very high speed and the cannon only recoils modestly?
(b) If the cannon is bolted to the ground, it does not recoil at all. How is this possible if the interaction with the cannonball exerts a force on the cannon?
Please explain carefully in both cases.

N1B.2 The moon's mass is about 1/81 times that of the earth. How does the gravitational force that the moon exerts on the earth compare to the gravitational force the earth exerts on the moon? Explain your answer.

N1B.3 A car with a mass of 1300 kg decreases its forward speed from 22 m/s to 14 m/s in 4.0 s. What are the magnitude and direction of the net force on the car? Show how you arrived at your result.

N1B.4 The magnitude of the net external force on a flying 0.25-kg ball is 2.5 N. What is the magnitude of the ball's acceleration? Explain how you arrived at your result.

N1B.5 What are the derivatives of the following functions of time (where a, b, and c are constants)?
(a) $at + c$
(b) $at^2 + (bt)^{-2}$
(c) $(at + b)^2$

N1B.6 What are the derivatives of the following functions of time (where a, b, and c are constants)?
(a) $ct(at^2 + b)$
(b) $(at + b)^{-2}$
(c) $(at^3 + b)/t$

N1B.7 An object's position as a function of time is given by $\vec{r}(t) = [-a/t, bt^2 + c, qt]$, where a, b, c, and q are constants with values $a = 2.0$ m·s, $b = 1.0$ m/s², $c = 5.0$ m, and $q = 2.0$ m/s. Calculate the components of this object's instantaneous velocity at time $t = 2.0$ s.

N1B.8 Suppose that an object moves along the x axis. Its x-position as a function of time is $x(t) = A \sin \omega t$, where A and ω are constants. What are the x, y, and z components of its velocity as a function of time?

N1B.9 An object travels exactly once around a circle of radius 5.0 m at a constant instantaneous speed of 3.0 m/s. What is its average velocity during this interval?

N1B.10 The driver of a car moving at 24 m/s due east sees something on the road and applies the brakes. The car comes to rest 4.0 s later. What are the magnitude and direction of the car's average acceleration during this interval?

N1B.11 A person hits a trampoline while moving downward with a speed of 5.0 m/s and rebounds a short time later with roughly the same speed upward. If the person is in contact with the trampoline for about 1.8 s, what are the magnitude and direction of the person's average acceleration during this time interval?

N1B.12 A spaceship accelerates from rest with a constant average acceleration of 15 m/s² (about the maximum that a human being can tolerate indefinitely). About how long will it take the ship to reach one-half the speed of light? (*Hint:* 1 Ms = 10^6 s ≈ 12 days.)

N1B.13 A car travels at a constant speed of 30 m/s around a circular track with a radius of 300 m. What is the magnitude of the car's acceleration?

N1B.14 Suppose an object travels at a constant speed $|\vec{v}|$ exactly halfway around a circle of radius r during a time T. What is the magnitude and direction (relative to the direction of the object's initial velocity) of the object's average acceleration $\vec{a}$ during this time?

Modeling

N1M.1 If one looks at the planet Venus through even a small telescope, one sees that sometimes it looks like a tiny crescent moon and at other times like a tiny full moon. Galileo noted that whenever Venus looks like a crescent, it also looks significantly *larger* than it does when it looks like a full moon. Argue (with the help of some diagrams) that this makes sense if Venus goes around the *sun*, but not if it goes around the earth. Contrast this situation with that of the moon, which *does* go around the earth.

N1M.2 Suppose you want to push on a 1500-kg car hard enough to get it to walking speed in no longer than four seconds. How massive a barbell must you be able to press upward if you can do this? (Make and describe appropriate estimates. Note that 1.0 m/s = 2.2 mi/h.)

N1M.3 A spaceship decreases its speed smoothly from $0.1c$ (where $c = 3.0 \times 10^8$ m/s = the speed of light) to rest in 30 min as it approaches a space station. Assume that the ship travels in a straight line. Compare the ship's acceleration to $6|\vec{g}|$, which is roughly the maximum acceleration that a human being can tolerate for a short period of time.

N1M.4 Two people, one with a mass of 70 kg and one with a mass of 45 kg, sit on 5-kg frictionless carts that are initially at rest on a flat surface. The lighter person pushes on the more massive person, causing the latter to accelerate at a rate of 1.0 m/s² toward the east. What are the magnitude and direction of the lighter person's acceleration vector?

N1M.5 Suppose an object of mass m hanging at the end of a spring oscillates vertically in such a way that its vertical position is $z(t) = A \sin \omega t$, where A and ω are positive constants and the $+z$ axis is upward.
(a) What are the SI units of A and ω?
(b) At what time t is the object's acceleration maximum in the upward direction? [*Hint:* First evaluate $v_z(t)$.]
(c) What is the magnitude of the maximum net force acting on the object? (Be sure to check your units!)
(d) What is the magnitude of the maximum force that the *spring* exerts on the object? [*Hint:* This is *not* the same as the answer to part (c).]

N1M.6 Suppose an object of mass m moves in the $+x$ direction with an x-velocity $v_x = Ae^{-bt}$, where A and b are positive constants.
(a) What are the SI units of A and b?
(b) What is the magnitude and direction of this object's acceleration?
(c) What is the net force (magnitude and direction) acting on the object according to Newton's second law?
(d) Check the units of your answer for part (c).
(e) Find a vector expression for the relationship between the object's velocity vector and this net force vector.

N1M.7 An astronaut hangs on to a handrail on the rim of a disk-shaped space station that has a radius of 15 m and rotates once in 30 s.
(a) What force must the rail exert on the astronaut? (Express this force as a fraction of the astronaut's normal weight on the earth's surface.)
(b) What is the *net* force on the astronaut (expressed in the same way)? (*Hint:* Remember Newton's second law.)

N1M.8 A 1500-kg car traveling at a steady speed of 20 m/s (about 44 mi/h) initially due northwest rounds a corner so that after 10 s, it is traveling due northeast. What is the magnitude and direction of the net force that must be acting on the car at the instant it is traveling due north? (*Hint:* It really helps if you draw a picture.)

Derivation

N1D.1 Here is how we can prove that equation N1.17 is fully correct as it stands. Consider an object that moves with a constant speed $|\vec{v}|$ in a circle of radius r.
(a) Argue that $|d\theta/dt| \equiv \lim_{\Delta t \to 0} |\Delta\theta/\Delta t|$, the object's angular speed as it goes around the circle, must equal $|\vec{v}|/r$.
(b) The angle between the object's velocity vectors for a given interval Δt will therefore be $|\Delta\theta| = |\vec{v}| \, \Delta t/r$. The change $\Delta\vec{v}$ in the object's velocity is therefore as shown in the diagram below.

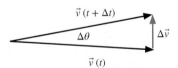

Since these vectors have the same magnitude $|\vec{v}|$, if we put them tail-to-tail like this, their tips mark out two points on a circular arc of radius $|\vec{v}|$. Therefore, the length of the arrow $\Delta\vec{v}$ will be approximately equal to the arclength $|\vec{v}| \, |\Delta\theta|$ along that circular arc. This approximation will become exact in the limit that $|\Delta\theta| \to 0$. Use this to argue that in the limit that $\Delta t \to 0$,

$$|\vec{a}| = \lim_{\Delta t \to 0} \frac{|\Delta\vec{v}|}{\Delta t} = \frac{|\vec{v}|^2}{r} \qquad \text{(N1.18)}$$

N1D.2 Suppose that an object goes at a constant speed once around a circle of radius R in time T. What is the magnitude of the object's acceleration in terms of R and T?

Rich-Context

N1R.1 One of the arguments against the concept that the earth is moving that is described (and refuted) by Galileo is this: Imagine dropping an object from a tall tower. If the earth is moving in a certain direction through space at a speed of many kilometers per second (as is required by the Copernican model), then it will move forward a considerable distance even during the short time before the object hits the ground. This means that since the earth moves forward under the object as it falls, the object will hit the ground at a point kilometers away from the tower in the opposite direction to the earth's motion. Since observations clearly show that an object dropped from a tower hits the ground almost exactly below the drop point, the earth must be at rest. Carefully construct an argument based on momentum flow in a reference frame at rest with respect to the sun to explain why, according to Newton's model, the object will hit the ground directly below the drop point even if the earth *does* move.

N1R.2 Suppose you have discovered a way to create a repulsive force field and you are trying to use it to make a device to protect against car crashes. Your device is mounted in a 1000-kg car and is capable of exerting a constant repulsive force of 40,000 N on another car until the cars come to rest with respect to each other. Generating the force field creates a lot of electric power, though, and you are trying to decide how big a battery you need to use. You want the battery to be able to supply power long enough to bring the cars to rest with respect to each other (not necessarily the road) when the protected car is involved in a head-on collision with a 2000-kg car when both are traveling at 30 m/s on frictionless ice. How long must your battery be able to supply power to the force field? (*Hint:* What is the velocity of the center of mass of the system consisting of the two cars? How is this relevant?)

N1R.3 Suppose you are driving on a large, dark parking lot in the middle of the night. Your headlights suddenly illuminate a wall directly ahead of you that is perpendicular to your direction of travel and stretching away on both sides as far as you can see. To avoid hitting the wall, is it better to turn your car to the right or left without braking or to brake as hard as you can while moving in a straight line toward the wall? (*Hints:* There is a limit to the force the interaction between your car's tires and the road can exert on your car before you begin to skid, and this limit is the same whether you are turning or braking. Also remember Newton's second law.)

ANSWERS TO EXERCISES

N1X.1 At an arbitrary time t, the components of this object's velocity are

$$\vec{v} = \begin{bmatrix} dx/dt \\ dy/dt \\ dz/dt \end{bmatrix} = \begin{bmatrix} 2at \\ -c \\ -2at + c \end{bmatrix} \qquad (N1.19)$$

At $t = 0$, the velocity is $\vec{v}(0) = [0, -4\text{ m/s}, 4\text{ m/s}]$, and the speed is $|\vec{v}| = 5.7$ m/s. At $t = 2.0$ s, the velocity is $\vec{v}(2\text{ s}) = [6\text{ m/s}, -4\text{ m/s}, -2\text{ m/s}]$, and the speed is $|\vec{v}| = 7.5$ m/s.

N1X.2 The ball's change in velocity during this interval is 30 m/s downward, so the ball's average acceleration in this case must be 10 m/s^2 downward.

N2 Forces from Motion

CORE

Chapter Overview

Section N2.1 The Kinematic Chain

According to chapter N2, an object's position, velocity, and acceleration are linked in the following chain of relationships:

$$\vec{r}(t) \; -\begin{bmatrix} \text{time} \\ \text{derivative} \end{bmatrix} \longrightarrow \; \vec{v}(t) \; -\begin{bmatrix} \text{time} \\ \text{derivative} \end{bmatrix} \longrightarrow \; \vec{a}(t) \qquad (\text{N2.1}a)$$

$$\vec{a}(t) \; -\begin{bmatrix} \text{anti} \\ \text{derivative} \end{bmatrix} \longrightarrow \; \vec{v}(t) \; -\begin{bmatrix} \text{anti} \\ \text{derivative} \end{bmatrix} \longrightarrow \; \vec{r}(t) \qquad (\text{N2.1}b)$$

We call this the **kinematic chain**.

Equation N2.1a implies that if we know either $\vec{r}(t)$ or $\vec{v}(t)$, we can determine the object's acceleration. We can then use Newton's second law to determine the net force acting on the object, which in turn allows us to infer things about the characteristics of the individual forces acting on that object. We will focus on this use of Newton's second law in this chapter and chapters N4 through N8. Alternatively, if we know all the forces that act on the object, we can use Newton's second law and equation N2.1b to determine the object's trajectory. We will focus on this approach in chapters N3 and N9 through N12.

Section N2.2: Classification of Forces

We begin by discussing a few tools that are useful for both approaches. This section further develops the force classification scheme first discussed in chapter C1, so that we can talk more precisely about the external forces acting on an object. Figure N2.1 pictorially describes the scheme. Throughout this text, I will denote a force vector by using $\vec{F}$ with a descriptive subscript specified by this classification scheme.

Section N2.3: Force Laws

The long-range gravitational, electrostatic, and ideal spring interactions between two objects exert forces on the objects that obey the following **force laws**:

$$\vec{F}_g = \frac{GMm}{r^2} \;\; \text{(toward the other object)} \qquad (\text{N2.4}a)$$

$$\vec{F}_e = \frac{1}{4\pi\varepsilon_0}\frac{Qq}{r^2} \;\; \text{(away from the other object)} \qquad (\text{N2.4}b)$$

$$\vec{F}_{Sp} = k_s(r - r_0) \;\; \text{(toward the other object)} \qquad (\text{N2.4}c)$$

- **Purpose:** Each of these force laws describes how the force exerted on an object by a certain interaction varies with the separation r between the interacting objects. $\vec{F}_g$ is the gravitational force, $G = 6.67 \times 10^{-11}$ N·m^2/kg^2 is the **universal gravitational constant**, and M and m are the objects' masses. $\vec{F}_e$ is the electrostatic force, $1/4\pi\varepsilon_0 = 8.99 \times 10^9$ N·m^2/C^2 is the **Coulomb constant**, and Q and q are the objects' charges. $\vec{F}_{Sp}$ is the spring force, k_s is the spring's **spring constant**, and r_0 is the separation between objects when the spring is relaxed.

- **Limitations:** The first two formulas apply only if each object is either a particle or spherically symmetric object.
- **Notes:** Physicists call these force laws **Newton's law of universal gravitation**, **Coulomb's law**, and **Hooke's law**, respectively.

One can derive these laws from the interactions' potential energy formulas as discussed in the section.

Section N2.4: Free-Body Diagrams

A **free-body diagram** is a tool for visually representing the external forces acting on an object. The diagram displays the object in question (abstracted from its surroundings), the external forces acting on it (depicted as labeled arrows), perhaps some coordinate axes, and (to avoid confusion) *nothing else*. The force arrows are drawn so their tails start at the object's center of mass. As we will see throughout the unit, such a visual representation makes it much easier to understand what goes into the left side of Newton's second law.

Section N2.5: Motion Diagrams

Now we turn to two tools that will specifically help us determine an object's acceleration from some kind of a description of its motion. A **motion diagram** is a tool for doing this that includes

1. A single image of the object in question.
2. A set of dots that represent the positions of the object's center of mass at sequential instants of time a constant interval Δt apart.
3. Arrows between adjacent dots that represent average velocities $\vec{v}\Delta t$.
4. Acceleration arrows centered on each dot (except the first and last).

One constructs an acceleration arrow by constructing the difference between two adjacent average velocity arrows. This arrow closely approximates $\vec{a}\Delta t^2$ as the object passes the dot between the velocity arrows. As long as Δt is fixed, the arrows $\vec{v}\Delta t$ and $\vec{a}\Delta t^2$ are proportional to, and thus visually represent, the object's velocity and acceleration at selected instants. We can use such diagrams in cases of one-dimensional motion but they are especially useful in cases of two-dimensional motion.

Section N2.6: Graphs of One-Dimensional Motion

In most cases where an object moves in one dimension, it is easiest to analyze *graphs* of the object's motion. When an object moves along a straight line (which we can define to be the x axis), its position, velocity, and acceleration are completely described by its x components $x(t)$, $v_x(t)$, and $a_x(t)$ (which are simple signed numbers). From a graph of $a_x(t)$, we can infer how the net force on an object changes with time.

Conventionally, we stack such graphs vertically so that we always place a graph of $x(t)$ above a graph of $v_x(t)$, which we in turn always place above a graph of $a_x(t)$. Given this convention, we can always construct a given graph from the graph above it by remembering that at any given instant of time, the *slope* on the upper graph in the pair corresponds to the value displayed on the lower graph: *the slope above equals the value below*.

Section N2.7: A Quantitative Example

The example in this section pulls many of the tools in this chapter together to analyze the forces acting on a car as it goes over the top of a hill.

N2.1   The Kinematic Chain

As we saw in chapter N1, an object's position $\vec{r}(t)$, its velocity $\vec{v}(t)$, and its acceleration $\vec{a}(t)$ as functions of time are connected by the following chain of derivative relationships:

The kinematic chain

$$\vec{r}(t) - \left[\begin{array}{c} \text{time} \\ \text{derivative} \end{array}\right] \longrightarrow \vec{v}(t) - \left[\begin{array}{c} \text{time} \\ \text{derivative} \end{array}\right] \longrightarrow \vec{a}(t) \qquad (\text{N2.1}a)$$

We can also follow the chain in the reverse direction:

$$\vec{a}(t) - \left[\begin{array}{c} \text{anti} \\ \text{derivative} \end{array}\right] \longrightarrow \vec{v}(t) - \left[\begin{array}{c} \text{anti} \\ \text{derivative} \end{array}\right] \longrightarrow \vec{r}(t) \qquad (\text{N2.1}b)$$

We call this chain of relationships the **kinematic chain** (note that **kinematics** is the mathematical study of motion).

Finding forces from motion: an overview

The forward and reverse directions of this chain correspond to two fundamentally different ways to apply Newton's second law. In many situations, one or more forces may act on an object whose magnitude and/or direction we do not know. But if we know how an object *moves* (that is, we know either its position or velocity as a function of time), we can follow the chain to compute the object's *acceleration* as a function of time. Newton's second law links that acceleration to the *net force* acting on the object, and knowing the net force is often sufficient to determine the unknown forces. This is how we determine *forces from motion*. We will talk about this in general terms in the remainder of this chapter and in more detail in chapters N4 through N8.

Finding motion from forces

On the other hand, note that if we know *all* the forces acting on an object, then Newton's second law tells us that object's acceleration, and following the kinematic chain in the reverse direction allows us to determine the object's trajectory. This is how we determine *motion from forces*. We will talk about this in fairly general terms in chapter N3, and explore more detailed applications in chapters N9 through N12.

N2.2   Classification of Forces

Review of the basic macroscopic force categories

In both approaches, having a clear and agreed-upon scheme for describing and naming forces that act on an object will help us greatly.

Forces ultimately reflect *interactions*. As discussed in chapter C1, it is helpful at the macroscopic level to divide interactions into **long-range interactions** (gravitational and electromagnetic interactions) that can act between objects even when they are physically separated, and **contact interactions** that arise from the microscopic electromagnetic interactions between the objects' atoms when the objects are in physical contact.

Subcategories of long-range forces

Figure N2.1 shows a useful scheme for further subdividing these categories. Let me make a few brief comments about this scheme. First, it is important to understand that *all* known long-range interactions are either gravitational or electromagnetic: there are no other possibilities. While *every* object exerts a gravitational force on every other object, unless at least one object in the interacting pair has a mass of planetary proportions, gravitational forces will be utterly negligible. Loosely speaking, an electromagnetic interaction is *electrostatic* if it involves at least one electrically charged object, and *magnetic* if it involves at least one magnet or flowing electric current. We will explore the distinction between these categories more fully in unit E.

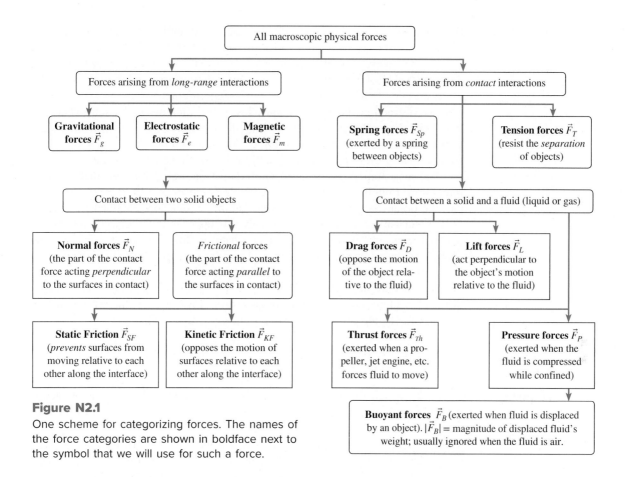

Figure N2.1
One scheme for categorizing forces. The names of the force categories are shown in boldface next to the symbol that we will use for such a force.

All forces that are *not* long-range gravitational, electrostatic, or magnetic forces arise from contact interactions. The subcategories of contact forces, however, are neither totally exhaustive nor even completely logical: these are simply categories that people have found *useful* in common situations. Broadly speaking, contact forces can be divided into two subcategories: *tension* forces, where the atoms (for example, in a string, cable, or chain) resist being pulled apart, and *other* contact forces. I have set aside the force exerted by a spring as a third subcategory because it can be either a compression force (if the spring is squeezed) *or* a tension force (if it is stretched).

The primary distinction between *normal* and *frictional* compression forces between two solid objects is the direction in which they act. In a certain sense, the basic compression force exerted by the contact interaction on either solid is divided somewhat artificially into a part that is *perpendicular* to the interface between the solids (the normal force) and a part that is *parallel* to this interface (the frictional force). (Remember that *normal* here does not mean "typical" but "perpendicular.") A similar distinction distinguishes *lift* forces from *drag* forces on an object moving relative to a fluid: a lift force is the part of the force on the object that acts *perpendicular* to its motion relative to the fluid, and a drag force is the part that acts *opposite* the object's motion.

One may also conveniently divide the frictional forces acting on solid objects in contact into two further categories depending on whether the interacting surfaces are actually sliding relative to each other (kinetic friction) or not (static friction). A *static* friction force acts to *prevent* the surfaces from starting to slide. For example, suppose you push horizontally on a crate, but it doesn't move. An object at rest is not accelerating, so by Newton's second law it must have zero net force on it, and therefore *some* force must be opposing your push. What is happening is that the crate and

Subcategories of contact forces

The distinction between normal and frictional forces

floor atoms are intertwined and locked together, and until you supply enough force to break these bonds, they will oppose any sideward push you might apply.

If it bothers you to call such a force a "friction" force (because friction to you connotes "rubbing" and thus relative motion of the surfaces), think of the subscript SF as meaning "sticking force" instead of "static friction."

The distinctions between the varieties of fluid forces can be somewhat artificial. (For example, should we think of the force exerted by the air on a helicopter rotor a *lift* force or a *thrust* force? A case could be made for either!) Even so, these categories are reasonably helpful in most common contexts.

The most important things to know about *buoyant forces* are that (1) such forces arise from height-dependent pressure *differences* in a fluid body (for example, the ocean or the atmosphere) that has been somehow confined by gravity; (2) the buoyant force on an object usually acts opposite to the object's weight and is equal in magnitude to the weight of whatever fluid the object displaces; and (3) buoyant forces on an object in air are typically so small compared to the object's weight that we ignore them unless *explicitly* stated.

Note that in this text, I will *always* use the symbol $\vec{F}$ for a force (this sharply distinguishes forces from other vector quantities). I will indicate the various categories of force by attaching the subscripts shown in figure N2.1.

The categories can be somewhat fuzzy

Buoyancy

Conventional notation for forces

N2.3   Force Laws

How to infer a *force law* from a potential energy formula

In unit C, we saw that the gravitational interaction and the electrostatic interaction have associated potential energy functions $V(r)$. We can infer a **force law** for each interaction from its potential energy function as follows.

First, note that in chapter C7, we saw that the principle that the laws of physics are independent of one's orientation in space implies that the forces that a long-range interaction between two particles can exert on those particles must be directed along the line between them, that is, the force on either must point either directly toward or directly away from the other particle. (If you did not read chapter C7, the first three paragraphs of section C7.7 provide this argument, which does not depend on anything else in chapter C7.)

In chapter C9, we also saw that force exerted on a particle moving subject to a one-dimensional potential energy function $V(x)$ was given by

$$F_x = -\frac{dV}{dx} \tag{N2.2}$$

So, consider a particle of mass m that is interacting gravitationally with another particle of mass M. At any instant, we can define the x axis so that the two particles lie on the x axis, with the other particle at the origin and the particle we are interested in at some x-position $x > 0$. This means that the particles' separation is $r = x$, and the potential energy for the interaction between the particles is therefore $V(x) = -GMm/r = -GMm/x$. This in turn means that the x component of the force acting on that particle must be

$$F_{g,x} = -\frac{dV_g}{dx} = -\frac{d}{dx}\left(-\frac{GMm}{x}\right) = -\frac{GMm}{x^2} \tag{N2.3}$$

Note that this force points in the $-x$ direction, which is *toward* the other particle. We know that the other components of this force must be zero by the argument in chapter C7, so we can say somewhat more abstractly that the gravitational force vector on our particle of interest must be

$$\vec{F}_g = \frac{GMm}{r^2} \text{ (toward the other particle)}$$

This is **Newton's law of universal gravitation**.

We similarly derive force laws for the electrostatic interaction and the ideal spring "interaction" (where we treat the spring connecting two objects as if it were a long-range interaction between them). To summarize,

$$\vec{F}_g = \frac{GMm}{r^2} \text{ (toward the other object)} \qquad (N2.4a)$$

$$\vec{F}_e = \frac{1}{4\pi\varepsilon_0} \frac{Qq}{r^2} \text{ (away from the other object)} \qquad (N2.4b)$$

$$\vec{F}_{Sp} = k_s(r - r_0) \text{ (toward the other object)} \qquad (N2.4c)$$

- **Purpose:** Each of these force laws describes how the force exerted on an object by a given interaction varies with the separation r between the interacting objects. $\vec{F}_g$ is the gravitational force, $G = 6.67 \times 10^{-11}$ N·m²/kg² is the universal gravitational constant, and M and m are the objects' masses. $\vec{F}_e$ is the electrostatic force, $1/4\pi\varepsilon_0 = 8.99 \times 10^9$ N·m²/C² is the Coulomb constant, and Q and q are the objects' charges. $\vec{F}_{Sp}$ is the spring force, k_s is the spring's spring constant, and r_0 is the separation between objects when the spring is relaxed.
- **Limitations:** The first two formulas apply only if each object is either a particle or spherically symmetric object.
- **Notes:** Physicists call these force laws **Newton's law of universal gravitation**, **Coulomb's law**, and **Hooke's law**, respectively.

Note that the right side of each equation is indeed a vector, because the right side specifies a magnitude and a direction. Note also that if Qq is negative (meaning that the charges have opposite signs), the direction of $\vec{F}_e$ is −(away from the other object) = toward the other object. Similarly, when $r < r_0$, the direction of $\vec{F}_{Sp}$ is −(toward the other object) = away from the other object.

Since contact interactions (with the exception of the spring interaction) are not described by potential energy functions, we cannot find nice force laws for contact interactions. Determining the forces exerted by such interactions is one of the main goals of the methods we will develop in this chapter!

Exercise N2X.1

Derive equations N2.4b and N2.4c from the potential energy formulas $V_e(r) = (1/4\pi\varepsilon_0)(Qq/r)$ and $V_{Sp}(r) = \frac{1}{2}k_s(r - r_0)^2$, respectively. (We introduced these potential energy formulas in unit C.)

N2.4   Free-Body Diagrams

Another essential tool, *especially* when we seek to infer forces by observing motions, is the *free-body diagram*. Newton's second law says that an object's acceleration is determined by external forces acting on it. A **free-body diagram** is a graphical tool that helps us name and display these external forces. Drawing such a diagram is a very useful first step in almost any problem involving Newton's second law: it helps us clarify for ourselves exactly what forces act on the object and how they are oriented relative to one another.

To draw a free-body diagram of an object, we do the following:

1. First, ***imagine the object in its context***, thinking of the things that interact with it. Be especially careful to look along the object's boundary in your mental image for places where it touches things outside itself.

2. Then ***draw a sketch of the object alone***, as if it were floating in space. Part of the diagram's purpose is to direct our attention to the object itself, reducing its surroundings merely to abstract forces that act on it.

3. ***Draw a dot*** representing the object's center of mass and then (for each external interaction acting on the object) draw an arrow (starting at the dot) that shows the magnitude and direction of force exerted by that interaction. ***Label each arrow*** with an appropriate symbol.

General do's and don'ts

Here are some general "do's and don'ts" about drawing free-body diagrams:

1. *Do not draw any arrows on the diagram depicting quantities that are* NOT *forces*. The main point of a free-body diagram is to display the *forces* that act on an object. Do not confuse the issue by also drawing arrows displaying the object's velocity or acceleration.

2. *Draw arrows for* ONLY *those forces that act directly on the object in question.* As stated above, part of the purpose of a free-body diagram is to direct our attention to a *single* object and display the forces acting on it. Do not confuse the issue by drawing forces that act on other objects.

3. *Make sure that every force arrow you draw reflects an interaction with some other object.* A common error is to draw arrows on a free-body diagram for alleged forces associated with an object's velocity and/or acceleration that have no place in the Newtonian model. *Every* force on an object must arise from an interaction that involves some *other* object.

Example N2.1

Problem: Draw a free-body diagram of a skier sliding at a constant velocity down a hill.

Solution A diagram of the skier in context (a) and the corresponding free-body diagram (b) appear below.

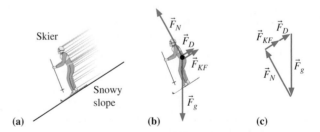

Comments: The skier participates in a gravitational interaction with the earth, which exerts a downward force $\vec{F}_g$ on the skier. The skier also participates in contact interactions with the snow and the air. The force due to the interaction with the snow can be divided into normal and frictional parts $\vec{F}_N$ and $\vec{F}_{KF}$ (the friction force is a *kinetic* friction force because the skier is moving relative to the slope).

The interaction with the air exerts a buoyant force $\vec{F}_B$ and a drag force $\vec{F}_D$ (we will ignore the former). There is nothing else that the skier is in contact with, so these will be the only forces that act.

Note that if the skier's velocity is *constant,* the skier's acceleration is zero. Newton's second law then implies that the sum of the external forces acting on the skier must be zero. Therefore, we should draw the force arrows on the diagram in such a way that they add to zero, as demonstrated in part (c).

Exercise N2X.2

Draw a free-body diagram of a boat moving at a constant velocity on a lake.

N2.5   Motion Diagrams

When we use Newton's second law to infer something about the forces acting on an object, knowing the magnitude and direction of the object's acceleration is essential. A *motion diagram* is a powerful tool that not only vividly illustrates an object's motion in one or two dimensions, but also helps us determine both the magnitude and direction of the object's acceleration.

A **motion diagram** is (1) a sequence of dots that represents the object's position at successive, equally spaced instants of time, and (2) a set of arrows constructed from those dots that represent the object's velocity and acceleration at successive instants of time. We already implicitly used motion diagrams in chapter N1, but here I want to discuss this tool in greater depth.

Before drawing such a diagram, I recommend *visualizing* what a "multiflash photograph" of the object's motion might look like. A **multiflash photograph** shows images of the object at successive equally spaced instants of time (as if the camera's flash fired at those instants). Figure N2.2a shows what a multiflash photograph of a braking car might look like. Note as the car slows down, its displacement in a given interval of time decreases, so the distances between successive images decrease.

The first step in drawing the actual diagram is to draw one image of the object (to indicate what object we are considering and how it is oriented) and then a sequence of dots representing the positions of the object's center of mass. Start the sequence with the first dot at the object's center of mass if you can (see figure N2.2b), but if this will be messy or too hard to interpret, you can put the dots below or to the side (figure N2.2c). Number the first dot 0 (because it occurs at time $t = 0$) and then number the others in sequence.

Why motion diagrams are useful

Steps in drawing a motion diagram

Visualize this:

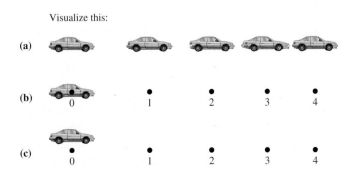

Figure N2.2

The first step in drawing a motion diagram of a braking car: (a) *Visualize* how a "multiflash photograph" of the object's motion would look in this particular situation.
(b) Actually draw one image of the object and then draw dots to indicate the position of the object's center of mass at successive equally spaced instants of time.
(c) One might place the first dot *below* the center of mass (and subsequent dots aligned with the first) if that improves the diagram's clarity for some reason.

Figure N2.3

(a) How to construct and label the velocity arrows on our motion diagram for a braking automobile.

(b) We can construct acceleration arrows at each point (except the end points) by constructing the *differences* between adjacent pairs of velocity arrows; for example, $\vec{a}_3\Delta t^2 = \vec{v}_{34}\Delta t - \vec{v}_{23}\Delta t$. (The lower labels on the transported velocity arrows are optional.)

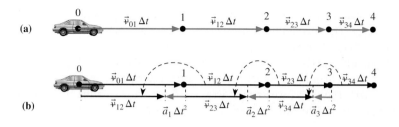

Drawing velocity arrows

Drawing acceleration arrows

The next major step is to construct velocity and acceleration arrows. Figure N2.3 illustrates the process as a sequence of smaller steps. We first draw an arrow from each dot to the next dot, as shown in figure N2.3a. These arrows represent the object's displacement during the time intervals between dots. But by the definition of average velocity, if the time interval between dots is Δt, then the displacement $\Delta\vec{r}_{01}$ in meters between dots 0 and 1 (for example) is related to the average velocity $\vec{v}_{01}$ during the interval between instants 1 and 2 by the expression $\vec{v}_{01} = \Delta\vec{r}_{01}/\Delta t$, implying that $\Delta\vec{r}_{01} = \vec{v}_{01}\Delta t$. The same reasoning applies to other pairs of dots. Therefore, we label the arrows we have drawn $\vec{v}_{01}\Delta t$, $\vec{v}_{12}\Delta t$, and so on, as shown. Since each arrow is proportional to the average velocity between dots, the sequence of arrows vividly depicts the object's changing velocity as it moves.

Indeed, the object's average velocity $\vec{v}_{01}$ between instants 0 and 1 best approximates the object's *instantaneous* velocity halfway between those instants, at an instant we might label "0.5." Similarly, the object's average velocity $\vec{v}_{12}$ is a good approximation for the instantaneous velocity at instant "1.5." So the velocity arrows we have drawn also approximately depict the directions and relative magnitudes of the object's *instantaneous* velocities at instants 0.5, 1.5, and so on.

Now, since instants 0.5 and 1.5 are also separated by a time interval Δt, the object's average *acceleration* between instants 0.5 and 1.5 is

$$\vec{a}_{0.5,1.5} = \frac{\Delta\vec{v}}{\Delta t} = \frac{\vec{v}(1.5) - \vec{v}(0.5)}{\Delta t} \approx \frac{\vec{v}_{12} - \vec{v}_{01}}{\Delta t} \tag{N2.5}$$

The average acceleration during this interval is closest to the object's *instantaneous* acceleration at the interval's midpoint, that is, at instant 1. So if we substitute this into the above and multiply both sides by Δt^2, we see that

$$\vec{a}_1\Delta t^2 \approx \vec{v}_{12}\Delta t - \vec{v}_{01}\Delta t \tag{N2.6}$$

This equation says that we can construct an arrow equal to $\vec{a}_1\Delta t^2$, where $\vec{a}_1$ is the object's *instantaneous* acceleration as it passes point 1 by constructing the difference between the average velocity arrows $\vec{v}_{12}\Delta t$ and $\vec{v}_{01}\Delta t$ that we drew for the flanking intervals. Similarly, $\vec{a}_2\Delta t^2 \approx \vec{v}_{23}\Delta t - \vec{v}_{12}\Delta t$ and so on. We can straightforwardly construct these acceleration arrows by moving the second velocity arrow back so its tail coincides with the first velocity arrow and then constructing the arrow representing their difference, as shown in figure N2.3b (note, for example, that the construction ensures that $\vec{v}_{12}\Delta t = \vec{a}_1\Delta t^2 + \vec{v}_{01}\Delta t$). This construction method naturally puts the tail of the arrow for $\vec{a}_1\Delta t^2$ at point 1, the tail for arrow $\vec{a}_2\Delta t^2$ at point 2 and so on. Drawing these acceleration arrows completes the motion diagram.

Note that such a diagram starts with a straightforward depiction of the object's motion and ends with arrows that (though approximate) both qualitatively and quantitatively represent the object's acceleration at successive instants of time. Knowing the object's acceleration, we can use Newton's second law to infer what the net force acting on the object must be. In the particular case of the braking car shown in figure N2.3, we can see from the motion diagram that the car's acceleration is toward the left (as we might expect for a car that is slowing down). This means that whatever forces act on the car, the *net* force must act toward the left.

So, to summarize, a motion diagram consists of the following elements:

A summary of items to appear in a motion diagram

1. A single image of the object in question.
2. A set of dots representing the position of the object's center of mass at successive, equally spaced instants of time.
3. (Average) velocity arrows drawn between adjacent dots.
4. Acceleration arrows attached to each dot (except the first and last).

Labels on these arrows are not strictly necessary, but can help with clarity.

Motion diagrams for objects that move in two dimensions are often even more helpful than those for objects moving in one dimension, because the magnitude and direction of the object's acceleration is often harder to intuit in the two-dimensional case. The example below illustrates how to construct a two-dimensional motion diagram and how to use that diagram to make *quantitative* statements about an object's acceleration.

Problem: Suppose the dots shown below show the actual-size center-of-mass positions at 0.1-s intervals for a puck sliding on a tilted surface. Construct the acceleration arrows at points 2, 3, and 4 and calculate the magnitude of the puck's acceleration at each of those points.

Example N2.2

2
•
                                                3
                                                •
1
•

                                                        4
                                                        •

0
•

Solution The drawing below shows the completed construction. The dashed arrows are velocity arrows that have been transported to new positions.

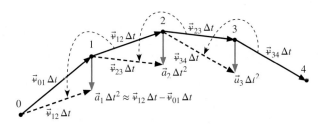

The constructed acceleration arrows (the colored arrows) all point downward, and I measure them to have a common length of 1.0 cm. Therefore $|\vec{a}_i|\Delta t^2 = 1.0$ cm $= 0.010$ m (where $i = 1, 2,$ or 3), meaning that

$$|\vec{a}_i| = \frac{0.010 \text{ m}}{\Delta t^2} = \frac{0.010 \text{ m}}{(0.1 \text{ s})^2} = 1.0 \frac{\text{m}}{\text{s}^2} \qquad (\text{N2.7})$$

Therefore, the puck's acceleration is 1.0 m/s^2 downward at each point.

N2.6   Graphs of One-Dimensional Motion

In the case of one-dimensional motion, vectors are represented by signed numbers

In cases where an object's motion is confined to a line, a different tool can be more useful. We can generally choose to orient our reference frame so that the line of motion coincides with the x axis (or we might choose the z axis if the motion is vertical). If we do this, then only one component (say, the x component) of an object's position vector $\vec{r}(t)$ is ever nonzero, and this in turn means that only the x component of its velocity vector $\vec{v}(t)$ is nonzero:

$$\vec{r}(t) = \begin{bmatrix} x(t) \\ y(t) \\ z(t) \end{bmatrix} \Rightarrow \vec{v}(t) \equiv \frac{d\vec{r}}{dt} = \begin{bmatrix} dx/dt \\ d(0)/dt \\ d(0)/dt \end{bmatrix} = \begin{bmatrix} dx/dt \\ 0 \\ 0 \end{bmatrix} \Rightarrow \vec{a}(t) = \begin{bmatrix} dv_x/dt \\ 0 \\ 0 \end{bmatrix} \qquad \text{(N2.8)}$$

Therefore, in this very special case of one-dimensional motion (and *only* in this case), we can represent an object's position, velocity, and acceleration vectors each by a *single signed number* we take to be a function of time (these numbers are $x(t)$, $v_x(t)$, and $a_x(t)$, respectively). These signed numbers provide all we need to know about the vectors $\vec{r}(t)$, $\vec{v}(t)$, and $\vec{a}(t)$ in this case: the *absolute value* of the number is the same as the *magnitude* of the vector (for example, $|\vec{a}| = |a_x|$), and the *sign* of the number tells us the *direction* of the corresponding vector (a plus or minus sign indicates that the vector points in the $+x$ or $-x$ direction, respectively).

Constructing a graph of $a_x(t)$ from one of $v_x(t)$

Now, the fact that $a_x(t) = dv_x/dt$ means that (as discussed in appendix NA), the *value* of the object's x-acceleration at every instant of time will correspond to the *slope* of $v_x(t)$ when the latter is plotted versus t on a graph. So, given a graph of $v_x(t)$, in principle we can construct a graph of $a_x(t)$. Figure N2.4a illustrates the process of constructing such a graph.

Constructing a graph of $v_x(t)$ from one of $x(t)$

Similarly, because $v_x = dx/dt$, the *value* of v_x at every instant of time will correspond to the *slope* of a graph of $x(t)$ versus t (see figure N2.4b). It is conventional when drawing pairs of graphs like these to put the graph of $x(t)$ above the paired graph of $v_x(t)$ and a graph of $v_x(t)$ above a paired graph of $a_x(t)$ so that the slope of the upper graph always corresponds to the value on the lower graph (always remember: *slope* above equals *value* below).

Keep components and magnitudes distinct!

In this text, to keep the distinction between the components and the magnitude of a vector clear, I will *always* use the symbols x, v_x, and a_x to refer to these signed numbers (because they really are just components of the corresponding vectors), and $|\vec{r}|, |\vec{v}|$, and $|\vec{a}|$ for the magnitudes of the corresponding vectors, even though

Figure N2.4

(a) How to construct a graph of $a_x(t)$ from a graph of $v_x(t)$ for a car speeding up to and then holding at a certain cruising speed. Each little black line segment on the top graph indicates the *slope* of that graph of $v_x(t)$ at a certain instant, while the circle on the bottom graph indicates the value of a_x at that instant. Note that at first the car's velocity is steadily increasing, so its slope is a constant positive number. When the car reaches its cruising speed, the slope quickly falls to zero and remains zero. (b) How to construct a graph of $v_x(t)$ from a graph of $x(t)$ using the same technique.

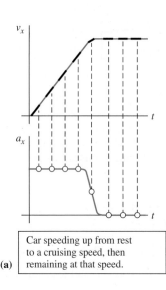

(a) | Car speeding up from rest to a cruising speed, then remaining at that speed.

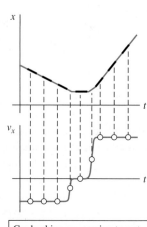

(b) | Car backing up, coming to rest for a moment, then moving forward at a faster rate.

physicists commonly use the simpler symbols r, v, and a as meaning either components or vector magnitudes. I think this is confusing, so I am going to use the more unambiguous notations described above, and I urge you to do likewise.

According to Newton's second law, an object's acceleration is proportional to the net force acting on the object, so a graph of $a_x(t)$ essentially tells us how $F_{net,x}$ depends on time. This can give us insight into the forces acting on the object, as example N2.3 illustrates.

Inferring forces

Problem: Suppose you hold a basketball some distance above the floor and then release it from rest. It falls toward the floor, hits the floor, and then rebounds upward at about the same speed. Draw graphs of the ball's vertical velocity and acceleration components as functions of time.

Solution The first thing we have to do is define our reference frame. Let us define our z axis to be vertically upward and take $z = 0$ to correspond to the floor. (In the case of vertical motion, it is more natural to align the z axis with the object's motions than to use the x axis.) Let us also define $t = 0$ to be the instant when the ball is dropped.

According to the description of the situation, the ball is released from rest ($v_z = 0$). As the ball falls, its unopposed gravitational interaction with the earth transfers downward momentum to the ball at a constant rate, and thus the ball's z-velocity should steadily become more negative as time increases. When the ball hits the floor, its z-velocity rapidly changes from being negative to positive (downward to upward). After the ball leaves the floor, its z-velocity again becomes steadily more negative.

From this information, we can construct the graph of v_z versus t shown in figure N2.5a. Then, we can use the methods discussed in this section to construct the graphs of $a_z(t)$ below this graph (figure N2.5b). Note how in each case the slope of the upper graph at a given t is equal to the value of the graph below it (as indicated at selected places).

We see from figure N2.5b that the net force on the ball when it is *not* in contact with the floor must be constant and downward: this is the constant force of gravity. When the ball is in contact with the floor, however, the graph indicates that the net force on the ball must become very large and upward. The gravitational force on the ball always has a constant (downward) value, so we must have a new upward force acting on the ball. This must be a normal force arising from the ball's interaction with the ground. The graph clearly indicates that this force must be larger than the gravitational force, and indeed, the briefer the bounce is, the larger this normal force has to be.

Example N2.3

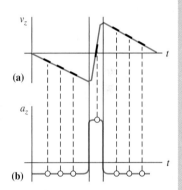

Figure N2.5
(a) A graph of $v_z(t)$ and (b) a graph of $a_z(t)$ for a basketball dropped on the floor. The ball is in contact with the floor during the time interval bracketed by the thin vertical lines.

Drawing such sets of graphs is a good way to visualize (and thus think more carefully about) the time dependence of v_x and a_x, so this is a skill you should practice and master. Note that your graphs need not be *quantitatively* accurate: the important thing is that they be *qualitatively* accurate.

Exercise N2X.3

A car traveling along a straight road at a constant speed suddenly brakes to a stop to avoid an animal and remains at rest thereafter. Draw graphs of $v_x(t)$ and $a_x(t)$ for this situation.

N2.7   A Quantitative Example

The following example illustrates how we might combine some of the tools we have discussed in this chapter (and also chapter N1) to answer some basic questions about a car going over a hill.

Example N2.4

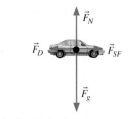

(a) Free-body diagram

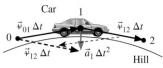

(b) Motion diagram

Problem: Suppose that a car traveling at a constant speed of 30 m/s goes over the top of a hill, and suppose that (near the top of the hill, anyway) the road curves vertically as if it were a part of a circle whose radius is 450 m. Is there a forward force on the car, and if so what is the source of this force? (Do not neglect air friction.) Also, what is the ratio of the magnitudes of the total normal force and gravitational forces acting on the car in this situation?

Solution The car touches the road and the air and also interacts gravitationally with the earth. The free-body diagram to the left illustrates that the car must experience a downward gravitational force from the earth, an upward normal force from the road, and a backward drag force from the air. Now, one only needs to quickly sketch the car's motion diagram (see the lower left) to see that the car's acceleration must be exactly downward at the top of the hill if the car's speed is constant (because $|\vec{v}_{01}| = |\vec{v}_{12}|$). Therefore, the total external force on the car must also be directly downward according to Newton's second law. Since we can have no backward component to the total force, there *must* be a forward external force acting on the car that cancels the rearward force drag force $\vec{F}_D$. What could this force be? We know that this has something to do with the car's engine, but the engine is *inside* the car, so it cannot exert an *external* force on the car! The only interaction that can exert an external forward force on the car is the car's contact interaction with the *road*.

What happens is this: While the engine cannot directly exert a forward force on the car, it *can* turn the wheels against the road, and doing this exerts a *backward* force on the road. Newton's third law then tells us that if the contact interaction between the tires and the road exerts a backward force on the road, it must also exert a forward force of equal magnitude on the tires and thus the car. Since the tires and road are at rest with respect to each other at the point where they touch (if we assume that the car is not skidding), this must be a *static friction force*, as I have labeled it in the free-body diagram.

Now let's consider the vertical aspect of the free-body diagram. To explain the car's downward acceleration, we must have a net downward force on the car. The diagram makes it clear that the magnitude of this force must be $|\vec{F}_{net}| = |\vec{F}_g| - |\vec{F}_N| = m|\vec{g}| - |\vec{F}_N|$. If the car's motion is nearly circular in this case, the magnitude of the car's acceleration is $|\vec{a}| = |\vec{v}|^2/R$, where $|\vec{v}| = 30$ m/s and $R = 450$ m. Newton's second law tells that $\vec{F}_{net} = m\vec{a}$, so we have

$$m|\vec{g}| - |\vec{F}_N| = |\vec{F}_{net}| = m|\vec{a}| \Rightarrow |\vec{F}_N| = m|\vec{g}| - m\vec{a} \Rightarrow \frac{|\vec{F}_N|}{|\vec{F}_g|} = \frac{m|\vec{g}| - m|\vec{a}|}{m|\vec{g}|}$$

$$\Rightarrow \frac{|\vec{F}_N|}{|\vec{F}_g|} = 1 - \frac{|\vec{a}|}{|\vec{g}|} = 1 - \frac{|\vec{v}|^2}{R|\vec{g}|} = 1 - \frac{(30 \text{ m/s})^2}{(450 \text{ m})(9.8 \text{ m/s}^2)} = 0.80 \qquad \text{(N2.9)}$$

So the normal force acting on the car must be only 80% as strong as the gravitational force on the car as it goes over the hill. Note that the units work out and the answer is plausible (not greater than 1, for example). Note also that a downward acceleration is consistent with the formula for uniform circular motion, which specifies an acceleration toward the circle's center.

TWO-MINUTE PROBLEMS

N2T.1 When a baseball pitcher throws a curveball, he or she puts top-spin on the ball. This causes the ball to interact with the passing air in such a way as to make the air push the ball downward perpendicular to the ball's velocity. (This causes the ball to dive sharply as it approaches the plate, making it quite difficult to hit.) We would classify this force as a "lift" force on the ball. T or F?

N2T.2 Suppose that a comet goes around the sun in an elliptical orbit such that the comet at the farthest point in its orbit is 3 times as far from the sun as it is when it is at its closest point. How many times stronger is the gravitational force on the comet due to the sun when it is at its closest point than when it is at its farthest?
A. $|\vec{F}_g|$ is the same at both positions.
B. $\sqrt{3}$ times stronger
C. 3 times stronger
D. 9 times stronger
E. 27 times stronger
F. The answer depends on the relative masses of the comet and the sun.

N2T.3 Suppose an object is hanging from the end of a spring. Let the spring's length when the object is hanging at rest be L. Now suppose we pull the object downward until the spring has been stretched a length of $2L$. By what factor has the force that the spring is exerting on the object increased?
A. The spring force is the same in both cases.
B. $\sqrt{2}$ times stronger
C. 2 times stronger
D. 4 times stronger
E. The answer depends on the spring's relaxed length.

N2T.4 A crate sits on the ground. You push as hard as you can on it, but you cannot move it. At any given time when you are pushing, what is the magnitude of the static friction force exerted on the crate by its contact interaction with the ground compared to the magnitude of your push (which is a *normal* force)? (*Hints:* What is the crate's acceleration? Draw a free-body diagram.)
A. $|\vec{F}_{SF}| < |\vec{F}_N|$
B. $|\vec{F}_{SF}| = |\vec{F}_N|$
C. $|\vec{F}_{SF}| > |\vec{F}_N|$
D. $|\vec{F}_{SF}| = 0$!
E. We do not have enough information to answer.

N2T.5 A box sits at rest on an inclined plank. How do the magnitudes of the normal force and the gravitational force exerted on the box compare? (*Hints:* What is the box's acceleration? Draw a free-body diagram!)
A. $|\vec{F}_N| < |\vec{F}_g|$
B. $|\vec{F}_N| = |\vec{F}_g|$
C. $|\vec{F}_N| > |\vec{F}_g|$
D. $|\vec{F}_N| = 0$!
E. We do not have enough information to answer.

N2T.6 A boat hits a sandbar and slides some distance before coming to rest. Which of the arrows shown below best represents the direction of the boat's acceleration as it is sliding? (*Hint:* Draw a motion diagram.)

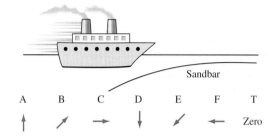

N2T.7 A car moving at a constant speed travels past a valley in the road, as shown below. Which of the arrows shown most closely approximates the direction of the car's acceleration at the instant it is at the position shown? (*Hint:* Draw a motion diagram.)

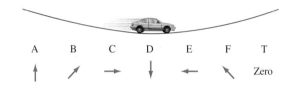

N2T.8 A bike (shown in a top view in the diagram) travels around a curve with its brakes on, so that it is constantly slowing down. Which of the arrows shown most closely approximates the direction of the bike's acceleration at the instant it is at the position shown? (*Hint:* Draw a motion diagram.)

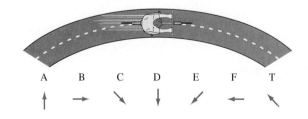

N2T.9 A car passes a dip in the road, going first down, then up. At the very bottom of the dip, when the car's instantaneous velocity is passing through horizontal, how does the magnitude of the normal force on the car compare to the magnitude of the car's weight?
A. $|\vec{F}_N| < |\vec{F}_g|$
B. $|\vec{F}_N| = |\vec{F}_g|$
C. $|\vec{F}_N| > |\vec{F}_g|$
D. $|\vec{F}_N| = 0$!
E. We do not have enough information to answer.

N2T.10 The drawing below is supposed to be a free-body diagram of a box that sits without slipping on the back of a truck that is moving to the right but is slowing down. Is the diagram correct?

A. Yes.
B. No, $\vec{F}_{SF}$ should point leftward.
C. No, the $\vec{F}_{SF}$ label should be $\vec{F}_{KF}$.
D. No, there should be a leftward drag force.
E. No, $|\vec{F}_N|$ should not be equal to $|\vec{F}_g|$.
F. No, there is some other problem (specify).

N2T.11 The drawing shown is supposed to be a free-body diagram of a crate that is being lowered by a crane and is speeding up as it is being lowered. Is the diagram correct? (Ignore air resistance.)

A. Yes.
B. No, $\vec{F}_T$ should be labeled $\vec{F}_N$.
C. No, $|\vec{F}_T|$ should be equal to $|\vec{F}_g|$.
D. No, $|\vec{F}_T|$ should be greater than $|\vec{F}_g|$.
E. No, there should be an upward drag force $\vec{F}_D$.
F. No, there is some other problem (specify).

N2T.12 A crate is sliding down a 30° incline, picking up speed as it slides. Ignore air friction. **(a)** Which of the forces listed below that acts on the crate has the greatest magnitude? **(b)** Which force has the smallest magnitude? **(c)** Which force listed does not actually act on the crate?
A. $\vec{F}_N$
B. $\vec{F}_g$
C. Both $\vec{F}_g$ and $\vec{F}_N$ (which are equal)
D. $\vec{F}_{SF}$
E. $\vec{F}_{KF}$

N2T.13 An object's x-position is shown in the boxed graph of the following set of graphs. Which of the other graphs in the set most correctly describes its x-velocity?

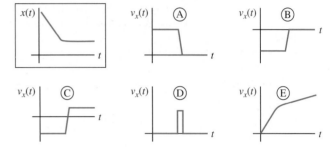

N2T.14 Which graph best describes the x-acceleration of the object described in problem N2T.13?

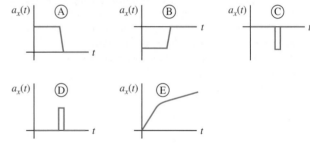

HOMEWORK PROBLEMS

Basic Skills

N2B.1 Draw a free-body diagram for each of the objects described below. Be careful to draw the force arrows with comparatively correct lengths (at least qualitatively).
(a) A baseball moving vertically upward just after it has left the bat (do not ignore air friction)
(b) A crate lifted by a crane at a constant upward velocity
(c) An airplane flying horizontally at a constant velocity (do not ignore air friction)

N2B.2 Draw a free-body diagram for each of the situations described below. Be careful to draw the force arrows with comparatively correct lengths (at least qualitatively).
(a) A box sliding down an incline at a constant velocity
(b) A box being hauled up an incline at a constant velocity by a rope

N2B.3 Draw a free-body diagram for each of the objects described below. Be careful to draw the force arrows with comparatively correct lengths (at least qualitatively).
(a) A baseball player sliding into second base
(b) A motorboat just after its motor has just run out of gas, but while the boat is still moving rapidly forward
(c) A rocket accelerating upward (do not ignore drag)

N2B.4 Draw a free-body diagram for each of the objects described below. Be careful to draw the force arrows with comparatively correct lengths (at least qualitatively)
(a) A child bouncing on a trampoline, at the instant the child is at rest at the lowest point of the bounce
(b) A baseball traveling vertically in the air, at the highest point in its trajectory
(c) A box sitting in an elevator whose downward speed is increasing as the elevator begins its descent

N2B.5   Draw a free-body diagram for a child holding on for dear life (with feet in the air!) to a rapidly spinning merry-go-round. (*Hint:* You might find a motion diagram useful.)

N2B.6   A car travels at a constant speed through a dip in the road that takes the car first down and then up. Draw a free-body diagram for the car as it passes the *bottom* of the dip, paying special attention to correctly indicating the relative magnitudes of the vertical forces on the car.

N2B.7   Supply the requested diagrams and explanations.
(a) Draw a free-body diagram for a motorboat at rest on a still lake. How should we classify the vertically upward force in this case? Explain.
(b) Now draw a free-body diagram for the boat when it is moving at a constant velocity. There is an *additional* upward force in this case (that causes the boat to rise somewhat in the water). How should we classify this force? Explain.

N2B.8   Consider the unfinished motion diagram shown. What is the direction of the object's acceleration at the instant it passes point 3? Show how you arrived at your answer.

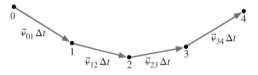

N2B.9   Consider the unfinished motion diagram shown. What is the direction of the object's acceleration in this case? Explain your reasoning.

N2B.10   Figure N2.6 shows a set of dots indicating a certain object's position every 0.10 s. After tracing or xeroxing these dots to your own sheet of paper, draw a complete and quantitatively accurate motion diagram and compute the magnitude of the object's acceleration at point 3.

N2B.11   Figure N2.7 shows a set of dots indicating a certain object's position every 0.05 s. After tracing or xeroxing these dots to your own sheet of paper, draw a complete and quantitatively accurate motion diagram and compute the magnitude of the object's acceleration at point 4.

N2B.12   Figure N2.8 shows a set of dots indicating a certain object's position every 0.10 s. Note that this is a scale diagram, where 1.0 cm on the diagram corresponds to 0.1 m of actual distance. After tracing or xeroxing the dots to your own sheet of paper, draw a complete and quantitatively accurate motion diagram for this object, and compute the magnitude of the object's acceleration at points 1, 2, and 3. (*Hint:* You can do the construction exactly as if it were an actual-size diagram, but when you calculate the acceleration, use a unit operator, that is,

$$\left(\frac{0.1 \text{ m [actual]}}{0.01 \text{ m [diagram]}} \right) \qquad \text{(N2.10)}$$

to convert diagram lengths to actual lengths.)

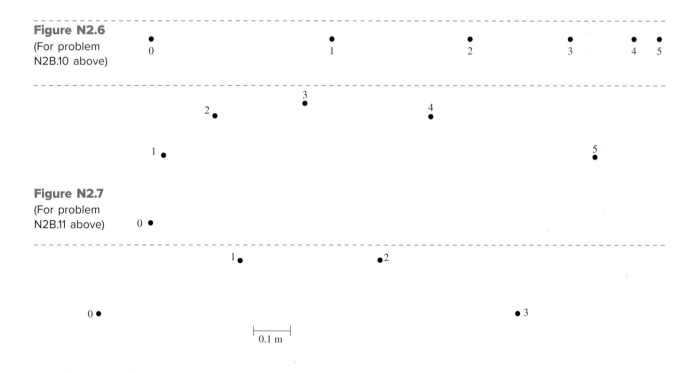

Figure N2.6
(For problem N2B.10 above)

Figure N2.7
(For problem N2B.11 above)

0.1 m

Figure N2.8
(For problem N2B.12 above)

N2B.13 Construct a graph of *x*-acceleration as a function of time for an object whose *x*-velocity is as shown.

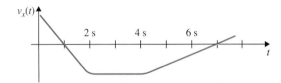

N2B.14 Construct a graph of *x*-velocity as a function of time for an object whose *x*-position is as shown.

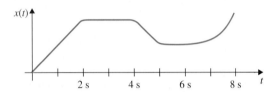

Modeling

N2M.1 An airplane travels at a constant speed in a horizontal circular path around an airport. Draw a *top-view* motion diagram and a *rear-view* free-body diagram of the plane. Also explain why the plane has to "bank" (that is, tilt so as to lower its inner wing and raise its outer wing) when it is flying in a circle.

N2M.2 A box sitting on the floor of a van slides toward the front of the van when the van suddenly brakes to a stop.
(a) Draw a motion diagram (as viewed from the *ground*, not the van) and a free-body diagram of the box.
(b) Explain qualitatively why the box moves forward relative to the van.

N2M.3 Consider an unpowered roller-coaster car at the instant it passes the peak of a vertical loop (that is, when the car is upside down; see figure N2.9). Do not ignore air resistance, and assume that the car is moving rapidly enough to remain in firm contact with the rails.
(a) Draw a motion diagram for the car, showing the car's acceleration at the peak of the loop.
(b) Draw a free-body diagram for the car at that peak.
(c) Explain why there is no *outward* force on the car.

N2M.4 Suppose an unpowered go-cart (with a child rider) traveling at an initial speed of 5 m/s coasts on a level road for about 50 m before coming to rest. If the cart and rider have a mass of 40 kg, about what force would you have to apply to keep the cart moving at a constant speed? Assume that the friction forces acting on the cart are approximately independent of its speed. (*Hint:* Use the "momentum requirement" from unit C to compute the magnitude of the total friction/drag force acting on the cart, then link this force to the force you must apply.)

N2M.5 Suppose you are driving your 1500-kg car along a straight road at a constant speed of 30 m/s. If you suddenly lift your foot

Figure N2.9
One of the two vertical loops on the Big Loop roller-coaster ride at Heide Park in Lower Saxony, Germany. (Credit: Jupiterimages/Comstock Images/Getty Images)

from the gas pedal, the car decelerates (that is, accelerates backward) with $|\vec{a}| = \frac{1}{20}|\vec{g}|$.
(a) What is the drag force on your car?
(b) At what rate (in horsepower) must the car's engine convert chemical energy to mechanical energy to keep the car's speed constant at 30 m/s? (*Hint:* Use energy concepts from unit C. In particular, argue that only the drag force does work on the car.)

N2M.6 Suppose a car rounds a curve at a constant speed of 10 m/s. The curve is a circular arc with a radius of 100 m. Ignore air friction initially.
(a) Draw free-body diagrams of the car, as viewed from the car's rear, the car's side, and the car's top.
(b) What is the direction and magnitude of the horizontal force that the road must exert on the car to keep it moving along the curve? Express this magnitude as a fraction of the car's weight.
(c) Now suppose that air friction is not zero, but is about half the result that you calculated for part (b). Redraw the three free-body diagrams you drew in part (a), considering the implications of drag (still assuming that the car's speed is constant, though!).
(d) Must the force that the road exerts on the car increase in magnitude compared to the no-drag case? Must it change its direction? Explain.

N2M.7 A balloon of mass M is floating at rest a distance H above the earth's surface. The rider throws out some ballast, and the balloon begins to rise with an initial acceleration of magnitude $|\vec{a}|$.
(a) What is the mass m of the ballast the rider throws out?
(b) Why is the word *initial* important in the description of the problem? What will cause this acceleration to change as time passes?

N2M.8 Suppose you are tracking a star that is orbiting the black hole at the center of our galaxy. You have taken three careful measurements of the star's position, and find its positions (separated by 10-day intervals) to have the coordinates

$$\vec{r}_1 = \begin{bmatrix} -2.10 \\ 19.65 \\ 0 \end{bmatrix} \text{Tm}, \; \vec{r}_2 = \begin{bmatrix} 0 \\ 20.11 \\ 0 \end{bmatrix} \text{Tm}, \; \vec{r}_3 = \begin{bmatrix} 2.10 \\ 19.52 \\ 0 \end{bmatrix} \text{Tm} \quad \text{(N2.11)}$$

in some suitable coordinate system (1 Tm = 10^{12} m) where the black hole is at the origin. Using these measurements alone, estimate the mass of the black hole. (*Hint:* Imagine constructing a motion diagram for this situation, but instead of graphically constructing the vector difference you will need, do the equivalent calculation using components. You should *not* assume that the star's orbit is circular.)

N2M.9 Imagine that during the time interval $0 \leq t \leq 2$ s, the x-velocity of a car with a mass $m = 1200$ kg is given by $v_x(t) = v_0 - bt^2$, where $v_0 = 20$ m/s and $b = 5$ m/s^3.
(a) Sketch graphs of $v_x(t)$ and $a_x(t)$ for this situation.
(b) Find an expression (in terms of m, b, t, and whatever else you need) for the x component of the combined static friction and drag forces acting on the car during that time interval.
(c) Calculate the magnitude of this force at time $t = 0.5$ s.

Rich-Context

N2R.1 The top of a small hill in a certain highway has a circular (vertical) cross section with an approximate radius of 57 m. A car going over this hill too fast might leave the ground and thus lose control. What speed limit should be posted? (*Hint:* Note that the magnitude of the normal force acting on the car as it passes over the crest of the hill is *not* equal to the car's weight. What must it be, approximately, if the car's tires just barely maintain contact with the road?)

N2R.2 You are designing an ejector seat for an automobile. Your seat contains two rocket engines that will burn for no more than 0.5 s (so as not to fry passers-by) and yet must throw the seat and occupant at least 150 m in the air after the engines shut off (to allow the parachute to deploy). Estimate the combined thrust your rockets will have to exert on the seat, and check whether the seat's acceleration will exceed the safe limit of about $10|\vec{g}|$. (*Hint:* You may assume that the rocket engines' thrust is constant. You may need to consider a range of passenger weights, or alternatively express the thrust as a function of the passenger's weight. Do not neglect the seat's weight.)

N2R.3 Consider a vertical loop in the roller coaster ride shown in figure N2.9. Suppose you are designing such a loop and you want the normal force that the track exerts on the cars to be at least 1/4 their weight at the instant the car is at the top of the loop. Let point A be a point on the roller-coaster track just before the track begins to rise into the loop, and suppose the vertical coordinate of a car's center of mass as it passes the top of the loop is 15 m above the vertical coordinate of the car's center of mass as it passes point A. What *minimum* speed should the roller-coaster car have as it passes point A?

ANSWERS TO EXERCISES

N2X.1 Again, if we define our x axis so the interacting particles both lie on the x axis, with the particle of interest at some positive x coordinate and the other particle at the origin, then we have $r = x$ and

$$F_{e,x} = -\frac{dV_e}{dx} = -\frac{d}{dx}\left(\frac{1}{4\pi\varepsilon_0}\frac{Qq}{x}\right) = +\frac{1}{4\pi\varepsilon_0}\frac{Qq}{x^2} \quad \text{(N2.12a)}$$

$$F_{Sp,x} = -\frac{dV_{Sp}}{dx} = -\frac{d}{dx}[\tfrac{1}{2}k_s(x-x_0)^2] = -\tfrac{1}{2}k_s 2(x-x_0) \quad \text{(N2.12b)}$$

In the first case, the force points *away* from the other particle (for positive Qq). In the second case, the force is *toward* the other particle. So general descriptions of these forces detached from the coordinate system would be

$$\vec{F}_e = \frac{1}{4\pi\varepsilon_0}\frac{Qq}{r^2} \text{ (away from the other object)} \quad \text{(N2.12c)}$$

$$\vec{F}_{Sp} = k_s(r-r_0) \text{ (toward the other object)} \quad \text{(N2.12d)}$$

as given in equations N2.4b and N2.4c.

N2X.2 A free-body diagram for the boat looks as shown in figure N2.10. (If the boat is moving quickly, we should probably add an upward "lift" force.) Since the boat is moving at a constant velocity, the net force should be zero.

N2X.3 Graphs of $v_x(t)$ and $a_x(t)$ in this situation will look qualitatively as shown in figure N2.11.

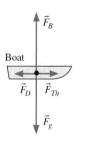

Figure N2.10

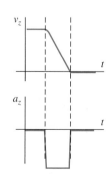

Figure N2.11

N3

CORE

Motion from Forces

Chapter Overview

Introduction

In chapter N2, we explored a number of examples in which we used information about an object's motion to learn about the forces acting on that object. In this chapter, we will discuss examples in which we use knowledge of the net force acting on an object to determine the object's motion. This will lay important foundations for exploring more complicated problems of this type in chapters N9 through N12.

Section N3.1: The Reverse Kinematic Chain

We can reverse the kinematic chain discussed in chapter N2 as follows:

$$\vec{a}(t) \longrightarrow \begin{bmatrix} \text{time} \\ \text{antiderivative} \end{bmatrix} \longrightarrow \vec{v}(t) \longrightarrow \begin{bmatrix} \text{time} \\ \text{antiderivative} \end{bmatrix} \longrightarrow \vec{r}(t) \tag{N3.1b}$$

If we know the net force acting on an object, we can use Newton's second law to determine the object's acceleration. The chain of relationships in equation N3.1b means that we can then predict the object's motion by computing **antiderivatives** of $\vec{a}(t)$. The only problem with doing this is that computing antiderivatives is *hard*. In this chapter, we will look at four different approaches to computing these antiderivatives.

Section N3.2: Graphical Antiderivatives

If the object moves only along the x axis, then we can use graphs of $a_x(t)$, $v_x(t)$, and $x(t)$ to describe an object's motion. If we stack such graphs vertically so a graph of $x(t)$ is above a graph of $v_x(t)$, which is above a graph of $a_x(t)$, then we can construct an antiderivative graph from the graph below it using the idea that the *slope above equals the value below*. To apply this idea:

1. Draw a short line segment on the upper graph of the pair whose *slope* reflects the *value* on the lower graph at a given instant of time.
2. Align successively drawn segments so they sketch out a continuous curve.

In either method, we must use separately stated **initial conditions** $x(0)$ and $v_x(0)$ to determine where to start on the upper graph.

Section N3.3: Integrals for One-Dimensional Motion

The mathematical definitions of velocity and acceleration imply that

$$v_x(t) = \int a_x(t)\,dt + C_1 \quad \text{and} \quad x(t) = \int v_x(t)\,dt + C_2 \tag{N3.2}$$

where $\int f(t)\,dt$ is any antiderivative of $f(t)$ and C_1 and C_2 are constants of integration. We can determine the values of C_1 and C_2 by plugging $t = 0$ into the expressions above and setting the values C_1 and C_2 so that the values of $v_x(0)$ and $x(0)$ match the specified initial conditions.

We can evaluate these constants of integration automatically (without separate steps) by using definite integrals:

$$v_x(t) - v_x(0) = \int_0^t a_x(t)\,dt \quad \text{and} \quad x(t) - x(0) = \int_0^t v_x(t)\,dt \tag{N3.8}$$

where the **definite integral** of $f(t)$ is *any one* of its possible antiderivative functions evaluated at the integral's upper limit minus the same function evaluated at the integral's lower limit.

When an object's x-acceleration is *constant*, these integrals imply that

$$v_x(t) = a_x t + v_{0x} \quad \text{and} \quad x(t) = \tfrac{1}{2} a_x t^2 + v_{0x} t + x_0 \qquad \text{(N3.6)}$$

- **Purpose:** These equations specify an object's x-velocity $v_x(t)$ and x-position $x(t)$ as functions of time t when its x-acceleration a_x is constant, where $v_{0x} \equiv v_x(0)$ is the object's initial x-velocity, and $x_0 \equiv x(0)$ is its initial x-position.
- **Limitations:** (Important!) This equation applies *only* if the object's x-acceleration is constant (or we can reasonably model it as constant).

Section N3.4: Free-fall in One Dimension

An object is freely falling if the only significant force acting on it is its weight. Newton's second law then says that $m\vec{a} = \vec{F}_{\text{net}} = m\vec{g}$ which implies that the object's acceleration is $\vec{a} = \vec{g}$, where $\vec{g}$ is the local **gravitational field vector** (sometimes called the **acceleration of gravity**). If the object moves only vertically, we can adapt equations N3.6 by replacing the x subscripts with z subscripts and using $a_z = -|\vec{g}|$.

Section N3.5: Integrals in Three Dimensions

In general, the definitions of velocity and acceleration imply that

$$\vec{v}(t) - \vec{v}(0) = \int_0^t \vec{a}(t)\,dt \quad \text{and} \quad \vec{r}(t) - \vec{r}(0) = \int_0^t \vec{v}(t)\,dt \qquad \text{(N3.20)}$$

- **Purpose:** These equations describe how to calculate an object's velocity $\vec{v}(t)$ and its position $\vec{r}(t)$ as functions of time t, given its acceleration $\vec{a}(t)$, its initial velocity $\vec{v}(0)$, and its initial position $\vec{r}(0)$.
- **Limitations:** The acceleration and velocity functions must be well-defined functions of time.
- **Note:** Each of these equations compactly expresses three independent component equations (see equations N3.18 and N3.19).

Section N3.6: Constructing Trajectory Diagrams

When an object's motion is two-dimensional, one can calculate its motion graphically by constructing a **trajectory diagram** as follows:

1. Draw an arrow $\vec{v}_0 \Delta t$ with its tip at the object's initial position $\vec{r}_0$ (point 0). Also draw the arrow $\vec{a}_0 \Delta t^2$ *centered* on point 0.
2. Construct an arrow $\vec{v}_{01} \Delta t$ that goes from the tail end of the $\vec{v}_0 \Delta t$ arrow to the tip of the $\vec{a}_0 \Delta t^2$ arrow.
3. Move this $\vec{v}_{01} \Delta t$ arrow so its tail is at point 0. The point at its tip is now point 1 (the object's position at time $t_1 = 1 \cdot \Delta t$).
4. Draw the arrow $\vec{a}_1 \Delta t^2$ with its tail attached to point 1.
5. Construct the arrow $\vec{v}_{12} \Delta t = \vec{v}_{01} \Delta t + \vec{a}_2 \Delta t^2$ from the tail of $\vec{v}_{01} \Delta t$ to the tip of $\vec{a}_2 \Delta t^2$.
6. Move this $\vec{v}_{12} \Delta t$ arrow so its tail is at point 1. Its tip is now point 2 (the object's position at time $t_2 = 2 \cdot \Delta t$).
7. Repeat steps 4 through 6 for points 2, 3, … to find the object's later positions.

This process works in *all* circumstances (as long as Δt is sufficiently small).

Trajectory diagrams are very useful, but tedious to construct by hand. The *Newton* app (sixideas.pomona.edu/resources.html) can construct trajectory diagrams very rapidly.

N3.1   The Reverse Kinematic Chain

The kinematic chain

In chapter N2, we discussed the *kinematic chain* of relationships that link an object's position to its velocity and its acceleration:

$$\vec{r}(t) \; \underline{\left[\begin{matrix} \text{time} \\ \text{derivative} \end{matrix}\right]} \longrightarrow \; \vec{v}(t) \; \underline{\left[\begin{matrix} \text{time} \\ \text{derivative} \end{matrix}\right]} \longrightarrow \; \vec{a}(t) \tag{N3.1a}$$

$$\vec{a}(t) \; \underline{\left[\begin{matrix} \text{time} \\ \text{antiderivative} \end{matrix}\right]} \longrightarrow \; \vec{v}(t) \; \underline{\left[\begin{matrix} \text{time} \\ \text{antiderivative} \end{matrix}\right]} \longrightarrow \; \vec{r}(t) \tag{N3.1b}$$

In chapter N2, we focused on using equation N3.1a to work from an object's observed motion to its acceleration, so that we could then use Newton's second law to answer questions about the forces acting on the object. In this chapter, we will focus on using the kinematic chain in the reverse direction given in equation N3.1b. If we know the forces acting on an object at all times, Newton's second law tells us the object's acceleration as a function of time. If we know how to compute that function's antiderivative, we can find its velocity and its position as functions of time as well. This is how we determine *motion from forces* (at least in principle).

The problem here is not the physics but the mathematics. One only needs a handful of rules to calculate the derivative of almost any function. Finding the *antiderivative* of an arbitrary function is generally not nearly so easy.

Four different approaches to solving the reversed chain

In this chapter, we will discuss *four* different approaches to finding the antiderivatives of acceleration—two approaches that work for one-dimensional motion and two for multidimensional motion. When an object moves in only one dimension, we can essentially reverse the graphical construction techniques we learned in chapter N2 to construct graphs of $v_x(t)$ and $x(t)$ from a graph of $a_x(t)$. Alternatively, we can *sometimes* use the techniques of integral calculus to find these functions mathematically. (For a review of the basic integral calculus we will need, see appendix NB at the end of this volume.)

If the object moves in three dimensions, we can still use the mathematical approach: we simply have to keep track of three vector components instead of one. We can also reverse the approach we used to construct motion diagrams in chapter N2: whereas before we used information about the object's position at equally spaced instants of time to construct arrows representing the object's acceleration, we now use known acceleration arrows to construct the object's trajectory. This latter technique turns out to be especially easy to turn into a computer algorithm that computes trajectories. (It is also essentially the approach that Newton himself used in his *Principia*.)

N3.2   Graphical Antiderivatives

We begin with the graphical approach to one-dimensional motion. In chapter N2, we learned how to use the "slope above equals value below" method to construct a graph of $v_x(t)$ below a graph of $x(t)$ and/or a graph of $a_x(t)$ below a graph of $v_x(t)$. In this section, we will learn how to do the process in reverse, constructing graphs of $v_x(t)$ and $x(t)$ from a graph of $a_x(t)$.

The slope method

The easiest way to do this simply reverses the approach we used in chapter N2. For example, suppose we are trying to construct a graph of $v_x(t)$ from a graph of $a_x(t)$. If we were instead constructing the graph of $a_x(t)$ from $v_x(t)$, we would look at the slope of the upper graph and plot its value on the lower graph. To do this in reverse, we should, for each of a set of specific instants of time, draw a little line segment on the upper graph whose slope reflects the *value* we see at that instant on the lower graph. If you arrange these line segments so that they

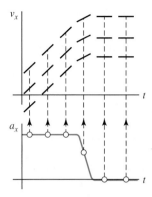

Figure N3.1
We can construct a graph of $v_x(t)$ from a graph of $a_x(t)$ by drawing short line segments on the upper graph whose *slopes* are consistent with the *value* of the lower graph at the same time. We can choose between the various possible upper curves we generate in this way only by actually choosing a value for, say, $v_x(0)$.

nearly touch end to end, as shown in figure N3.1, then they will sketch out the desired upper curve.

Unfortunately, this process is ambiguous, because we can draw *several* upper graphs that satisfy this criterion, as shown. These different graphs all have the same *shape* but are offset vertically from one another by different constant values. The only way to resolve this ambiguity is to *choose* the value that the upper graph should have at a certain time (generally at $t = 0$). In some cases, the problem statement may suggest the value (for example, it might say that the object starts at *rest*), but otherwise we simply have to *pick* an arbitrary value. Once we have chosen or found a value for $v_x(t)$ at any given instant of time, the rest of the graph is completely determined.

Initial conditions resolve the ambiguities

Problem: Imagine a car (initially at rest at $x = 0$) powered by a rocket engine that pivots so we can direct its thrust either forward or backward. The engine ignites at $t = 0$, subsequently exerting a constant forward thrust on the car between $t = 0$ and $t = T$. Then, the engine is turned around so it exerts a rearward thrust on the car of the same magnitude from $t = T$ to $t = 2T$. Draw graphs of $a_x(t)$, $v_x(t)$, and $x(t)$ for this car. (Ignore drag or friction.)

Solution Assuming that no significant drag or friction forces oppose the thrust on the car, and assuming that the car moves along a level road (so that the vertical normal and gravitational forces on the car cancel out), then the net force on the car is the same as that applied by the rocket engine. If we take the $+x$ direction to be forward, the problem description implies that the car's x-acceleration is some positive constant $|\vec{a}|$ for a certain interval of time (call it $t = 0$ to $t = T$) and then $-|\vec{a}|$ for $t = T$ to $t = 2T$. This means the graph of $a_x(t)$ should look as shown in the bottom graph in figure N3.2.

Working upward, we can use the slope method to show that the car's velocity first increases linearly and then decreases linearly. The problem description states that $v_x(0) = 0$, so a graph of the car's $v_x(t)$ must look as shown in the middle graph.

To create the position graph, we start with $x = 0$ at $t = 0$. Note that as the velocity increases, the slope of the $x(t)$ graph increases; and as the velocity decreases, the slope of $x(t)$ decreases, as shown in the top graph.

Note that we have (qualitatively at least) completely described the car's motion, given the forces acting on it!

Example N3.1

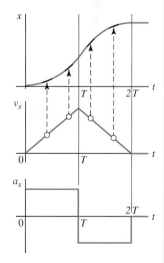

Figure N3.2
Graphs of $a_x(t)$, $v_x(t)$, and $x(t)$ for the rocket car.

N3.3   Integrals for One-Dimensional Motion

Using the *antiderivatives* to do the same thing mathematically

Another whole approach to constructing the functions $v_x(t)$ and $x(t)$ from $a_x(t)$ is to use the techniques of integral calculus to determine these quantities mathematically. In this section, we will explore this method in the context of one-dimensional motion.

Since $dx/dt = v_x(t)$ and $dv_x/dt = a_x(t)$, we have

$$v_x(t) = \int a_x(t)\,dt + C_1 \qquad \text{and} \qquad x(t) = \int v_x(t)\,dt + C_2 \qquad (N3.2)$$

where C_1 and C_2 are unknown **constants of integration** and $\int f(t)\,dt$ is the conventional notation for a particular antiderivative of $f(t)$, that is, any function $F(t)$ whose time derivative is $f(t)$. The constants C_1 and C_2 express mathematically what figure N3.1 expresses graphically: in the absence of additional information, *we can only determine a function's antiderivative up to an overall constant.* (See appendix NB for more discussion of this issue.)

Initial conditions determine the constants of integration

We can, however, determine C_1 if we know the object's initial velocity $v_x(0)$: we simply substitute $t = 0$ into the expression for $v_x(t)$ and choose the value of C_1 that gives the correct initial velocity $v_x(0)$. We can similarly determine C_2 by choosing it so we get the right value of $x(0)$. We call $v_x(0)$ and $x(0)$ **initial conditions**; we can completely determine an object's position and velocity from its acceleration *only* if we also know these quantities.

Example N3.2

Problem: After a stoplight turns green at time $t = 0$, a car accelerates with $\vec{a} = 2.0$ m/s^2 forward along a straight stretch of road (which we will take to define our x axis). **(a)** Find a general expression for the car's x-velocity and x-position as a function of time t in terms of the car's x-acceleration a_x and its initial conditions $v_{0x} \equiv v_x(0)$ and $x_0 \equiv x(0)$ for as long as a_x is constant. **(b)** Assuming that the car starts from rest and we define $x = 0$ to be the car's initial position, what are the car's x-velocity and x-position after 5.0 s?

Solution **(a)** According to the first expression in equation N3.2,

$$v_x(t) = \int a_x(t)\,dt + C_1 = \int a_x\,dt + C_1 = a_x t + C_1 \qquad (N3.3a)$$

since $d(a_x t)/dt = a_x t$. Plugging $t = 0$ into this result, we find that

$$v_x(0) = a_x \cdot 0 + C_1 \quad \Rightarrow \quad C_1 = v_x(0) \equiv v_{0x} \quad \Rightarrow \quad v_x(t) = a_x t + v_{0x} \qquad (N3.3b)$$

We can take the antiderivative again to find $x(t)$:

$$x(t) = \int v_x(t)\,dt + C_2 = \int a_x t\,dt + \int v_{0x}\,dt + C_2 = \tfrac{1}{2}a_x t^2 + v_{0x}t + C_2 \qquad (N3.4a)$$

since $d(\tfrac{1}{2}a_x t^2)/dt = a_x t$ and $d(v_{0x}t)/dt = v_{0x}$. At time $t = 0$, we have

$$x(0) = \tfrac{1}{2}a_x \cdot 0^2 + v_{0x} \cdot 0 + C_2 \quad \Rightarrow \quad C_2 = x(0) \equiv x_0$$

$$\Rightarrow \quad x(t) = \tfrac{1}{2}a_x t^2 + v_{0x}t + x_0 \qquad (N3.4b)$$

(b) Because the car's acceleration is forward (that is, in the $+x$ direction), we have $a_x = +|\vec{a}| = 2.0$ m/s^2. We also know that $x_0 = 0$ and $v_{0x} = 0$. Therefore,

$$v_x(t) = a_x t = v_{0x} = (+2.0 \text{ m/s}^2)(5.0 \text{ s}) + 0 = 10 \text{ m/s} \quad \text{at } t = 5 \text{ s} \qquad (N3.5a)$$

$$x(t) = \tfrac{1}{2}a_x t^2 + v_{0x}t + x_0 = \tfrac{1}{2}(+2.0 \text{ m/s}^2)(5.0 \text{ s})^2 + 0 + 0 = 25 \text{ m} \qquad (N3.5b)$$

Equations N3.3*b* and N3.4*b* are useful whenever an object's *x*-acceleration is constant:

$$v_x(t) = a_x t + v_{0x} \quad \text{and} \quad x(t) = \tfrac{1}{2} a_x t^2 + v_{0x} t + x_0 \qquad (\text{N3.6})$$

- **Purpose:** These equations specify an object's *x*-velocity $v_x(t)$ and *x*-position $x(t)$ as functions of time *t* when its *x*-acceleration a_x is constant, where $v_{0x} \equiv v_x(0)$ is the object's initial *x*-velocity, and $x_0 \equiv x(0)$ is its initial *x*-position.
- **Limitations:** (Important!) This equation applies *only* if the object's *x*-acceleration is constant (or we can reasonably model it as constant).

These equations are important for at least two reasons. First, this is a relatively simple case that illustrates the integral approach nicely. Second, these equations are useful because the assumption that $a_x =$ constant is a reasonably good *model* for a variety of situations. Physicists often use this model to make quick estimates even when it is clear that a_x is *not* particularly constant.

On the other hand, one of the most common student errors I have seen is thinking that these equations are more general than they actually are. These equations really apply only to the very *specific* case in which a_x is *constant*. While this is occasionally an excellent approximation to a realistic acceleration and more often an adequate first approximation, there are many more situations where equations N3.6 are not even close to being correct.

It is also worth noting that in example N3.2, the constants of integration and C_1 and C_2 happen to be equal to the initial conditions $v_x(0)$ and $x(0)$. This is *not* generally true! In every problem where an object's acceleration is *not* constant, you *must* carefully evaluate these constants of integration by using the specified initial conditions.

Exercise N3X.1

A car whose initial *x*-position and *x*-velocity are $x(0) = 0$ and $v_x(0) = +12$ m/s experiences forces that give it an *x*-acceleration of $a_x(t) = -|\vec{a}|$, with $|\vec{a}| = 2.0$ m/s². Find $v_x(t)$ and $x(t)$ and evaluate these quantities at $t = 5.0$ s.

One can, however, *automatically*, *correctly*, and *implicitly* evaluate these constants of integration as follows. If we take the definitions $a_x \equiv dv_x/dt$ and $v_x \equiv dx/dt$ and "integrate both sides from t_A to t_B," the fundamental theorem of calculus implies that

$$\int_{t_A}^{t_B} \frac{dv_x}{dt}\, dt = \int_{t_A}^{t_B} a_x(t)\, dt \quad \Rightarrow \quad v_x(t_B) - v_x(t_A) = \int_{t_A}^{t_B} a_x(t)\, dt \qquad (\text{N3.7}a)$$

$$\int_{t_A}^{t_B} \frac{dx}{dt}\, dt = \int_{t_A}^{t_B} v_x(t)\, dt \quad \Rightarrow \quad x(t_B) - x(t_A) = \int_{t_A}^{t_B} v_x(t)\, dt \qquad (\text{N3.7}b)$$

where $\int_{t_A}^{t_B} f(t)\, dt$ is the **definite integral** of $f(t)$, which we calculate by finding any function $F(t)$ whose derivative is $f(t)$ and evaluating the difference $F(t_B) - F(t_A)$ between that function's values at the limits of the integration. In particular, if $t_A = 0$ and t_B is some arbitrary time *t*, then these equations read

$$v_x(t) - v_x(0) = \int_0^t a_x(t)\, dt \quad \text{and} \quad x(t) - x(0) = \int_0^t v_x(t)\, dt \qquad (\text{N3.8})$$

Note how these equations *automatically* include the initial conditions. Also, as evaluating the definite integrals involves computing the *difference* of antiderivatives evaluated at two different times, the antiderivative's unknown constant of integration will cancel out of the difference, so we never even need to think about it! I think that once you get used to this method, you will find it reduces the likelihood of error and is simpler in complicated problems.

Example N3.3

Problem: Consider a similar situation as discussed in example N3.2 where a car accelerates from rest when a stoplight turns green. But in this case, let us assume that a drag force that increases with speed opposes the constant forward force applied to the car, so that the car's x-acceleration ends up being $a_x(t) = b/(t + T)^3$, where $b = 2000$ m·s, and $T = 10$ s (both are constants). What are the car's x-velocity and x-position as a function of time in this case? What are the car's x-velocity and x-position at $t = 5.0$ s?

Solution Note that the opposing drag force makes $a_x(t)$ *decrease* as t increases: $a_x(t)$ decreases to one-eighth its original value by time $t = T$. Note also that at $t = 0$, $a_x = (2000$ m·s$)/(10$s$)^3 = 2.0$ m/s^2, which is the same as the *constant* acceleration value in example N3.2. Since $v_x(0) = 0$ in this case, equation N3.8 implies that

$$v_x(t) = \int_0^t a_x(t)\,dt = \int_0^t \frac{b}{(t+T)^3}\,dt = b\int_0^t \frac{dt}{(t+T)^3}$$

$$= \frac{b}{-2}\left[\frac{1}{(t+T)^2} - \frac{1}{(0+T)^2}\right] = \frac{b}{2T^2} - \frac{b}{2(t+T)^2} \tag{N3.9}$$

since the derivative of $-\frac{1}{2}b(t+T)^{-2}$ is $+b(t+T)^{-3}$. Plugging this into equation N3.8a, again defining the car's initial position to be $x(0) \equiv 0$, and using the constant and sum rules for integration (see appendix NB), we get

$$x(t) = \int_0^t v_x(t)\,dt = \int_0^t \left[\frac{b}{2T^2} - \frac{b}{2(t+T)^2}\right]dt = \frac{b}{2T^2}\int_0^t dt - \frac{b}{2}\int_0^t \frac{dt}{(t+T)^2}$$

$$= \frac{b}{2T^2}(t-0) - \frac{b}{2}\left(\frac{-1}{t+T} - \frac{-1}{T}\right) = \frac{b}{2}\left(\frac{t}{T^2} + \frac{1}{t+T} - \frac{1}{T}\right) \tag{N3.10}$$

since the integral of $(t+T)^{-2}$ is $-(t+T)^{-1}$. Evaluating $v_x(t)$ and $x(t)$ at time $t = 5.0$ s, we get

$$v_x(5.0\text{ s}) = \frac{2000\text{ m·s}}{2(10\text{ s})^2} - \frac{2000\text{ m·s}}{2(15\text{ s})^2} = 5.6\,\frac{\text{m}}{\text{s}} \tag{N3.11a}$$

$$x(5.0\text{ s}) = \frac{2000\text{ m·s}}{2}\left[\frac{5.0\text{ s}}{(10\text{ s})^2} + \frac{1}{15\text{ s}} - \frac{1}{10\text{ s}}\right] = 17\text{ m} \tag{N3.11b}$$

Note that these values are smaller than the results in example N3.2, as we might expect.

Exercise N3X.2

Why might we expect the results to be smaller? Also, check that the expressions for $v_x(t)$ and $x(t)$ yield the correct results at $t = 0$.

In closing this section, let me emphasize again that just knowing $a_x(t)$ is *not* sufficient to determine an object's motion: we also need to specify the object's initial x-position $x_0 \equiv x(0)$ and x-velocity $v_{0x} \equiv v_x(0)$ to completely determine its motion. This is *always* the case when we are trying to determine an object's trajectory from its acceleration, so dealing with initial conditions is going to be a recurring theme in chapters N9 through N12.

The importance of initial conditions

N3.4   Free-fall in One Dimension

An important application of equations N3.6 is the case of a *freely falling* object. If the only significant force acting on an object near the earth is its weight, we say that is **freely falling**. According to Newton's second law, we have

The definition of a freely falling object

$$m\vec{a} = \vec{F}_{net} = \vec{F}_g = m\vec{g} \quad \Rightarrow \quad \vec{a} = \vec{g} \qquad (N3.12)$$

where $\vec{g}$ is a vector whose magnitude is $|\vec{g}| = 9.8$ m/s^2 and whose direction is downward. Note that this equation implies that *all* objects fall with the same acceleration $\vec{g}$, independent of their mass! (The vector $\vec{g}$ is properly called the **gravitational field vector**, but people often call it the *acceleration of gravity* for this reason.) The non-obvious result that $\vec{a} = \vec{g}$ has important implications that we will explore in chapters N8 and N9.

If we define the z axis to be vertically upward, then $a_z = -|\vec{g}|$. So, if the falling object only moves along the z axis, we can adapt equations N3.6 for this situation by switching the x subscripts to z subscripts, yielding

$$v_z(t) = v_{0z} - |\vec{g}|t \qquad (N3.13a)$$

$$z(t) = -\tfrac{1}{2}|\vec{g}|t^2 + v_{0z}t + z_0 \qquad (N3.13b)$$

Problem: Suppose that at $t = 0$ we drop a ball from rest at an initial z-position $z_0 = 25$ m above the ground (which we define to have z-position $z = 0$). If we ignore air drag, then how fast is the ball moving after 2.2 s? How far above the ground is it at that time?

Example N3.4

Solution Plugging $v_{0z} = 0$ and $t = 2.2$ s into equation N3.13a yields

$$v_z(2.2 \text{ s}) = 0 - (9.8 \text{ m/s}^2)(2.2 \text{ s}) = -21.6 \text{ m/s} \qquad (N3.14)$$

So the ball is moving *downward* with a speed of 21.6 m/s = 48.3 mi/h after falling for 2.2 s. Similarly, plugging numbers into equation N3.13b yields

$$z(2.2 \text{ s}) = -\tfrac{1}{2}(9.8 \text{ m/s}^2)(2.2 \text{ s})^2 + 0 + 25 \text{ m} = +1.3 \text{ m} \qquad (N3.15)$$

N3.5   Integrals in Three Dimensions

We can easily generalize equation N3.8 to handle cases where the object in question is *not* constrained to move in one dimension. According to the kinematic chain, we have $d\vec{v}/dt \equiv \vec{a}(t)$ and $d\vec{r}/dt = \vec{v}(t)$. If we integrate both sides of this expression from 0 to t and treat these vector functions as if they were ordinary numerical functions, we get

The integral of the vector
acceleration (in abstract)

$$\vec{v}(t) - \vec{v}(0) = \int_0^t \vec{a}(t)\,dt \quad \text{and} \quad \vec{r}(t) - \vec{r}(0) = \int_0^t \vec{v}(t)\,dt \qquad \text{(N3.16)}$$

But can we really do this when the functions are *vector* functions? Yes! Remember that the vector equation $d\vec{v}/dt \equiv \vec{a}(t)$ is equivalent to three independent component equations

$$\begin{bmatrix} dv_x/dt \\ dv_y/dt \\ dv_z/dt \end{bmatrix} \equiv \begin{bmatrix} a_x(t) \\ a_y(t) \\ a_z(t) \end{bmatrix} \quad \Rightarrow \quad \begin{array}{l} dv_x/dt \equiv a_x(t) \\ dv_y/dt \equiv a_y(t) \\ dv_z/dt \equiv a_z(t) \end{array} \qquad \text{(N3.17)}$$

Each one of these three equations involves only ordinary functions of time. If we integrate both sides of each of these scalar equations from 0 to t, we get

What this means in terms of
components

$$\begin{array}{l} v_x(t) - v_x(0) = \int_0^t a_x(t)\,dt \\[2mm] v_y(t) - v_y(0) = \int_0^t a_y(t)\,dt \\[2mm] v_z(t) - v_z(0) = \int_0^t a_z(t)\,dt \end{array} \quad \Rightarrow \quad \begin{bmatrix} v_x(t) \\ v_y(t) \\ v_z(t) \end{bmatrix} = \begin{bmatrix} \int_0^t a_x(t)\,dt \\ \int_0^t a_y(t)\,dt \\ \int_0^t a_z(t)\,dt \end{bmatrix} + \begin{bmatrix} v_{0x} \\ v_{0y} \\ v_{0z} \end{bmatrix} \qquad \text{(N3.18)}$$

where $v_{0x} \equiv v_x(0)$, $v_{0y} \equiv v_y(0)$, and so on. The equation on the right shows what the first of equations N3.16 really means at the component level.

Similarly, integrating both sides of $d\vec{r}/dt \equiv \vec{v}(t)$ from 0 to t, we get

The integral of the vector velocity

$$\vec{r}(t) - \vec{r}(0) = \int_0^t \vec{v}(t)\,dt \quad \Rightarrow \quad \begin{bmatrix} x(t) \\ y(t) \\ z(t) \end{bmatrix} = \begin{bmatrix} \int_0^t v_x(t)\,dt \\ \int_0^t v_y(t)\,dt \\ \int_0^t v_z(t)\,dt \end{bmatrix} + \begin{bmatrix} x_0 \\ y_0 \\ z_0 \end{bmatrix} \qquad \text{(N3.19)}$$

Equations N3.18 and N3.19 provide a complete and general description of how we can compute an object's velocity components and position components as a function of time if we know the object's initial velocity, its initial position, and its acceleration at all times as a function of time.

So we see that when it comes to integrating and taking derivatives, we *can* treat vector functions (symbolically) as if they were ordinary functions:

$$\vec{v}(t) - \vec{v}(0) = \int_0^t \vec{a}(t)\,dt \quad \text{and} \quad \vec{r}(t) - \vec{r}(0) = \int_0^t \vec{v}(t)\,dt \qquad \text{(N3.20)}$$

- **Purpose:** These equations describe how to calculate an object's velocity $\vec{v}(t)$ and its position $\vec{r}(t)$ as functions of time t, given its acceleration $\vec{a}(t)$, its initial velocity $\vec{v}(0)$, and its initial position $\vec{r}(0)$.
- **Limitations:** The acceleration and velocity functions must be well-defined functions of time.
- **Note:** Each of these equations compactly expresses three independent component equations (see equations N3.18 and N3.19).

N3.6   Constructing Trajectory Diagrams

Equations N3.20 provide a very general method for calculating trajectories, but actually computing the integrals in these equations can be *very* difficult in many situations of interest. In this section, we will see that we can reverse the procedure that we used to construct motion diagrams to construct an object's trajectory from its acceleration under *any* circumstances.

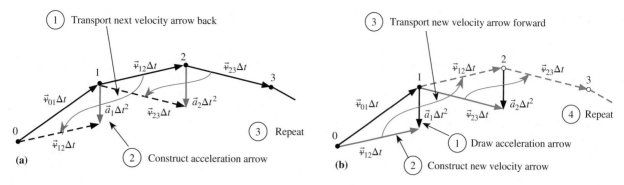

Figure N3.3

(a) How to construct acceleration arrows from a given trajectory. (b) How to construct a trajectory from given acceleration arrows. In both figures, black arrows represent known vectors, colored arrows represent constructed vectors, dashed arrows represent transported vectors, and white dots represent constructed positions.

Figure N3.3a reviews how we construct a motion diagram. We first plot the object's position at equally spaced instants of time and draw displacement arrows $\Delta \vec{r}_{01} = \vec{\mathbb{v}}_{01}\,\Delta t$, $\Delta \vec{r}_{12} = \vec{\mathbb{v}}_{12}\Delta t$, and so on between the dots. Since Δt is the same for all these arrows, these arrows depict both the direction and the relative magnitudes of the average velocities $\vec{\mathbb{v}}_{01}$, $\vec{\mathbb{v}}_{12}$, ... between the dots. To construct an arrow representing the object's acceleration at, say, point 1, we move the velocity arrow $\vec{\mathbb{v}}_{12}\Delta t$ just after the point back so that its tail end coincides with the tail end of the velocity arrow $\vec{\mathbb{v}}_{01}\Delta t$ just before the point, and we construct the vector difference $\Delta\vec{\mathbb{v}}\,\Delta t \approx \vec{a}_1\Delta t^2$. As long as Δt is small compared to the time it takes the object's acceleration to change significantly, this process yields reasonably accurate acceleration arrows.

Review of motion diagrams

To construct a **trajectory diagram** of the object's motion, we simply do this process backward, as illustrated in figure N3.3b. Assume that we know the object's initial position $\vec{r}_0$ at time $t = 0$ (represented by dot 0) and the first average velocity arrow $\vec{\mathbb{v}}_{01}\Delta t$ (which is related to the object's initial velocity $\vec{v}_0$ in a way that we will discuss in a minute). This vector's tip locates dot 1, the object's position at time $t_1 = \Delta t$.

To draw a trajectory diagram, reverse the procedure for constructing a motion diagram

Now, to find the object's position at time $t_2 = 2\Delta t$, we carefully draw the object's acceleration arrow $\vec{a}_1\Delta t^2$ (which we can calculate by using Newton's second law) so its tail end coincides with point 1. We then construct the arrow $\vec{\mathbb{v}}_{12}$ so that the difference $\vec{\mathbb{v}}_{12}\Delta t - \vec{\mathbb{v}}_{01}\Delta t = \vec{a}_1\Delta t^2$, that is, so that

$$\vec{\mathbb{v}}_{12}\Delta t = \vec{a}_1\Delta t^2 + \vec{\mathbb{v}}_{01}\Delta t \qquad (N3.21)$$

We then transport the arrow $\vec{\mathbb{v}}_{12}\Delta t$ so its tail end coincides with point 1; its tip now indicates where to draw dot 2, the object's position at time t_2.

We can then repeat the process indefinitely, using the velocity arrow $\vec{\mathbb{v}}_{12}\Delta t$ and the computed acceleration arrow $\vec{a}_2\Delta t^2$ to find the object's position at time t_3, and so on. Figure N3.3b shows how this works. If you carefully compare figure N3.3a and N3.3b, you will see that the construction processes are simple inverses of each other.

The only minor complication is that we are usually given the object's initial position $\vec{r}_0$ and the object's instantaneous velocity $\vec{v}_0$ *at the same instant* $t = 0$. Note that the first average velocity vector $\vec{\mathbb{v}}_{01}$ is *not* the same as $\vec{v}_0 \equiv \vec{v}(0)$: the average velocity $\vec{\mathbb{v}}_{01}$ most closely approximates the object's instantaneous velocity at the *midpoint* of the time interval between instants $t = 0$ and $t = \Delta t$ (that is, at $t = \frac{1}{2}\Delta t$), *not* $t = 0$. This might seem like a small issue if Δt is small, but if we blithely set $\vec{\mathbb{v}}_{01} = \vec{v}_0$, then for practical sizes of Δt in many situations, we end up creating diagrams that are noticeably wrong.

Figure N3.4

Here is how we can more accurately locate point 1 on a trajectory diagram. Construct the arrows $\vec{v}_0 \Delta t$ with its *tip* at point 0 and $\vec{a}_0 \Delta t^2$ *centered* on point 0. Since $\vec{v}_0 \Delta t$ represents the velocity a *half* step to the past of $\vec{v}_{01} \Delta t$'s center value instead of a full step as usual, adding *half* of $\vec{a}_0 \Delta t^2$ (which automatically happens when we center it on 0) yields the correct arrow $\vec{v}_{01} \Delta t$. Transporting $\vec{v}_{01} \Delta t$ as usual then accurately locates point 1.

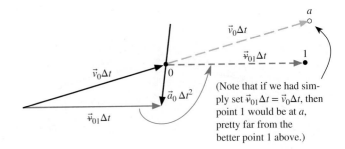

(Note that if we had simply set $\vec{v}_{01} \Delta t = \vec{v}_0 \Delta t$, then point 1 would be at a, pretty far from the better point 1 above.)

The special procedure for correctly locating point 1

Figure N3.4 shows a method for more accurately constructing the first velocity arrow $\vec{v}_{01} \Delta t$ and locating point 1. Draw the initial velocity arrow $\vec{v}_0 \Delta t$ so its *tip* is at point 0. Normally, the velocity arrow whose tip is at a given point most closely approximates the object's instantaneous velocity a half time step *before* that point, but in this special case the arrow $\vec{v}_0 \Delta t$ is the object's *exact* instantaneous velocity *at* point 0 by definition. The velocity $\vec{v}_{01} \Delta t$ thus best represents the object's instantaneous velocity at a time only $\frac{1}{2} \Delta t$ later than the instant where the previous velocity arrow is most accurate instead of a full time step Δt that is usually the case. Therefore, the difference between $\vec{v}_0 \Delta t$ and $\vec{v}_{01} \Delta t$ should be only *half* of what it usually would be, that is, $\frac{1}{2} \vec{a}_0 \Delta t^2$ instead of $\vec{a}_0 \Delta t^2$. An easy way to do this is to draw the arrow $\vec{a}_0 \Delta t^2$ the usual length but then *center* it on point 0.

So here is the process for *accurately* constructing $\vec{v}_{01} \Delta t$ from $\vec{v}_0 \Delta t$:

1. Draw the arrow $\vec{v}_0 \Delta t$ with its *tip* at point 0.
2. Calculate and draw the acceleration arrow $\vec{a}_0 \Delta t^2$ *centered* on point 0.
3. Draw $\vec{v}_{01} \Delta t$ from the tail of $\vec{v}_0 \Delta t$ to the tip of the centered arrow $\vec{a}_0 \Delta t^2$.
4. Move $\vec{v}_{01} \Delta t$ so its tail is at point 0; its tip is then point 1. One can then go on from here as before.

This process works in *all cases* (if Δt is small)

We can use this construction process *anytime* we know the object's velocity and position at time $t = 0$ and the object's acceleration at all times. This is great, because it empowers us to calculate trajectories in realistic situations where solving Newton's second law mathematically would be difficult or impossible. However, the construction method assumes that the object's average velocity during an interval is actually *equal* to the instantaneous velocity at the center of that interval, and the same holds for the acceleration. In general, this is true only in the limit that Δt goes to zero, so the constructed trajectory is only an *approximation* if $\Delta t \neq 0$. Still, as long as Δt is fairly small compared to the time it takes the acceleration to change appreciably (and we use the method above to construct the first velocity arrow), the constructed trajectory will be reasonably accurate.

Example N3.5

Problem: Suppose we launch a marble with an initial velocity of 1.0 m/s at an angle of 30° above the horizontal, and it subsequently falls freely. Use the graphical construction method to predict its future trajectory. (Suggestion: Set $\Delta t = 0.03$ s; this yields a diagram with a convenient size.)

Solution With $\Delta t = 0.03$ s, the marble's acceleration arrow at all points on the diagram should have a length of $|\vec{g}| \Delta t^2 = (9.8 \text{ m/s}^2)(0.03 \text{ s})^2 = 0.0088$ m long and always points downward. The length of the marble's initial velocity vector arrow should be $\vec{v}_0 \Delta t = (1.0 \text{ m/s})(0.03 \text{ s}) = 0.03 \text{ m} = 3.0$ cm. Figure N3.5

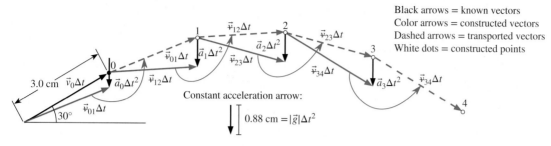

Figure N3.5
An actual-size trajectory diagram for the freely falling marble discussed in example N3.5.

shows the resulting constructed trajectory. Note how I have handled the construction of the first average velocity arrow using the method described on the previous page. You may already know that a freely falling object generally follows a parabolic path: note that our constructed trajectory looks plausibly parabolic.

Exercise N3X.3
Construct the first few steps of the trajectory diagram for a marble whose initial velocity is 0.80 m/s and horizontal. Use $\Delta t = \frac{1}{30}\,\text{s} = 0.033\,\text{s}$.

Constructing trajectory diagrams by hand is not very accurate and quickly becomes tedious. However, because the process is simple, repetitive, and universally applicable, it provides an excellent foundation for a computer algorithm to compute trajectories.

This process provides a good basis for a computer calculation

The web app Newton (posted at sixideas.pomona.edu/resources.html) uses precisely this algorithm to construct two-dimensional trajectories rapidly and accurately whenever you can specify the object's acceleration as a function of $\vec{r}$ (position relative to the origin), $\vec{v}$ (the object's instantaneous velocity), and/or time t. In chapters N9 through N11, we will find the Newton app to be a very powerful tool for computing trajectories in cases where symbolic integration is difficult or impossible.

The Newton web app

As the app gets more frequent updates than this volume, I won't describe the details of its operation here (such instructions might well be out of date by the time this volume goes to press!). I strongly urge you to go to sixideas.pomona.edu/resources.html and read the "About" information for the app. Suffice it to say here that the Newton app simply automates the construction procedure described in this section. You have the option of displaying the constructed arrows if you want to see what the application is doing, or simply displaying the trajectory dots if you are more interested in results. Because the application does all the constructing far more accurately than humanly possible, you can also choose smaller time steps than would be practical when constructing a trajectory diagram by hand, which will make the computer's trajectory diagram a better approximation to the object's actual trajectory.

TWO-MINUTE PROBLEMS

N3T.1   An object's x-velocity $v_x(t)$ is shown in the boxed graph at the top left. Which of the other graphs in the set most correctly describes its x-position?

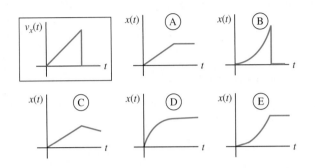

N3T.2   An object's x-acceleration $a_x(t)$ is shown in the boxed graph at the top left. Which of the other graphs in the set most correctly describes its x-velocity?

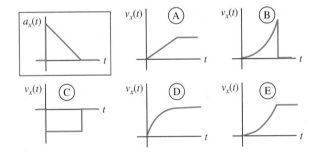

N3T.3   If a car has an x-acceleration of $a_x(t) = -bt + c$, and its initial x-velocity at time $t = 0$ is $v_x(0) = v_0$, which function below best describes $v_x(t)$?
A. $-b$
B. $-b + v_0$
C. $\frac{1}{2}bt^2 + ct + v_0$
D. $-\frac{1}{2}bt^2 + ct + v_0$
E. $-2bt^2 + v_0$
F. $-\frac{1}{2}bt^2 + v_0$

N3T.4   If a car's x-position at time $t = 0$ is $x(0) = 0$ and it has an x-velocity of $v_x(t) = b(t - T)^2$, where b and T are constants, which function below best describes $x(t)$?
A. $x(t) = 2b(t - T)$
B. $x(t) = 3b(t - T)^3$
C. $x(t) = \frac{1}{3}b(t - T)^3$
D. $x(t) = \frac{1}{2}b(t - T)$
E. $x(t) = \frac{1}{3}b[(t - T)^3 + T^3]$
F. Other (specify)

N3T.5   Suppose you are preparing an actual-size trajectory diagram of a freely falling object. The time interval between positions is 0.02 s. How long should you draw the acceleration arrows on your diagram?
A. 9.8 m
B. 0.20 m
C. 0.04 m
D. 3.9 cm
E. 0.39 cm
F. Other (specify)

N3T.6   At time $t = 0$ a person is sliding due east on a flat, frictionless plane of ice. The net force on this person is due to a battery-powered fan the person holds that exerts a southward thrust force on the person. Assuming that drag is negligible, the eastward component of the person's velocity is unaffected by this force. True (T) or false (F)?

N3T.7   Consider the person described in problem N3T.6. The person's trajectory will look most like which of the following? (The dot shows the person's position at $t = 0$ and east is to the right and north to the top.)

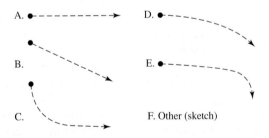

HOMEWORK PROBLEMS

Basic Skills

N3B.1   Construct a graph of the object's x-velocity as a function of time for an object whose x-acceleration is as given below. Assume that $v_x(0) = 0$.

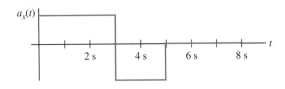

N3B.2   Construct a graph of x-position as a function of time for an object whose x-velocity is as given below. Assume that $x_0 = (0)$.

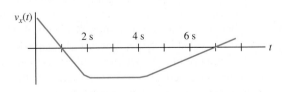

N3B.3   A stone dropped from rest from the middle of a bridge hits the water below 2.5 s later. How far is the bridge above the water? (Ignore air resistance.)

N3B.4   An object freely falling from rest for 5.0 s will have what final speed? (Express your result in both meters per second and miles per hour, and ignore air resistance.)

N3B.5   Suppose you are constructing an actual-size motion diagram of a falling object with an initial horizontal velocity of 2.0 m/s. Take the time interval between position points to be 0.05 s.
(a) How long should you draw the object's initial velocity arrow on the diagram?
(b) Its acceleration arrows?

N3B.6   Suppose you are trying to construct an actual-size motion diagram of a falling object with an initial horizontal velocity of 4.0 m/s. Take the time interval between position points to be 0.15 s.
(a) How long should you draw the object's initial velocity arrow on the diagram?
(b) Its acceleration arrows?

Modeling

N3M.1   Imagine that the net force on a car moving in one dimension is initially large and forward, but subsequently decreases linearly to zero and then remains zero thereafter. Draw graphs of $a_x(t)$, $v_x(t)$, and $x(t)$ for this car, assuming that it starts from rest at $x = 0$.

N3M.2   A car starts at $t = 0$ from rest at $x = 0$ and accelerates at a constant rate until it reaches a cruising speed of 15 m/s. After maintaining that speed for a while, the driver of the car, seeing that a bridge ahead is out, brakes suddenly, and the car comes to rest in a relatively short time. After remaining at rest for a few seconds, the driver then backs up at about 3 m/s. Draw qualitatively accurate graphs of the car's x-position, x-velocity, and x-acceleration as functions of time. Assume that the road is straight and the car initially travels in the positive x direction.

N3M.3   Suppose that the x-position of one car is given by the function $x_1(t) = bt^2$ (with $b = 2.5$ m/s^2), whereas the position of another car is given by the function $x_2(t) = ct$ (where $c = 18$ m/s). Note that both cars have a position equal to zero at $t = 0$. At time $t = 5.0$ s, which car is moving faster? Which one is farther ahead?

N3M.4   Suppose that a car's x-acceleration during a certain time period is given by the function $a_x(t) = -bt$. If the car's x-velocity at $t = 0$ is 32 m/s, and $b = 1.0$ m/s^3, how long will it take the car to come to rest?

N3M.5   The z-velocity of a deep-sea probe descending into the ocean is computer controlled to be $v_z(t) = c + bt^2$ (where c is a constant and $b = 0.040$ m/min^3) until the probe comes to rest. At $t = 0$, $z(0) = -22$ m (that is, 22 m below the ocean surface) and $v_z(0) = -120$ m/min.

(a) What are the units of c? What value must c have?
(b) Find the probe's z-acceleration as a function of time.
(c) At what time and position will the probe come to rest?

N3M.6   A spaceship is approaching Starbase Beta at an initial velocity of 130 km/s in the +x direction. The ensign sets its computer to initiate a braking program starting at time $t = 0$. After this time, the computer controls the ship until the ship docks at time $t = T$ so that its x-acceleration is $a_x(t) = b(t - T)$, where $b = 0.26$ m/s^3. Note that $a_x < 0$ for $t < T$, but gets smaller as t approaches the docking time T.
(a) What must T be so that the ship's x-velocity is also zero at the time of docking?
(b) How far is the ship from the station at $t = 0$? (*Hint:* Use equation NB.14 in appendix NB.)

N3M.7   A mad scientist invents an antigravity device that shields the region of space above a large metal plate from the earth's gravitational field. Specifically, if the device is turned on at time $t = 0$, the effective magnitude of the gravitational acceleration in the region above the plate decreases exponentially with time according to $|\vec{a}| = |\vec{g}|e^{-qt}$, where $|\vec{g}|$ is the usual gravitational field strength and q is a constant. When enough time has passed that qt becomes large, the effective gravitational field above the plate becomes very small compared to $|\vec{g}|$. Assume that the value of q is 3.0 when expressed in the appropriate SI units.
(a) What are the SI units of q?
(b) The mad scientist keeps a 7-kg bowling ball on a shelf above the plate. When the mad scientist throws the gigantic wall switch to turn on the apparatus one day, the vibrations jostle the bowling ball loose, and it rolls off the shelf. The scientist watches in horror as the ball smashes into the plate 1.0 s later. If the ball leaves the shelf with essentially zero vertical velocity at time $t = 0$, what is its speed when it smashes into the delicate apparatus? (*Hints:* See equation NB.17 in appendix NB, and set $b = -q$. Note that $e^0 = 1$.)
(c) How high was the shelf above the plate?

N3M.8   Suppose that if you step hard on a certain car's accelerator at $t = 0$ its forward acceleration is given by $a_x(t) = a_0 \sin \omega t$ for $0 \le \omega t \le \pi$ and becomes zero afterward, where $a_0 = 5.0$ m/s^2 and ω is a constant with a value of $\pi/5$ when expressed in the appropriate SI units. Note that the car's acceleration is initially zero when $t = 0$ but increases to a maximum of a_0 as the engine reaches its maximally efficient speed but then decreases as drag and friction forces begin to oppose the car's motion significantly. When the car reaches its cruising speed at a time t such that $\omega t = \pi$, the car's acceleration becomes zero. (This is not likely to be the actual acceleration function for any realistic car, but we can use it as a simple model.)
(a) What are the SI units of ω?
(b) Assume that the car starts from rest at $x = 0$. What are the car's x-velocity and x-position at time $t = 5.0$s? (*Hint:* The derivative of $\sin \omega t$ is $\omega \cos \omega t$ and the derivative of $\cos \omega t$ is $-\omega \sin \omega t$. What, therefore, are the antiderivatives of $\sin \omega t$ and $\cos \omega t$?)

N3M.9 Suppose we throw a marble with an initial speed of 2.2 m/s at an angle of 60° above the horizontal.
(a) Draw a trajectory diagram for the marble, using $\Delta t = \frac{1}{30}$ s = 0.033 s for at least 10 time steps, and estimate from your diagram how long it takes the marble to reach the peak of its trajectory.
(b) Use the Newton application to check your work.

N3M.10 Suppose an outfielder throws a baseball with an initial speed of 12 m/s in a direction 30° up from the horizontal.
(a) Use a trajectory diagram to determine the height of the peak of its trajectory. I suggest using a time step of $\Delta t = 0.2$ s and a scale of 1.0 m (in reality) = 2.0 cm (on diagram).
(b) Use the Newton application to check your work.

N3M.11 Suppose an air puck slides frictionlessly on a level air table. The puck is connected to a string going through a hole in the table's center. A clever student pulls on the string with a force having the right constant magnitude to cause the puck to accelerate with a constant magnitude of 1.0 m/s². Assume that at time $t = 0$, the puck is 4.0 cm from the center of the table in the $-x$ direction and is moving at a speed of 13 cm/s in the $+y$ direction.
(a) Construct a trajectory diagram that shows the puck's position at subsequent instants of time separated by $\Delta t = 0.1$ s until $t = 1.15$ s. [*Hint:* the main difference between figure N3.5 and what you will draw is that the acceleration arrow you should draw through each position dot during step 1 should point from the puck's position at that instant *toward the table's center,* since the string's tension force acts in that direction.]
(b) Check your work using the Newton program, and submit a printout to compare to your drawing.

N3M.12 In this problem, we will explore the necessary conditions for uniform circular motion.
(a) According to chapter N1, an object moving in a circle of radius r at constant speed $|\vec{v}|$ experiences an acceleration of constant magnitude $|\vec{a}| = |\vec{v}|^2/r$ directed toward the circle's center. But is the converse true? Does an object having a constant magnitude of acceleration toward the origin necessarily move with a constant speed in a circle if its initial velocity is perpendicular to its position? Set the time step to be 0.05 s, the object's initial position to be $[x, y] = [1 \text{ m}, 0 \text{ m}]$, its initial velocity to be $[v_x, v_y] = [0 \text{ m/s}, 1 \text{ m/s}]$ (note: $\vec{v} \perp \vec{r}$), and the acceleration to be -1 m/s² in the r direction (= 1 m/s² in the $-r$ direction). Note that the acceleration has a magnitude equal to $|\vec{v}|^2/r$ for the given initial conditions. Does this generate a circular trajectory with constant speed and the right radius? Submit a graph justifying your conclusions.
(b) Reset the time step size to 0.05 s but increase the velocity to 1.2 m/s in the y direction while leaving everything else the same. Now is the path circular? Does making the time step smaller change your result? Submit at least one graph in support of your claims.

(c) Instead of making the acceleration constant and directed toward a central point, make it constant and always perpendicular to the object's velocity. Is the orbit circular now? For all speeds? If so, are the trajectories you get consistent with $a = |\vec{v}^2|/r$? Submit at least one graph with a speed not equal to 1 m/s to support your case.

Derivations

N3D.1 Integrate the following functions from 0 to t.
(a) $f(t) = bt$
(b) $f(t) = b(t - T)$ (b and T are constants)

N3D.2 Integrate the following functions from 0 to t.
(a) $f(t) = bt^3$
(b) $f(t) = b/(t + T)^{1/2}$ (b and T are constants)

N3D.3 Check that if you use equation N3.2 to evaluate $v_x(t)$ in the situation in example N3.3 and set the constant of integration C_1 so that $v_x(0) = 0$, you get the same result as in equation N3.9.

N3D.4 An object's x-velocity is $v_x(t) = v_0 e^{-bt}$, where v_0 has units of m/s and b has units of s⁻¹. Find the object's x-position as a function of time if $x(0) = 0$.

Rich-Context

N3R.1 You are a police officer. Your squad car is at rest on the shoulder of an interstate highway when you notice a car fitting the description given in an all-points bulletin passing you at its top speed of 85 mi/h. You jump in your car, start the engine, and find a break in the traffic, a process which takes 25 s. You know from the squad car's manual that when it starts from rest with its accelerator pressed to the floor, the magnitude of its acceleration is $|\vec{a}| = a_0 - bt^2$ (where $a_0 = 2.5$ m/s² and $b = 0.0028$ m/s⁴) until $a_0 = bt^2$, and then it remains zero thereafter. Can you catch the car before it reaches the next exit 5.3 mi away?

N3R.2 You are traveling in the fabled Mines of Moria when you come across an open water well in the floor of a tunnel. To find out how deep the well is, you drop a stone into the well and count the seconds until you hear the splash. If you hear the splash a time $t_h = 5.0$ s after you drop the stone, roughly what is the depth D of the well? Be sure you describe any approximations you have to make to do the problem. [*Hints:* The speed of sound is $|\vec{v}_s| = 340$ m/s. Let the actual time of the splash be t_s. This is different from t_h: why? I strongly recommend you solve for D symbolically before plugging in numbers (or you will likely become totally lost), but pay very careful attention to signs: in particular, make sure that D is positive. If you end up with two possible answers, select the answer that leads to a positive value for t_s. Why is a negative t_s absurd?]

ANSWERS TO EXERCISES

N3X.1 In this case, equations N3.6 imply that

$$v_x(t) = v_{0x} - at \qquad \text{and} \qquad x(t) = v_{0x}t - \tfrac{1}{2}at^2 \quad \text{(N3.22)}$$

At $t = 5.0$ s, $v_x(5.0 \text{ s}) = 2.0$ m/s, $x(5.0 \text{ m}) = +3.5$ m.

N3X.2 The car's x-acceleration at $t = 0$ is the same for both cases, but in the second example the acceleration decreases with time and thus is generally less than in the first case. Therefore, it makes sense that it should reach a smaller final x-velocity and not go as far in the same time interval.

$$v_x(0) = \frac{b}{2T^2} - \frac{b}{2(0 + T)^2} = \frac{b}{2T^2} - \frac{b}{2T^2} = 0 \qquad \text{(N3.23a)}$$

$$x(0) = \frac{b}{2}\left(\frac{0}{T^2} + \frac{1}{0 + T} - \frac{1}{T}\right) = \frac{b}{2}\left(\frac{1}{T} - \frac{1}{T}\right) = 0 \qquad \text{(N3.23b)}$$

N3X.3 The trajectory diagram should look like this:

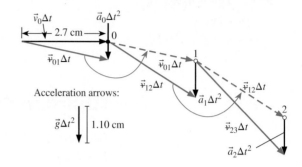

N4 Statics

CORE

Chapter Overview

Section N4.1: Forces from Motion: An Overview

This chapter begins the "Forces from Motion" subdivision of the unit. In this subdivision, we will consider progressively more complicated cases in which we use Newton's second law and our knowledge about an object's motion to infer things about the forces acting on the object. This chapter kicks off the subdivision by considering the simplest kind of motion: *no* motion.

Section N4.2: Introduction to Statics

We call any problem involving an extended object at rest a **statics problem**. Since the acceleration of the center of mass of an object at rest is zero, Newton's second law implies that *the net external force on the object is zero:*

$$\vec{F}_1 + \vec{F}_2 + \cdots = m\vec{a} = 0 \tag{N4.1}$$

The simplest statics problems involve only this principle.

This section carefully discusses the seemingly simple case of determining the force that a hanging mass m exerts on the hook from which it is suspended. While we all know that the force is $m|\vec{g}|$ downward, carefully arguing this displays how even simple cases can entail a multistep argument involving both Newton's second and third laws. This section also notes that we generally call interactions mediated by strings *tension* interactions even when the actual physical connection between the string and the object involves a compression interaction.

Section N4.3: Statics Problems Involving Torque

Some statics problems involve cases in which forces act at different places on an extended object at rest. The angular momentum of an object at rest is not changing; the net torque on the object (around any origin O you might choose) *must also be zero:*

$$\vec{\tau}_1 + \vec{\tau}_2 + \cdots = \frac{d\vec{L}}{dt} = 0 \tag{N4.2}$$

An extended object at rest must satisfy equations N4.1 and N4.2 simultaneously. As the examples illustrate, these equations taken together provide a very powerful tool for calculating forces that intuitively look impossible to determine.

Section N4.4: Solving Force-from-Motion Problems

This section describes a problem-solving framework for *any* problem in which we use Newton's second law and information about an object's motion to determine forces. The *translation* step of such a problem should do the following:

1. Display the object in its context.
2. Define necessary *symbols*.
3. Provide a list of known symbols.
4. Define an appropriate coordinate system.

The *modeling* step of such a problem must answer these questions: (1) What is the object's acceleration $\vec{a}$? (2) What are *all* the forces acting on the object? (3) How can we write these force and/or acceleration vectors as column vectors? (4) What equations do we need in addition to Newton's second law? (5) What approximations and/or assumptions must we make? To address these needs efficiently:

1. Draw an acceleration arrow (or write $\vec{a} = 0$) on your main diagram.
2. Draw a separate free-body diagram of the object (this also defines force symbols).
3. Write Newton's second law (the master equation) in column-vector form, explicitly indicating signs (when possible).
4. Link in enough other equations and/or cancel terms until you have enough equations to solve for the unknowns.
5. Explain any assumptions and approximations.

Following these steps will produce an adequate conceptual model for the problem. You should use our usual guidelines for doing the *math* and *check* parts of the solution (for example, do algebra with symbols, track units, check your results, and so on).

Section N4.5: Solving Statics Problems

If a problem also requires that we balance torques on an object, we should modify the first three modeling steps above as follows:

1. Write $\vec{a} = 0$ *and* $\vec{\tau}_{net} = 0$ on your diagram.
2. Instead of a free-body diagram, we draw a **torque diagram** that shows (a) a labeled arrow for each external force on the object, with its tail starting at the point where the force acts, (b) coordinate axes, (c) an origin O, and (d) a labeled arrow for the position vector for each point where a force acts, relative to O.
3. Write Newton's second law in column vector form *and* the relevant component of the torque equation (explicitly indicating signs when possible).

One can use the checklists below for such statics problems (commented examples in the section illustrate how).

Statics Checklist (no torque)
☐ Show coordinate axes in all drawings.
☐ Draw a *free-body diagram* (*in addition* to a general picture of the situation).
☐ Write Newton's second law (with $\vec{a}$ set to zero) in *column-vector* form.
☐ Make all signs *explicit* when possible.

General Checklist
☐ Define symbols.
☐ Draw a picture.
☐ Describe model & assumptions.
☐ Use a master equation.
☐ List knowns and unknowns.
☐ Do algebra symbolically.
☐ Track units. ☐ Check result.

Statics Checklist (with torque)
☐ Show coordinate axes in all drawings.
☐ Draw a *torque diagram* with each force arrow starting at the point where the force acts, and position arrows from a specified origin O to each such point.
☐ Write Newton's second law (with $\vec{a}$ set to zero) in *column-vector* form.
☐ Write the relevant component of the torque equation (note signs!)
☐ Specify which component this is.
☐ Make all signs *explicit* when possible.

General Checklist
☐ Define symbols.
☐ Draw a picture.
☐ Describe model & assumptions.
☐ Use a master equation.
☐ List knowns and unknowns.
☐ Do algebra symbolically.
☐ Track units.
☐ Check result.

N4.1   Forces from Motion: An Overview

An overview of the chapters in this subdivision

This chapter opens the "Forces from Motion" subdivision of the unit. In this subdivision, we will explore a variety of situations in which we use an object's known motion and Newton's second law to determine one or more of the forces acting on the object.

Knowing how to infer forces from motion is useful in many practical physical contexts. For example, knowing the force that a rope must exert in a given circumstance can help us ensure that we choose a rope that will not break. In other cases, the presence or absence of contact forces can tell us whether an object is touching another object. Knowing how strongly friction forces oppose an object's motion can tell us how hard we have to push on the object to get it moving or to keep it moving. These are just some of the many practical applications of knowing how to determine forces from motion.

The simplest kind of motion is *no* motion, so in *this* chapter we focus on what we can learn about the forces acting on an object that either we see does not move or we want to ensure does not move. In subsequent chapters, we consider progressively more complicated motions: linear motion in chapters N5 and N6 and circular motion in chapter N7. Chapter N8 closes the subdivision by considering cases in which observing an object's motion from an improper reference frame can tempt us to infer the existence of forces that are really not there (learning to recognize and handle such cases correctly is essential if one is to apply Newton's second law successfully in many realistic situations). Taken together, these chapters provide a solid introduction to this important use of Newton's second law.

N4.2   Introduction to Statics

Definition of a statics problem

With this big picture in mind, let's begin considering what we can learn from the fact that an object is at rest. Imagine an extended object that has various external forces ($\vec{F}_1$, $\vec{F}_2$, ...) acting on it. We call any problem involving an extended object at rest a **statics problem**.

If the object is *completely* at rest, it means its center of mass is at rest and thus is not accelerating. Newton's second law then implies that *the net external force on the object is zero*:

$$\vec{F}_1 + \vec{F}_2 + \cdots = M\vec{a} = 0 \qquad (N4.1)$$

The simplest statics problems involve only this principle.

Example N4.1

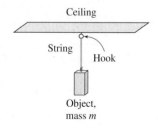

Ceiling

String

Hook

Object, mass m

Figure N4.1
An object hanging from the ceiling.

Problem: An object of mass m hangs in the earth's gravitational field from a string attached to a hook embedded in the ceiling. What is the magnitude of the downward force that the hanging mass exerts on the hook?

Solution Figure N4.1 shows a picture of the situation. We know intuitively that the answer is $m|\vec{g}|$, which is the magnitude of the gravitational force on the object. But this gravitational force acts on the hanging object, *not* on the hook directly! So how is the force on the hook connected to this gravitational force? This may seem like a trivial question, but thinking about this simple situation *carefully* illustrates issues and techniques that we will need to deal with more complicated problems.

The hanging object is not moving, so it is not accelerating. Therefore, by Newton's second law, the downward gravitational force on it must be canceled by some *upward* force of the same magnitude for the net force to be zero. Since the string is the only thing the object touches that could exert such a force, the string must exert an upward tension force of magnitude $m|\vec{g}|$ on the mass.

Now, by Newton's *third* law, this means that the mass exerts a *downward* force of magnitude $m|\vec{g}|$ on the string. Since the string is also at rest, its acceleration is zero, implying (by Newton's *second* law again) that the net external force on the *string* is zero. Therefore, to cancel the downward force on the string, an upward force of magnitude $m|\vec{g}|$ must act on the string, applied by the string's interaction with the hook. By Newton's *third* law, then, if the hook must exert an upward force of magnitude $m|\vec{g}|$ on the string, the string must exert a downward tension force of magnitude $m|\vec{g}|$ on the hook.

This is as we expected, but it is valuable to understand fully *why*.

Now that we know that we *can* analyze this situation fully, we will not do it again: you may safely assume in the future (as you may have already done in the past) that a hanging mass *at rest* exerts a force equal to its weight on whatever it is attached to. Note, however, that this simple statement is actually a summary of a rather involved line of reasoning!

Let's consider one last thing before we leave this case. In this example, I called the force exerted on the hook by its contact interaction with the string a *tension* force. But the interaction between the hook and the loop of string tied around it is technically a *compression* interaction, as shown in figure N4.2. (Think: if the atoms of the loop and hook could freely intermingle, the loop would simply slip *through* the hook as a wire would through pudding!) However, we *usually* consider a wire or string *not* as an object in its own right but rather simply as a mediator of an interaction between two other objects (the hanging weight and the hook in this case). Since this interaction opposes those objects' separation, it makes sense to think of it as a tension interaction. Even when we *do* consider the wire or string as an object in its own right (as we will in chapter N6), we usually ignore the exact nature of the connection and simply describe the interaction as being a tension interaction.

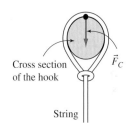

Figure N4.2
The top of the loop of the string connecting the hook to the hanging mass technically exerts a downward *compression* force on the hook. But we usually ignore the detailed nature of the connection between an object and a string and simply call this a *tension* force.

N4.3 Statics Problems Involving Torque

A more interesting set of statics problems involves situations in which multiple forces act at various *different positions* on an extended object at rest. Newton's second law still implies that the net external force on that object must be zero. But forces acting at different positions on an extended object also exert *torques* on that object. If the object is completely at rest, it is not rotating, so its angular momentum is not changing. This implies that *the net torque on the object* (around any origin you might choose) *must also be zero:*

The net torque on an object must also be zero if it is at rest

$$\vec{\tau}_1 + \vec{\tau}_2 + \cdots = \frac{d\vec{L}}{dt} = 0 \qquad (\text{N4.2})$$

Equations N4.1 and N4.2 must apply *simultaneously* to any object fully at rest.

In chapter C7, we saw that when a force $\vec{F}_1$ acts on the object at a specific position $\vec{r}_1$ relative to the origin, it exerts a torque $\vec{\tau}_1 = \vec{r}_1 \times \vec{F}_1$ about that origin. If you did not read that chapter, all you need to know here is that in such a case, $|\vec{\tau}_1| = |\vec{r}_1||\vec{F}_1| \sin\theta$, where θ is the angle between $\vec{r}_1$ and $\vec{F}_1$, and $\vec{\tau}_1$ points in the direction indicated by your right thumb when you curl your fingers in the direction that the torque *wants* to make the object rotate.

Example N4.2

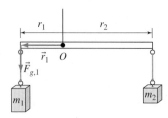

Figure N4.3

Where should we attach the wire to suspend this rod?

Problem: Two objects with masses m_1 and m_2 hang from the ends of a comparatively lightweight rod of length L. *(a)* At what position (relative to m_1) would you want to suspend the rod so the rod could hang horizontally at rest? *(b)* What upward force must we exert on the rod at that point to suspend the whole system at rest?

Solution **(a)** Figure N4.3 illustrates the situation. Suppose we suspend the rod from a point O that is a distance r_1 from where the object with mass m_1 is suspended. The distance to the point from which the object with mass m_2 is suspended is then $r_2 = L - r_1$. The rod will only remain at rest if the net torque on the rod around the suspension point O is zero. The supporting tension force $\vec{F}_3$ acts at O by definition, so it exerts zero torque around O. However, you can see that the tension force $\vec{F}_1$ exerted by mass m_1 will exert a counterclockwise torque on the rod (that is, $\vec{\tau}_1$ points directly out of the plane of the picture) while $\vec{F}_2$ from the other weight exerts a clockwise torque (that is, directly into the plane of the picture). Since these torques point in opposite directions, they will add to zero if they have the same magnitude. If the rod is horizontal, the force exerted on the rod by each hanging mass is perpendicular to the rod, and thus perpendicular to the position vector $\vec{r}_1$ or $\vec{r}_2$ from O to the place where the force is applied. This means that the sine of the angle θ between $\vec{r}$ and $\vec{F}$ in each case is $\sin 90° = 1$, so the magnitude of each torque is simply the product of the magnitude of the force and the distance $r/|\vec{r}|$ between O and the point where the force is applied. Since these torques must have equal magnitudes, we must have

$$r_1|\vec{F}_1| = r_2|\vec{F}_2| \tag{N4.3}$$

But the magnitudes of the applied forces here are just the magnitudes of the objects' weights ($|\vec{F}_1| = m_1|\vec{g}|$ and $|\vec{F}_2| = m_2|\vec{g}|$), so equation N4.3 becomes

$$r_1 m_1|\vec{g}| = r_2 m_2|\vec{g}| \tag{N4.4}$$

The problem says that we are given m_1, m_2, and L (and we know $|\vec{g}|$). We need to find either r_1 or r_2 to locate the hanging point. Equation N4.4 and the relationship $r_2 = L - r_1$ give us the two equations we need to solve for r_1 and r_2. Substituting $r_2 = L - r_1$ into equation N4.4 and dividing by $|\vec{g}|$ yield

$$r_1 m_1 = (L - r_1)m_2 \quad \Rightarrow \quad r_1(m_1 + m_2) = Lm_2 \quad \Rightarrow \quad r_1 = \frac{m_2 L}{m_1 + m_2} \tag{N4.5}$$

The value of r_1 here is actually the same as the distance between m_1 and the system's center of mass, implying that the suspension point O should be at the same horizontal position as the system's center of mass! (Perhaps this is no great surprise to you.)

(b) The external forces on the rod are the upward tension force $\vec{F}_3$ exerted by the supporting wire and the downward tension forces exerted by the hanging weights. Requiring that the total external force be zero means that the total upward force should balance the total downward force:

$$|\vec{F}_3| = |\vec{F}_1| + |\vec{F}_2| = (m_1 + m_2)|\vec{g}| \tag{N4.6}$$

that is, the upward force on the rod should be equal to the sum of the hanging objects' weights. Again, this should be no great surprise.

In example N4.2, Newton's second law and the law of zero torque were not linked, so we were able to work out the implications of these equations separately. In many problems, however, we have to combine both equations to solve the problem. Doing this allows us to calculate the magnitudes of contact forces that we might think at first glance are impossible to determine. Example N4.3 illustrates such a case.

Problems linking Newton's second law and the law of zero torque

Example N4.3

Problem: A uniform plank of mass m and length L sits on two supports, one a distance $2L/5$ and the other a distance $L/5$ from the plank's center. What are the magnitudes of the normal forces that each exerts on the plank?

Solution The drawing to the right shows the situation. We see that three forces act on the plank: the two supporting forces and the plank's weight (which we can model as acting on the plank's center of mass). If we take the origin O to be the plank's center of mass, then the torque that the weight force exerts around the origin is zero (since the distance between the point where the weight force is applied and the origin is zero by definition). The interaction with the left support exerts a torque into the plane of the drawing, while that with the right support exerts a torque out of the plane of the drawing. These torques will cancel if their magnitudes are equal. Since the normal forces exerted by these interactions are perpendicular to the plank, the magnitude of the torque due to each will be simply $r|\vec{F}|$, where $\vec{F}$ is the force that the interaction exerts and r is the distance between the origin and the point where the force acts. Since the magnitudes of the torques exerted by the supports must be equal, we have

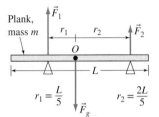

$$r_1|\vec{F}_1| = r_2|\vec{F}_2| \tag{N4.7}$$

We know that $r_1 = L/5$ and $r_2 = 2L/5$, but this is not enough information to find $|\vec{F}_1|$ and $|\vec{F}_2|$ separately. How can we proceed?

Well, since the center of mass of the plank is not moving, the three forces acting on the plank must add to zero as well: $\vec{F}_1 + \vec{F}_2 + \vec{F}_g = 0$. Since $\vec{F}_1$ and $\vec{F}_2$ both act upward and $\vec{F}_g$ acts downward, this will happen only if

$$|\vec{F}_1| + |\vec{F}_2| = |\vec{F}_g| = m|\vec{g}| \tag{N4.8}$$

This provides the second equation we need to solve for the two unknowns. Solving equation N4.8 for $|\vec{F}_2|$ and substituting the result into equation N4.7 allow us to solve for $|\vec{F}_1|$:

$$r_1|\vec{F}_1| = r_2\left(m|\vec{g}| - |\vec{F}_1|\right) \quad \Rightarrow \quad |\vec{F}_1|(r_1 + r_2) = r_2 m|\vec{g}|$$

$$|\vec{F}_1| = \frac{r_2}{r_1 + r_2} m|\vec{g}| = \frac{2\cancel{L}/5}{3\cancel{L}/5} m|\vec{g}| = \frac{2}{3} m|\vec{g}| \tag{N4.9}$$

Substituting this back into equation N4.8, we find that $|\vec{F}_2| = \frac{1}{3}m|\vec{g}|$.

Exercise N4X.1

We don't *have* to take the origin O to be the plank's center of mass. Redo the problem taking point O to be the position of the left-hand support. (You should get the same answer.)

N4.4   Solving Force-from-Motion Problems

A general framework for solving force-from-motion problems

In this section, I will describe a powerful framework for solving *any* problem in which we use Newton's second law and information about an object's motion to determine something about the forces acting on that object. In section N4.5, I will describe how we can extend and adapt this general framework to handle the special case of statics problems.

Recall from chapter C5 that solving virtually any physics problem involves four major steps: (1) You must first **translate** the problem from however it is initially stated into the language of mathematics. (2) You must then **model** the situation, a task that involves recognizing the physical principles that apply and describing whatever approximations and assumptions you need to make the problem solvable. (3) A good model will then provide a set of equations you can **solve** for the desired result. (4) Experienced physicists will then **check** the solution to ensure it makes sense.

What a good translation step should include

In the solution to any problem involving Newton's second law (including statics problems), the *translation* step should do the following:

1. Draw a picture of the object of interest in the context of its surroundings.
2. Define necessary *symbols* for masses, distances, angles, and so on.
3. Provide a list of known and unknown symbols.
4. Define an appropriate coordinate system.

We did not often need coordinate systems in conservation-of-energy problems, but as Newton's second law is a *vector* equation, a coordinate system is *essential* so that we can unambiguously define vector components in column vectors describing forces acting on the object and the object's acceleration.

Core questions that your model must resolve

The core questions that we must resolve in the *model* step of *any* force-from-motion problem solution are the following: What is the object's acceleration $\vec{a}$? What are *all* the forces acting on the object? How can we write these force and/or acceleration vectors in terms of symbols defined in the translation step? What equations do we need in addition to Newton's second law? What approximations and/or assumptions must we make?

The *modeling* step in a problem solution is not complete if it does not address each and every one of these concerns. To address these needs as efficiently as possible, I recommend following this sequence of sub-steps:

Sub-steps for constructing an efficient conceptual model

1. Draw an acceleration arrow (or write $\vec{a} = 0$) on your diagram.
2. Draw a separate free-body diagram of the object (isolated from its surroundings) that displays and labels *all* the forces acting on it. This drawing should also display your coordinate axes.
3. Your master equation in all such problems is Newton's second law. Write this equation in *column-vector form*, expressing the components in terms of the symbols you have defined earlier.
4. Link in additional equations and/or cancel common factors until you have enough equations to solve for all the unknowns.
5. Briefly explain any assumptions or approximations you make.

Important recommendations about executing these steps

When drawing a free-body diagram, think about *everything* that touches the object, and then draw arrows for all forces that might arise from those contact interactions. Then, add any forces that might arise from long-range interactions. Note that you should *never* draw arrows representing velocities or accelerations on a free-body diagram: the sole purpose of a free-body diagram is to display the *forces* that act on the object and where they act. Note also that $m\vec{a}$ is *not* a force acting on the object: it instead describes how the object *responds* to the vector sum of all forces acting on the object! Therefore, the quantity $m\vec{a}$ should *never* appear on a free-body diagram.

When writing Newton's second law in column-vector form, I personally find it valuable to express vector components in terms of symbols whose values we know to be *positive*. For example, if we know that a contact force on a certain object acts in the $-z$ direction, I would write

$$\begin{bmatrix} 0 \\ 0 \\ -|\vec{F}_C| \end{bmatrix} \quad \text{instead of} \quad \begin{bmatrix} 0 \\ 0 \\ F_{C,z} \end{bmatrix} \qquad \text{(N4.10)}$$

though either would be technically correct. The point is that this *explicitly* displays the sign of the last component instead of hiding it inside the $F_{C,z}$ symbol, and I find that displaying the sign in this way greatly reduces the probability that I will make a sign error. Of course, if we don't *know* whether $\vec{F}_C$ points in the $+z$ or $-z$ direction (or want to allow for both possibilities), then we must use $F_{C,z}$, but otherwise try to use positive symbols exclusively.

If you have trouble determining how many equations you need to solve a problem, you might put a check mark by each of the symbols you know in the master equation and any supporting equations as a way of helping you zero in on and count the unknown quantities.

Going through the five modeling sub-steps listed earlier will help you efficiently deal with the *modeling* step, which is the core of the solution. The steps where you *solve* the equations and *check* the result are similar to what you have done in previous problems. You should solve *symbolically* for any unknowns, *then* plug in numbers (keeping track of units), and check that your result has the correct units and a plausible sign and magnitude.

Following these steps also gives you a way to start a problem even when you are unsure about how to find a solution. Simply going through the steps will often help you work through any confusion you might have.

N4.5 Solving Statics Problems

The five modeling sub-steps described in section N4.4 are adequate as stated for any statics problem that does *not* involve torque. If a problem also requires that we balance torques on an object, we should modify the first three sub-steps for the modeling step as follows:

Modifications of the conceptual model steps for statics problems

1. Write both $\vec{a} = 0$ and $\vec{\tau}_{net} = 0$ on your basic diagram.
2. Draw a **torque diagram** (instead of a free-body diagram) as follows:
   a. Draw a picture of the object.
   b. Draw an arrow for each external force acting on the object, placing each arrow's tail at the point on the object *where the force acts*.
   c. Draw and label an origin O around which you will calculate torques as well as the coordinate axes you have defined.
   d. Draw and label the position vectors (relative to the origin O) of each point where a force acts.
3. In addition to writing Newton's second law in column-vector form, write down the *torque equation* $0 = \vec{\tau}_1 + \vec{\tau}_2 + \dots$. Typically, only one component of this vector torque equation will be relevant. These two equations *together* are your master equations: they need to be satisfied simultaneously by your solution.

In other regards, your solution for a statics problem will be the same as for any other force-from-motion problem involving Newton's second law.

In step 2, you can define the origin O to be *anywhere you like*. Defining O to be at the location where an unknown force acts is often helpful, because that unknown force will not then appear in the torque equation.

Comments on these modifications

Handling the torque equation

All the torque vectors in the torque equation will usually be either parallel to or opposite each other. If you define your coordinate system appropriately, these vectors will all have a single nonzero component along a common coordinate axis. When this is the case, I recommend writing only the relevant component of the torque equation (be sure to specify which component this is, though). You can express the relevant component of a torque $\vec{\tau}_1$ in terms of the torque's magnitude $|\vec{\tau}_1| = |\vec{r}_1||\vec{F}_1||\sin\theta|$ (where θ is the angle between $\vec{r}_1$ and $\vec{F}_1$), and an *explicit* sign, depending on whether the vector points in the positive or negative axis direction. Determine the torque's direction (and thus the component's sign) by using the right-hand rule. Don't use negative values for the angle θ, because this will bury the sign in the angle where one can't see it.

In unusual situations, the torques may *not* all lie along some convenient coordinate axis: in such cases, you will have to write the torque equation as a column-vector equation, like the force equation.

Statics-problem checklists

So, to summarize, here are checklists for solving statics problems:

Statics Checklist (no torque)	**General Checklist**
☐ Show coordinate axes in all drawings.	☐ Define symbols.
☐ Draw a *free-body diagram* (*in addition* to a general picture of the situation).	☐ Draw a picture.
	☐ Describe model & assumptions.
☐ Write Newton's second law (with $\vec{a}$ set to zero) in *column-vector* form.	☐ Use a master equation.
	☐ List knowns and unknowns.
☐ Make all signs *explicit* when possible.	☐ Do algebra symbolically.
	☐ Track units.
	☐ Check result.

Statics Checklist (with torque)	**General Checklist**
☐ Show coordinate axes in all drawings.	☐ Define symbols.
☐ Draw a *torque diagram* with each force arrow starting at the point where the force acts, and position arrows from a specified origin O to each such point.	☐ Draw a picture.
	☐ Describe model & assumptions.
	☐ Use a master equation.
☐ Write Newton's second law (with $\vec{a}$ set to zero) in *column-vector* form.	☐ List knowns and unknowns.
	☐ Do algebra symbolically.
☐ Write the relevant component of the torque equation (note signs!).	☐ Track units.
	☐ Check result.
☐ Specify which component this is.	
☐ Make all signs *explicit* when possible.	

The following commented examples display how one can use these problem-solving checklists to solve fairly complicated problems pretty easily. Example N4.4 does not involve torque, whereas example N4.5 does.

Example N4.4

Problem: A 65-kg rock climber hangs from two ropes while crossing a gap between two rocks. One end of each rope is tied to the climber's harness; the other ends are anchored 5.5 m to the left and 2.1 m vertically up and 7.2 m to the right and 1.5 m vertically up from the climber's harness, respectively. If the climber's full weight is to be supported by the ropes, what tension force must each rope exert? If the ropes can supply 2500 N of tension force without breaking, is the climber safe?

Solution Here is a diagram of the situation, a free-body diagram for the climber, and a list of knowns and unknowns:

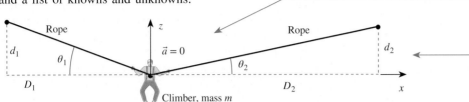

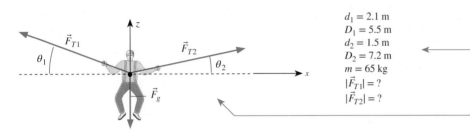

Statement that the object's acceleration is zero

Diagram of the situation that also defines symbols and coordinate axes

List of knowns and unknowns

Free-body diagram

$d_1 = 2.1$ m
$D_1 = 5.5$ m
$d_2 = 1.5$ m
$D_2 = 7.2$ m
$m = 65$ kg
$|\vec{F}_{T1}| = ?$
$|\vec{F}_{T2}| = ?$

Torque is not an issue in this problem: we are interested in only the motion (or lack thereof) of the climber's center of mass here. A climber "supported" by the ropes will be at rest, so $\vec{a} = 0$. The climber only touches the ropes and the air. The buoyant force exerted by the air on a person is negligible, and (unless a very stiff wind is blowing) any drag force exerted by the air will be negligible. Newton's second law therefore implies that $\vec{F}_{T1} + \vec{F}_{T2} + \vec{F}_g = 0$, or

$$\begin{bmatrix} 0 \\ 0 \\ 0 \end{bmatrix} = \begin{bmatrix} -|\vec{F}_{T1}|\cos\theta_1 \\ 0 \\ +|\vec{F}_{T1}|\sin\theta_1 \end{bmatrix} + \begin{bmatrix} +|\vec{F}_{T2}|\cos\theta_2 \\ 0 \\ +|\vec{F}_{T2}|\sin\theta_2 \end{bmatrix} + \begin{bmatrix} 0 \\ 0 \\ -m|\vec{g}| \end{bmatrix} \qquad \text{(N4.11)}$$

Statement that the object's acceleration is zero

Statement about the object touches (so we know what forces act)

Approximations and assumptions stated

Master equation: Newton's second law in column vector form

Note explicit signs!

The top and bottom components of this equation provide two equations in the two unknowns $|\vec{F}_{T1}|$ and $|\vec{F}_{T2}|$. The top component equation tells us that

$$|\vec{F}_{T1}|\cos\theta_1 = |\vec{F}_{T2}|\cos\theta_2 \quad \Rightarrow \quad |\vec{F}_{T2}| = |\vec{F}_{T1}|\frac{\cos\theta_1}{\cos\theta_2} \qquad \text{(N4.12)}$$

Substituting this into the bottom component equation yields

$$m|\vec{g}| = |\vec{F}_{T1}|\sin\theta_1 + |\vec{F}_{T1}|\frac{\cos\theta_1}{\cos\theta_2}\sin\theta_2 = |\vec{F}_{T1}|(\sin\theta_1 + \cos\theta_1\tan\theta_2)$$

$$|\vec{F}_{T1}| = \frac{m|\vec{g}|}{\sin\theta_1 + \cos\theta_1\tan\theta_2} = \frac{m|\vec{g}|}{\cos\theta_1(\tan\theta_1 + \tan\theta_2)} \qquad \text{(N4.13)}$$

Now note that

$$\tan\theta_1 = \frac{d_1}{D_1} = \frac{2.1 \text{ m}}{5.5 \text{ m}} = 0.382, \quad \tan\theta_2 = \frac{d_2}{D_2} = \frac{1.5 \text{ m}}{7.2 \text{ m}} = 0.208,$$

$$\Rightarrow \theta_1 = \tan^{-1}(0.382) = 20.9°, \quad \theta_2 = \tan^{-1}(0.208) = 11.8°$$

$$\Rightarrow |\vec{F}_{T1}| = \frac{(65 \text{ kg})(9.8 \text{ m/s}^2)}{\cos(20.9°)(0.382 + 0.208)}\left(\frac{1 \text{ N}}{1 \text{ kg} \cdot \text{m/s}^2}\right) = 1160 \text{ N} \qquad \text{(N4.14)}$$

Algebra with symbols

Calculation includes and tracks units

Substituting this into equation N4.12 yields

$$|\vec{F}_{T2}| = |\vec{F}_{T1}|\frac{\cos\theta_1}{\cos\theta_2} = (1160 \text{ N})\frac{\cos(20.9°)}{\cos(11.8°)} = 1110 \text{ N} \qquad \text{(N4.15)}$$

These both come out with the right units and signs (plus signs are appropriate for magnitudes) and seem reasonable. So the ropes *should* hold!

Final check of plausibility

Example N4.5

Problem: Suppose you are in charge of renovating a castle. A 580-kg drawbridge 6.6 m long spans a moat around the castle. One end of the drawbridge is connected by a hinge to the castle wall. The drawbridge is also held up by two chains, each of which is connected to a point on the drawbridge 4.4 m from the hinge. The bridge is raised by reeling in the chains, which enter the castle wall 4.4 m above the hinge. When the bridge is raised just above its support on the far side of the moat (but the bridge is still essentially horizontal), what magnitude of tension force must each chain be able to exert? Also find the magnitude of the force that the *hinge* must exert on the drawbridge. (The answers to these questions will help you determine what parts to procure to replace these rusted parts.)

Solution Here are the pictures we need.

List of knowns and unknowns

Diagram of the situation that also defines symbols and coordinate axes

Statement that the object's acceleration and net torque are zero

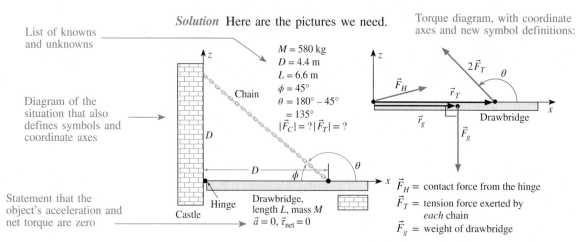

Torque diagram, with coordinate axes and new symbol definitions:

$\vec{F}_H$ = contact force from the hinge
$\vec{F}_T$ = tension force exerted by *each* chain
$\vec{F}_g$ = weight of drawbridge

Statement about the object touches (so we know what forces act)

Approximations and assumptions stated

Clarification of the y axis (which we can't easily see on the diagram)

Directions of all torques discussed

The drawbridge is at rest, so the drawbridge's acceleration and net torque must be zero. The drawbridge touches only the hinge, the air, and the two chains. We will treat contact forces with the air as negligible, and we will assume that each chain exerts the same tension force $\vec{F}_T$ on the drawbridge. Assuming that the drawbridge is uniform, its center of mass (which is the effective position where its weight acts) is located halfway along its length. We know nothing about either the magnitude or the direction of the contact force $\vec{F}_C$ the hinge exerts on the drawbridge, so it is convenient to define the origin O at the hinge so this force drops out of the torque equation. Note that in both diagrams, the y axis for a right-handed coordinate system points *into* the plane of the drawing. As one can show using the right-hand rule, the torque $\vec{\tau}_g$ due to the gravitational force $\vec{F}_g$ on the drawbridge points in this direction, and torque $\vec{\tau}_T$ due to the tension forces $\vec{F}_T$ points in the opposite direction, so only the y component of the torque equation $0 = \vec{\tau}_g + \vec{\tau}_T$, and the gravitational torque's y component is positive and the tension torque's y component is negative. Therefore, the y component of the torque equation is

First master equation

$$0 = |\vec{r}_g||\vec{F}_g||\sin 90°| - |\vec{r}_T|2|\vec{F}_T||\sin\theta| = (\tfrac{1}{2}L)\,M|\vec{g}| - 2D|\vec{F}_T||\sin\theta| \quad \text{(N4.16)}$$

Algebra with symbols

This equation implies that $2D|\vec{F}_T||\sin\theta| = \tfrac{1}{2}Lm|\vec{g}|$, or

Calculation includes and tracks units

$$|\vec{F}_T| = \frac{LM|\vec{g}|}{4D|\sin\theta|} = \frac{(6.6\text{ m})(580\text{ kg})(9.8\text{ m/s}^2)}{4(4.4\text{ m})|\sin 135°|}\left(\frac{1\text{ N}}{1\text{ kg}\cdot\text{m/s}^2}\right) = 3010\text{ N} \quad \text{(N4.17)}$$

Newton's second law in this case reads

Second master equation

$$\begin{bmatrix} 0 \\ 0 \\ 0 \end{bmatrix} = \begin{bmatrix} F_{H,x} \\ 0 \\ F_{H,z} \end{bmatrix} + \begin{bmatrix} -2|\vec{F}_T|\cos\phi \\ 0 \\ +2|\vec{F}_T|\sin\phi \end{bmatrix} + \begin{bmatrix} 0 \\ 0 \\ -m|\vec{g}| \end{bmatrix} \quad \text{(N4.18)}$$

If we substitute the result for $|\vec{F}_T|$ into the first and third component equations for Newton's second law, we find that

$$F_{H,x} = 2|\vec{F}_T|\cos\phi = 2(3010 \text{ N})\cos 45° = 4260 \text{ N} \qquad \text{(N4.19}a\text{)}$$

$$F_{H,z} = M|\vec{g}| - 2|\vec{F}_T|\sin\phi = (580 \text{ kg})\left(9.8 \frac{\text{m}}{\text{s}^2}\right)\left(\frac{1 \text{ N}}{1 \text{ kg} \cdot \text{m/s}^2}\right) - 2(3010 \text{ N})\sin 45°$$

$$= 5680 \text{ N} - 4260 \text{ N} = 1420 \text{ N} \qquad \text{(N4.19}b\text{)}$$

Symbolic algebra followed by a calculation that includes and tracks units

Therefore, the magnitude of the force on the hinge is

$$|\vec{F}_H| = \sqrt{F_{H,x}^2 + F_{H,y}^2} = \sqrt{(4260 \text{ N})^2 + (1420 \text{ N})^2} = 4490 \text{ N} \qquad \text{(N4.20)}$$

The units all come out right, and $\vec{F}_H$ is in the first quadrant (as we guessed in the drawing; this is the only way the vectors can add to zero). The magnitudes of these forces are comparable to the drawbridge's weight $m|\vec{g}| = 5680$ N, so everything seems plausible. Better cough up the money for the 10-kN-rated chains and hinge, just to be on the safe side (we don't want tourists falling into the moat, do we?).

Final check of plausibility

Exercise N4X.2

If the angle ϕ were larger than 45° when the drawbridge is horizontal, would the magnitude of the force exerted by the hinge increase, decrease, or remain the same? Explain. (*Hint:* Note that $\sin\phi = \sin\theta$.)

Exercise N4X.3

Suppose that people with a total mass of $\frac{1}{2}M = 290$ kg were standing on the end of the drawbridge farthest from the castle. Would 10-kN-rated chains still be able to support the drawbridge?

TWO-MINUTE PROBLEMS

N4T.1   A weight hangs from a string but is pulled to one side by a horizontal string, as shown. The tension force exerted by the angled string is
A. Less than the hanging object's weight.
B. Equal to the hanging object's weight.
C. Greater than the hanging object's weight.

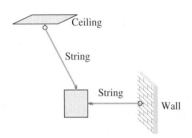

N4T.2   A person would like to pull a car out of a ditch. This person ties one end of a chain to the car's bumper and wraps the other end around a tree so that the chain is taut. The person then pulls on the chain perpendicular to its length, as shown in the picture. The magnitude of the force that the chain exerts on the car in this situation is
A. Much smaller than the force the person exerts on the chain.
B. About equal to the force the person exerts on the chain.
C. Much bigger than the force the person exerts on the chain.

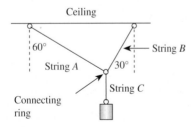

N4T.3   Suppose an object is suspended from strings as shown in the diagram below.

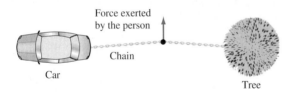

(a) Which exerts the largest force on the connecting ring?
(b) Which exerts the smallest force on that ring?
(*Hint:* Draw a free-body diagram of the ring.)
A. String *A*
B. String *B*
C. String *C*

N4T.4   The lid of a grand piano is propped open as shown. Which arrow most closely approximates the direction of the force that the hinge exerts on the lid? Assume that the prop exerts a force in parallel to the prop's length. (*Hint:* Draw a torque diagram of the lid.)

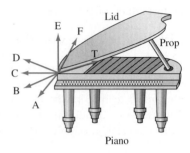

N4T.5   If we choose our origin to be the hinge in the situation discussed in problem N4T.4, the torque exerted by the lid's weight around the hinge is away from the viewer. T or F?

N4T.6   When viewed from above, suppose that a helicopter's rotor spins clockwise. The helicopter engine must continually exert a torque on the rotor to keep it spinning against the drag that the air exerts on the rotor. Note that a helicopter is usually designed so its center of mass is directly under the rotor. In order for the helicopter to hover motionless in the air, a small rotor at the helicopter's tail is necessary. As viewed by someone looking at the tail from the helicopter's front, the small rotor must blow air
A. To the left.
B. To the right.
C. Vertically upward.
D. Vertically downward.
E. In some combination of these directions.

N4T.7   A board of mass *m* lies on the ground. What is the magnitude of the force you would have to exert to lift *one end* of the board barely off the ground (assuming that the other end still touches the ground)?
A. $2m|\vec{g}|$
B. $m|\vec{g}|$
C. $\frac{1}{2}m|\vec{g}|$
D. The answer depends on the board's length.
E. Other (explain)

N4T.8   Suppose you continue to lift the board described in problem N4T.7. Assume that the force you exert is always perpendicular to the board and that one end of the board always remains on the ground. What happens to the magnitude of the force you exert on the end as the angle between the board and the ground increases? It
A. Increases.
B. Decreases.
C. Remains the same.
D. The answer depends on information not given.

HOMEWORK PROBLEMS

Basic Skills

N4B.1   Consider the situation discussed in problem N4T.1. If the diagonal string makes an angle of θ from the vertical, what is the magnitude of the tension force it exerts on the hanging mass, as a fraction or multiple of the magnitude $m|\vec{g}|$ of the hanging mass's weight? Explain carefully.

N4B.2   Imagine a glider with mass m on a frictionless air track that is inclined at an angle θ with respect to the horizontal. The glider is tied to the upper end of the track with a string that is parallel to the track. Find an expression for the tension force that this string exerts on the glider in terms of $m, |\vec{g}|$, and θ. Please explain your reasoning.

N4B.3   Suppose we place a 10-g weight at the 10-cm mark on a uniform meter stick, and we find that the meter stick now balances at the 45-cm mark. What is the mass of the meter stick? (*Hint:* Put the origin O at the balance point.)

N4B.4   Suppose a ladder leans against the house at an angle of 20° with respect to the vertical. A person with a weight of 650 N stands on the ladder at a point 1.0 m from its bottom. What are the magnitude and direction of the torque that this person exerts on the ladder around the point where the ladder touches the ground? (Express the direction as seen by the person standing on the ladder.)

N4B.5   A gymnast stands on a balance beam of length L that is supported above the gym floor by two supports at its ends, something like the situation shown below.

(Credit: Al Bello/Allsport/Getty Images)

Place the origin O at the point where the gymnast's foot touches the beam. Let's say that the gymnast is $\frac{1}{5}L$ from the right end, the mass of the gymnast is m, and the mass of the beam is M. Let the contact forces exerted on the beam by the right and left supports be $\vec{F}_{CR}$ and $\vec{F}_{CL}$, respectively.
(a) In terms of these quantities, what are the magnitude and direction of the torque exerted by the right support about the origin?
(b) What are the magnitude and direction of the torque exerted by the left support?
(c) What are the magnitude and direction of the torque exerted by the beam's weight?
(Express your directions relative to the picture.)

N4B.6   A skateboard rider with mass m stands on the middle of a skateboard of length L. Let's place the origin O at the skateboard's rear wheels. Let the upward forces exerted by the road on the skateboard's front and rear wheels be $\vec{F}_{NF}$ and $\vec{F}_{NR}$, respectively.
(a) In terms of these forces, what are the magnitude and direction of the torque exerted on the skateboard by the front wheels?
(b) What are the magnitude and direction of the torque exerted by the rear wheels?
(c) What are the magnitude and direction of the torque exerted by the skateboard rider? (Express your directions relative to the forward direction of the skateboard.)

N4B.7   A plank of length L and mass m leans against a wall. Assume that the point of contact between the plank and wall is nearly frictionless. Typically, the force of static friction $\vec{F}_{SF}$ between the ladder's bottom and the ground will have a maximum magnitude of $\mu_s|\vec{F}_N|$, where $\vec{F}_N$ is the normal force the ground exerts on the ladder and $\mu_s = 0.25$ to 0.6 or so (depending on the surface). Our eventual goal is to find the maximum angle θ that the plank can make with the wall before it starts slipping.
(a) The drawing below represents a partial torque diagram. Draw in the missing arrow for the force $\vec{F}_{SF}$.

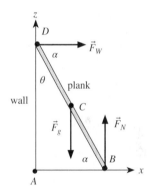

(b) Newton's second law in column-vector form for this situation is as shown below. Fill in the missing component and then write down the two relationships that this implies between the force magnitudes.

$$0 = \begin{bmatrix} |\vec{F}_W| \\ 0 \\ 0 \end{bmatrix} + \begin{bmatrix} 0 \\ 0 \\ -m|\vec{g}| \end{bmatrix} + \begin{bmatrix} \underline{} \\ 0 \\ 0 \end{bmatrix} + \begin{bmatrix} 0 \\ 0 \\ |\vec{F}_N| \end{bmatrix} \quad \text{(N4.21)}$$

(c) Define an origin. Which point is most convenient?
(d) Which component of the torques exerted about your origin by various forces are nonzero?
(e) Setting the net torque about B to zero requires that

$$0 = \underline{} + |\vec{F}_W|L|\sin\alpha| \quad \text{(N4.22)}$$

Fill in the missing item and solve for $|\vec{F}_W|$.
(f) Note that $\sin\alpha = \sin(90° - \theta) = \cos\theta$. Find $|\vec{F}_W|$ as a function of θ.
(g) We must have $|\vec{F}_{SF}| \le \mu_s|\vec{F}_N|$. Use this and the equations above to find the maximum θ as a function of μ_s.

N4B.8   Pole-vaulters hold their poles in front of them as they run at the beginning of their vault (see below).

(Credit: Clive Brunskill /Allsport/ Getty Images)

Explain qualitatively why the upward force that a vaulter exerts on the pole with his or her front hand must exceed the pole's weight.

Modeling

N4M.1   In the situation discussed in problem N4T.2, find the force that the chain can exert on the car if the distance between the car and the tree is 5.0 m and the length of the chain between the car and the tree is 5.2 m. (Estimate the force that a person can exert on the chain.)

N4M.2   A plank 3.0 m long and having a mass of 20 kg is supported at its ends. Imagine that a person with a mass of 50 kg stands 1.0 m from one end. What are the magnitudes of the forces exerted by the supports?

N4M.3   Imagine you have a plank 2.0 m long. You put a small piece of wood under the plank 60 cm from its far end, making the plank into a lever. A friend with a mass of 65 kg stands on the far (short) end of the plank. How much downward force do you have to exert on your end to lift your friend?

N4M.4   Suppose that a pole with a mass of 8 kg and a length of 1.8 m is connected to a wall so that the pole sticks out horizontally from the wall. One end of the pole is connected directly to the wall, while the other end is connected to a higher point on the wall by a chain that makes a 45° angle with respect to the pole. If a 65-kg person hangs from the center of the pole, what is the tension on the chain?

N4M.5   A certain board is 4.0 m long and rests horizontally and somewhat above the ground on two cylinders of wood, each supporting the board 0.5 m from the corresponding end of the board. (The axis of each cylinder is perpendicular to the length of the board.) The board has a mass of 21 kg. If a person with a mass of 68 kg steps on one end of the board, will it support that person?

N4M.6   A board has one end wedged under a rock having a mass of 380 kg and is supported by another rock that touches the bottom side of the board at a point 85 cm from the end under the rock. The board is 4.5 m long, has a mass of about 22 kg, and projects essentially horizontally out over a river. Is it safe for an adult with a mass of 62 kg to stand at the unsupported end of the board? If not, how far out on the board can one safely go? (*Hint:* The end of the board that is wedged under the rock is wedged *between* that rock and the ground, so the ground will exert an upward contact force on the board. The magnitude of that force can vary as needed. But what would happen if that force were to ever go to zero? What would that mean?)

N4M.7   The diagram below shows a 20-kg bucket of paint on top of a folding ladder that is much lighter.

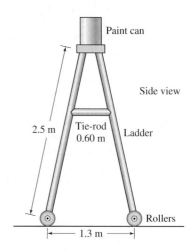

The ladder's feet have rollers to make it easy to move the ladder around the hard floor. What forces must the tie-rods exert on the ladder to keep the ladder from collapsing?

N4M.8   A ladder of length L and negligible mass leans against a wall, with its bottom $\frac{1}{2}L$ from the wall. Assume that the point of contact between the ladder and wall is nearly frictionless. The force of static friction $\vec{F}_{SF}$ between the ladder's bottom and the ground can have a maximum magnitude of $\mu_s|\vec{F}_N|$, where $\vec{F}_N$ is the normal force the ground exerts on the ladder and μ_s is a unitless constant that depends on the nature of the surface on which the ladder's bottom might slide. If the value of $|\vec{F}_{SF}|$ that is required to keep the ladder stationary is greater than $\mu_s|\vec{F}_N|$, the ladder will slip.
(a) What distance D along the ladder can a person of mass m safely climb up the ladder? (*Hint:* the answer is independent of the person's mass.)
(b) If $L = 5.0$ m and $\mu_s = 0.4$, what is D in meters?

N4M.9   A ladder of length L and mass m leans against the side of a house, making an angle of θ with the vertical. Assume that the ladder is free to slide at the point where it touches the side of the house (there is no significant friction) but that it is *not* free to slide on the ground. Find an expression for the normal force that the side of the house exerts on that end of the ladder in terms of $m, |\vec{g}|, L,$ and θ.

Rich-Context

N4R.1   A person's forearm consists of a bone (which we can model by a rigid rod) that is free to rotate around the elbow. When the forearm is held level with the elbow bent at a 90° angle (see below), the bone is supported by the biceps muscles, which are attached by a tendon to the bone a few centimeters away from the elbow.

(Credit: Courtesy of M. Oldstone- Moore/Jerry Martinez/Thomas Moore)

(a) By making measurements on your own forearm and appropriate approximations and estimations (be sure you describe them), estimate the tension force on the biceps tendon when you hold a 10-kg weight in your hand with your forearm level and elbow bent at 90°. (*Hint:* Feel where your biceps tendon is and measure its distance from the hinge of your elbow.)

(b) It turns out that the maximum magnitude of force that a mammalian muscle can exert is $|\vec{F}_{max}| = \sigma A$, where A is the cross-sectional area of the muscle and σ is a constant of proportionality $\approx$ 300,000 N/m². What minimum cross-sectional area must a person's biceps muscle have to hold the 10-kg weight?

(c) Does your result for part (b) seem plausible?

N4R.2   You are part of a research team in the deepest jungles of South America. At one point, the team has to cross a 20-m-wide gorge by stringing a cable across it and then crawling along the cable (see the photo below).

(Credit: O. Alamany & E. Vicens/The Image Bank/Getty Images)

After someone has managed to hook the cable on a rock on the other side of the gorge, a new team member connects the near end of the cable to a tree in such a way that the cable is very tight. A more experienced member makes the neophyte re-connect the cable so there is some slack.

(a) Explain why this is necessary. (Assume that the cable cannot stretch much, even when it is under tension.)

(b) Is the tension on the cable greater when a person is hanging near one side of the gorge or in the middle? (*Hint:* When the person is near one side, part of the cable is nearly vertical, while the other part is nearly horizontal. What must the tension force on the nearly vertical part be, roughly? On the horizontal part?)

(c) If the cable can exert a tension force of 3500 N without breaking and the heaviest member of the group has a mass of 110 kg, how much slack do you need? (That is, how much longer must the cable be than 20 m?)

ANSWERS TO EXERCISES

N4X.1   The torque exerted by the left-hand support is now zero, but gravity exerts a clockwise torque on the plank around the new O. Assuming that the gravitational force effectively acts as if it is applied to the object's center of mass, the distance from our new O to the center of mass is $r_{CM} \equiv \frac{1}{5}L$. Force $\vec{F}_2$, which is now applied at a distance of $r_2 = \frac{3}{5}L$ from the origin, exerts a counterclockwise torque on the plank. In each case, the forces are exerted perpendicular to the plank, so the torque magnitudes are simply $r_{CM}|\vec{F}_g|$ for the gravitational force and $r_2|\vec{F}_2|$ for the other force. These torques can cancel only if their magnitudes are equal, so we must have $r_{CM}m|\vec{g}| = r_2|\vec{F}_2|$

$$\Rightarrow \quad |\vec{F}_2| = \frac{r_{CM}}{r_2}m|\vec{g}| = \frac{\cancel{L}/5}{3\,\cancel{L}/5}m|\vec{g}| = \frac{1}{3}m|\vec{g}| \quad \text{(N4.23)}$$

as we found before. Substituting this into equation N4.8 yields the same result as in equation N4.9.

N4X.2   If ϕ increases, then $\sin\phi$ increases. Therefore, equation N4.17 implies that $|\vec{F}_T|$ decreases as $1/\sin\phi$. Moreover, $\cos\phi$ decreases, so $F_{H,x}$ in equation N4.19a must decrease. On the other hand, $F_{H,z}$ as given by equation N4.19b remains the same, because the quantity $|\vec{F}_T|\sin\phi$ does not change and the drawbridge's weight does not change. So $|\vec{F}_H|$ gets smaller, but only because F_{Hx} gets smaller.

N4X.3   In this case, equation N4.16 becomes

$$0 = (\tfrac{1}{2}L)M|\vec{g}| + L(\tfrac{1}{2}M)|\vec{g}| - 2D|\vec{F}_T||\sin\theta| \quad \text{(N4.24)}$$

This doubles the positive (clockwise) torque on the drawbridge, so $|\vec{F}_T|$ will have to double. But $2(3010\text{ N}) \approx 6$ kN is still comfortably less than 10 kN.

Linearly Constrained Motion

Chapter Overview

Introduction

This chapter continues our exploration of what we can learn about forces by observing motion. In this chapter, we will consider problems involving objects that move in straight lines with either constant or nonconstant speeds (these are **linearly constrained motion** problems).

Section N5.1: Motion at a Constant Velocity

If an object moves in a straight line at a constant velocity, its acceleration is zero, implying that the net force on the object must be zero, just as if the object were at rest. It is convenient in such problems to orient the reference frame so that as many forces as possible lie along coordinate axes.

Section N5.2: Static and Kinetic Friction Forces

A contact interaction between two solid objects can exert a force *parallel* to the surfaces in contact. If this force prevents the surfaces from sliding relative to each other, we call it a **static friction force**. This force arises because atoms in the surfaces become essentially "cold-welded" to one another. A static friction force automatically adjusts its magnitude to whatever value keeps the surfaces from sliding as long as it is less than a certain limit $\vec{F}_{SF,\text{max}}$ that depends on the characteristics of the surfaces and how strongly they are pressed together. Empirically, we find that

$$|\vec{F}_{SF}| \leq |\vec{F}_{SF,\text{max}}| \approx \mu_s |\vec{F}_N| \qquad \text{(N5.3)}$$

- **Purpose:** This equation describes the maximum magnitude of the static friction force $\vec{F}_{SF}$ that a contact interaction between solid objects exerts on those objects, where $\vec{F}_N$ is the normal force exerted by the same contact interaction and μ_s is the **coefficient of static friction**, a unitless constant that depends on the characteristics of the surfaces.
- **Limitations:** This is a simplified empirical model that is only approximately true.
- **Notes:** This equation specifies an *upper limit* $|\vec{F}_{SF,\text{max}}|$ for $|\vec{F}_{SF}|$, not its actual value at a given time.

If the surfaces are sliding relative to one another, then we call the part of the contact force that acts parallel to the surfaces a **kinetic friction force**. Empirically,

$$|\vec{F}_{KF}| \approx \mu_k |\vec{F}_N| \qquad \text{(N5.4)}$$

- **Purpose:** This equation describes the magnitude of the kinetic friction force $\vec{F}_{KF}$ exerted by a contact interaction between solid objects, where $\vec{F}_N$ is the normal

force exerted by the same contact interaction and μ_k is the **coefficient of kinetic friction**, a unitless constant that depends on the characteristics of the surfaces (and sometimes weakly on speed).
- **Limitations:** This is a simplified empirical model that is only approximately true.

Although equations N5.3 and N5.4 look similar, the static friction equation only specifies an *upper limit* to the static friction force while the other specifies the kinetic friction force's nearly steady value.

Section N5.3: Drag Forces

When a sufficiently large object moves through a fluid, its contact interaction with the fluid exerts an opposing drag force whose magnitude is roughly

$$|\vec{F}_D| \approx \tfrac{1}{2} C \rho A |\vec{v}|^2 \qquad (N5.9)$$

- **Purpose:** This equation describes the magnitude of the drag force $\vec{F}_D$ exerted by a fluid on an object moving through it with speed $|\vec{v}|$, where ρ is the fluid's density, A is the object's cross-sectional area, and C is a unitless **drag coefficient** that depends on the object's shape and surface characteristics.
- **Limitations:** This expression works only if A and $|\vec{v}|$ are sufficiently large, ρ is sufficiently small, and the fluid is not very viscous. This expression works for most sports projectiles in air, but $|\vec{F}_D| \propto |\vec{v}|$ for very tiny and slowly moving objects.

Section N5.4: Linearly Accelerated Motion

Solving problems in which the object accelerates while moving along a line is much easier if you *orient your reference system so one axis is parallel to* $\vec{a}$. Some such problems are hybrid problems in which we use information about motion (or lack thereof) in one component direction to determine characteristics of the object's motion in another direction.

Section N5.5: A Constrained-Motion Checklist

You can use the general framework discussed in section N4.4 for solving **constrained-motion problems**, except that you should draw an acceleration arrow $\vec{a}$ on your situation diagram and make *sure* it is right. Here is a checklist for such problems (which will be useful through chapter N7):

Constrained-Motion Problem Checklist
- ☐ Draw an arrow for $\vec{a}$ in your main drawing and align one axis with $\vec{a}$.
- ☐ Show coordinate axes in all drawings.
- ☐ Draw a *free-body diagram* (*in addition* to a general picture of the situation).
- ☐ Write down Newton's second law in *column-vector* form.
- ☐ Make all signs *explicit* when possible.

General Checklist
- ☐ Define symbols.
- ☐ Draw a picture.
- ☐ Describe model and assumptions.
- ☐ Use a master equation.
- ☐ List knowns and unknowns.
- ☐ Do algebra symbolically.
- ☐ Track units.
- ☐ Check result.

N5.1   Motion at a Constant Velocity

In this chapter, we continue in our exploration of what we can learn about the forces acting on an object by considering its motion. In the last chapter, we considered objects at rest. In this chapter, we will consider objects that are moving, but which are constrained to move in a straight line. Considering this common but simple case will help develop our skills for tackling more complicated types of motion.

Constant-$\vec{v}$ problems are like static problems

We begin by considering the especially simple case of an object moving at a constant velocity. An object moving in this manner has zero acceleration. Newton's second law then implies that the net force acting on that object must be *zero*, just as if the object were at rest. Indeed, we can consider *rest* to be simply a special case ($\vec{v} = 0$ = constant) of constant-velocity motion.

Example N5.1

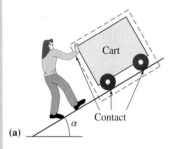

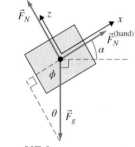

Figure N5.1
(a) A person pushing a cart up a ramp. (b) A free-body diagram for the cart in this situation.

Problem: Suppose you push a 50-kg cart with frictionless wheels at a constant speed up a ramp that makes an angle of 30° with the horizontal. If you push parallel to the ramp, how hard do you have to push?

Solution Figure N5.1a shows a picture of the situation. The dashed outline around the cart helps us focus on what touches the cart, which helps us draw the free-body diagram shown in figure N5.1b. The two normal forces are perpendicular to each other, so if we tilt our reference frame axes so the x axis is parallel to the incline, then each normal force has only one nonzero component.

Because the cart's velocity is constant, Newton's second law implies that $0 = \vec{F}_{net} = \vec{F}_g + \vec{F}_N + \vec{F}_N^{(hand)}$, which in component form reads

$$\begin{bmatrix} 0 \\ 0 \\ 0 \end{bmatrix} = \begin{bmatrix} -m|\vec{g}|\sin\theta \\ 0 \\ -m|\vec{g}|\cos\theta \end{bmatrix} + \begin{bmatrix} 0 \\ 0 \\ |\vec{F}_N| \end{bmatrix} + \begin{bmatrix} |\vec{F}_N^{(hand)}| \\ 0 \\ 0 \end{bmatrix}$$

$$\Rightarrow \quad \begin{array}{c} |\vec{F}_N^{(hand)}| = m|\vec{g}|\sin\theta \\ 0 = 0 \\ |\vec{F}_N| = m|\vec{g}|\cos\theta \end{array} \qquad \text{(N5.1)}$$

Note that since $\phi + 90° + \alpha = 180° = 90° + \phi + \theta$, θ must be the same as the angle of the ramp: $\theta = \alpha = 30°$. The top and bottom rows of equation N5.1 give us enough information to solve for our unknowns $|\vec{F}_N|$ and $|\vec{F}_N^{(hand)}|$.

Solving the top row (x component) of equation N5.1 for $|\vec{F}_N^{(hand)}|$ yields

$$|F_N^{(hand)}| = m|\vec{g}|\sin\theta = (50 \text{ kg})(9.8 \text{ m/s}^2)\sin 30° \left(\frac{1 \text{ N}}{1 \text{ kg} \cdot \text{m/s}^2}\right) = 245 \text{ N} \qquad \text{(N5.2)}$$

This has the right units and is positive, as a magnitude should be, and seems reasonable. Although we were not asked for the value of $|\vec{F}_N|$, note that in the bottom row (z component) of equation N5.1 $|\vec{F}_N| = m|\vec{g}|\cos\theta < m|\vec{g}|$. Thus, the magnitude of the normal force that the contact interaction with the incline exerts on the cart is *smaller* than the cart's weight here!

The key to solving force-from-motion problems

The most important thing to note in example N5.1 is how I expressed Newton's second law in column-vector form, expressing the components of each force vector in terms of that vector's *magnitude* (making signs explicit), and then reduced the vector equation to three separate *component* equations. These steps are the key to solving quantitative force-from-motion problems.

How to orient frame axes when $\vec{a} = 0$

Note also that when we solve a problem in which the object's acceleration is zero, it is usually most convenient to orient our reference frame so that as many force components as possible are zero. I did this in example N5.1 by tilting the reference frame so the x axis is parallel to the incline.

N5.2   Static and Kinetic Friction Forces

Many realistic situations involving objects at rest or objects moving at constant speeds involve *friction* forces. The purpose of this section is to explore the quantitative nature of static and kinetic friction forces exerted by contact interactions between two solid objects.

Static friction

Imagine the following situation. A large and heavy box sits on the floor of your room. Imagine that you push on the box in a direction parallel to the floor, as shown in figure N5.2a. What happens? You know from experience that if you don't push hard enough, you can exert a steady horizontal force on the box and yet it doesn't move.

How can we reconcile this behavior with Newton's second law? Since the box remains at rest, it is not accelerating, and therefore there must be *zero* net force on the box. Since you are pushing horizontally forward on the box, there must be another force acting backward on the box that exactly balances whatever force you exert, as shown in figure N5.2b. But what exerts this force, and why does it seem to automatically adjust its value to match however hard we push on the box?

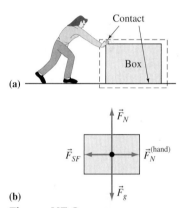

Figure N5.2
(a) Pushing on a box at rest.
(b) A free-body diagram of the same.

The following model helps us understand this force. When you place a box on the floor, the box's atoms and the floor's atoms come into contact, microscopic hills in the box's surface become interlocked with microscopic valleys in the floor and vice versa, and some of the box's atoms actually become "cold-welded" to floor atoms. When you push the box parallel to the floor, the interactions between the atoms of the box and the atoms of the floor under the box push the floor atoms forward, and thus (by Newton's third law) also push the box atoms backward. We call the total force exerted on the box due to this interaction between cold-welded atoms a **static friction force**.

This force adjusts its magnitude to keep the box at rest. If you push harder forward, the interaction intensifies, pushing the floor atoms farther forward and the box atoms more strongly backward. If you ease up, the interaction becomes less intense, allowing atoms on the surface of both the floor and the box to relax back closer to their original positions. Thus, however hard you push, the static friction force adjusts to exactly cancel the sum of horizontal components of the other forces acting on the box, keeping the box at rest.

Kinetic friction

However, this works only up to a point. If you push hard enough, the box will suddenly start to slide. Once it starts sliding, you can keep the box moving at a constant velocity by exerting a steady (and usually smaller) force.

Again, how do we reconcile this behavior with Newton's second law? Here we are exerting a steady horizontal force on the box, and yet it is only moving with a constant velocity, not accelerating (there may be only a brief moment of acceleration as the box began to move). This means that there *still* must be some horizontal force acting to cancel the effect of your push. This force is the **kinetic friction force** between the rubbing surfaces.

Here is a simplified model for understanding this force. When you finally get the box moving, the microscopic hills in the box's surface must push past microscopic hills in the floor's surface. When two hills come in contact, they momentarily weld together and deform as the surfaces continue to move before ultimately the welds break. As the hills on the box and floor surfaces meet and deform, their interaction exerts a force on the box's hills that opposes the deformation. The net result of the tiny forces exerted by untold numbers of interacting hills is a (roughly) steady force opposing the box's motion.

$|\vec{F}_{KF}|$ is usually less than the maximum value of $|\vec{F}_{SF}|$

It turns out that while the magnitude of this friction force can depend weakly on the speed of the box, it is generally *less* than the magnitude of force required to start the box moving at first (the welds that form on the fly are not as strong as ones that can form when surfaces are at rest relative to each other). This means you have to exert a lot of force to get the box going, but once it is moving, not as much force is required to *keep* it moving.

This phenomenon is illustrated by the graph shown in figure N5.3. This graph illustrates what happens when you slowly and steadily increase a manually applied horizontal force $\vec{F}$ to a motionless box. At first, the magnitude of the static friction force due to the box's interaction with the floor increases exactly in step with that of applied horizontal force, keeping the box at rest. But as the magnitude of the applied force continues to increase, there comes a point where the bonds between floor atoms and box atoms cannot stretch any farther without breaking. At this point, the atoms are exerting the maximum possible static friction force they can (we call the magnitude of this force $|\vec{F}_{SF,\max}|$). If you push still harder, these bonds break, and the horizontal force exerted by the contact interaction suddenly drops to the lower kinetic friction value. There is now a difference in magnitude between the horizontal applied force and the opposed friction force that allows the box to accelerate from rest. The box will continue to accelerate until either the kinetic friction force has increased (with increasing speed) to match your applied force or (more likely) you ease up automatically on the box so as to match your push to its new lower friction force.

This experimentally observed behavior has some interesting implications. *Anti-lock brakes* on cars prevent the brakes from locking the cars' wheels, so the tires stay locked to the road instead of skidding. Since the maximum static friction force the road can exert on the tires is greater than the kinetic friction force it can exert if the tires are skidding, anti-lock brakes help the driver to brake the car more rapidly (as well as keep the car under better control).

Figure N5.3

A graph illustrating the behavior of the static and kinetic friction forces. If we manually apply a horizontal force to an object sitting on a surface, the magnitude of the static friction force will at first increase in step with that of the applied force until a certain maximum is reached, after which the object suddenly begins to slide and the friction force drops to the smaller kinetic friction value for sliding surfaces. (Usually a person pushing will ease up after the object starts to move, which is why the applied force shown turns over.)

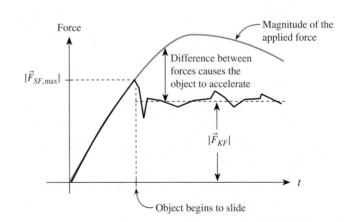

Empirically, the maximum possible magnitude of the static friction force that a contact interaction can exert seems to be roughly *proportional* to the magnitude of the normal force exerted by the same contact interaction:

$|\vec{F}_{SF,\max}|$ is proportional to $|\vec{F}_N|$ (for the same interaction)

$$|\vec{F}_{SF}| \leq |\vec{F}_{SF,\max}| \approx \mu_s|\vec{F}_N| \qquad \text{(N5.3)}$$

- **Purpose:** This equation describes the maximum magnitude of the static friction force $\vec{F}_{SF}$ that a contact interaction between solid objects exerts on those objects, where $\vec{F}_N$ is the normal force exerted by the same contact interaction and μ_s is the **coefficient of static friction**, a unitless constant that depends on the characteristics of the surfaces.
- **Limitations:** This is a simplified empirical model that is only approximately true.
- **Notes:** This equation specifies an *upper limit* $|\vec{F}_{SF,\max}|$ for $|\vec{F}_{SF}|$, not its actual value at a given time.

It makes sense that $|\vec{F}_{SF}|$ should depend on $|\vec{F}_N|$ because the normal force reflects how strongly the surfaces are pressed together and thus how extensively the surface atoms are likely to interpenetrate and bond together. Note that μ_s is unitless (since $\vec{F}_{SF}$ and $\vec{F}_N$ have the same units). Its value depends on the nature of the interacting surfaces.

Since the normal force on an object sitting on a surface typically (but not always!) cancels the object's weight, the applied force needed to get an object moving is often proportional to the object's weight. This is the Newtonian explanation for the intuitive idea that one has to exert a force strong enough to "overcome an object's inertia" before it starts to move. This Aristotelian notion thus works well enough in many cases, but it fails to explain why even a tiny force suffices to accelerate an object in space or in an otherwise frictionless environment. If a freight train car could be mounted on sufficiently frictionless wheels, even a toddler could cause it to accelerate!

A *similar* relation usefully models the relatively steady magnitude of the kinetic friction forces exerted by surfaces moving relative to one another:

$|\vec{F}_{KF}|$ is proportional to $|\vec{F}_N|$ for the same interaction

$$|\vec{F}_{KF}| \approx \mu_k|\vec{F}_N| \qquad \text{(N5.4)}$$

- **Purpose:** This equation describes the magnitude of the kinetic friction force $\vec{F}_{KF}$ exerted by a contact interaction between solid objects, where $\vec{F}_N$ is the normal force exerted by the same contact interaction, and μ_k is the **coefficient of kinetic friction**, a unitless constant that depends on the characteristics of the surfaces (and sometimes weakly on speed).
- **Limitations:** This is a simplified empirical model that is only approximately true.

Table N5.1 on the next page lists measured values of μ_s and μ_k for various surface pairs. Note that (almost always) $\mu_k < \mu_s$ for a given pair of surfaces.

Although they look very similar, there is an important difference between equations N5.3 and N5.4. Equation N5.4 expresses the *actual value* of the kinetic friction force (which has a fairly fixed value), while equation N5.3 expresses an *upper limit* on the adjustable static friction force.

Table N5.1   Some coefficients of friction (These values are approximate.)

	μ_s	μ_k
Climbing boots on rock	1.0	0.8
Tires on dry concrete	1.0	0.7
Rubber shoes on wood	0.9	0.7
Steel on steel (dry)	0.7	0.5
Tires on wet concrete	0.7	0.5
Tires on asphalt	0.6	0.4
Leather shoes on carpet	0.6	0.5
Rope or metal on wood	0.5	0.3
Wood on wood	0.4	0.2
Tires on icy concrete	0.3	0.2
Leather shoes on wood	0.3	0.2
Waxed wood on wet snow	0.14	0.1
Steel on steel (lubricated)	0.1	0.05
Shoes on ice	0.1	0.05
Ice on ice	0.1	0.03
Teflon on Teflon	0.04	0.04
Best human joints	0.01	0.003

Example N5.2

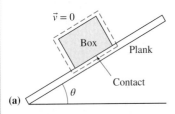

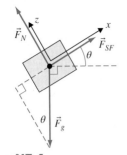

(a)

(b)

Figure N5.4
(a) A box sitting on an inclined plank. (b) A free-body diagram for the situation.

Problem: Suppose that a box sits on a plank of wood. If we gradually lift one end of the plank, we find that the box suddenly starts to slide down the plank when the plank's angle with the horizontal reaches 28°. What is the coefficient of static friction between the box and the plank?

Solution Figure N5.4 illustrates the situation in which the plank is inclined at an angle $\theta < 28°$ and the box is still at rest. Since the box is at rest, it is not accelerating, so the net force acting on it must be zero: $\vec{F}_{net} = 0$. If we orient the reference frame as shown in figure N5.4b, then writing $0 = \vec{F}_{net} = \vec{F}_g + \vec{F}_N + \vec{F}_{SF}$ in column-vector form yields

$$\begin{bmatrix} 0 \\ 0 \\ 0 \end{bmatrix} = \begin{bmatrix} -m|\vec{g}|\sin\theta \\ 0 \\ -m|\vec{g}|\cos\theta \end{bmatrix} + \begin{bmatrix} 0 \\ 0 \\ +|\vec{F}_N| \end{bmatrix} + \begin{bmatrix} +|\vec{F}_{SF}| \\ 0 \\ 0 \end{bmatrix} \qquad (N5.5)$$

We also know that

$$|\vec{F}_{SF}| \le \mu_s |\vec{F}_N| \qquad (N5.6)$$

This gives us only three useful equations in the four unknowns m, $|\vec{F}_{SF}|$, $|\vec{F}_N|$, and μ_s, but it turns out that the mass will cancel out when we solve for μ_s.

If we solve the top and bottom components (the x and z components) of this vector equation for the force magnitudes $|\vec{F}_{SF}|$ and $|\vec{F}_N|$, we get

$$|\vec{F}_{SF}| = m|\vec{g}|\sin\theta \qquad \text{and} \qquad |\vec{F}_N| = m|\vec{g}|\cos\theta \qquad (N5.7)$$

Using these equations to eliminate $|\vec{F}_{SF}|$ and $|\vec{F}_N|$ from equation N5.6 yields

$$m|\vec{g}|\sin\theta \le \mu_s m|\vec{g}|\cos\theta \quad \Rightarrow \quad \mu_s \ge \frac{\cancel{m}\cancel{|\vec{g}|}\sin\theta}{\cancel{m}\cancel{|\vec{g}|}\cos\theta} = \tan\theta \qquad (N5.8a)$$

So, for the box to remain at rest, $\tan \theta$ must be less than or equal to μ_s. Since this condition breaks down if θ exceeds $\theta_{max} = 28°$, we must have

$$\mu_s = \tan \theta_{max} = \tan(28°) = 0.53 \qquad (N5.8b)$$

Note that the value of μ_s is unitless (as we would expect) and reasonable.

Exercise N5X.1

When the plank in example N5.2 is inclined at 20°, what is the ratio of the magnitude of the actual static friction force on the box to the maximum possible value of $\mu_s |\vec{F}_N|$ for a box sitting on a plank at this particular angle?

N5.3   Drag Forces

While the kinetic friction force between two solid objects is fairly independent of relative speed, the **drag force** experienced by an object as it moves through a fluid depends strongly on speed. Empirically, the magnitude of the drag force on most objects moving through air is given by the formula

The drag on a large object moving quickly through air

$$|\vec{F}_D| \approx \tfrac{1}{2} C \rho A |\vec{v}|^2 \qquad (N5.9)$$

- **Purpose:** This equation describes the magnitude of the drag force $\vec{F}_D$ exerted by a fluid on an object moving through it with speed $|\vec{v}|$, where ρ is the fluid's density, A is the object's cross-sectional area, and C is a unitless **drag coefficient** that depends on the object's shape and surface characteristics.
- **Limitations:** This expression works only if A and $|\vec{v}|$ are sufficiently large, ρ is sufficiently small, and the fluid is not very viscous. This expression works for most sports projectiles in air but $|\vec{F}_D| \propto |\vec{v}|$ for very tiny and slowly moving objects.

The drag coefficient C is about 0.5 for a sphere and is smaller for smooth and streamlined shapes, but it can be as large as 2 for irregular shapes.

Exercise N5X.2

A certain streamlined car has a frontal area of about 3.0 m^2 and a drag coefficient of 0.3. What is the approximate magnitude of the drag force exerted on the car when it travels at a speed of 25 m/s (55 mi/h)?

The drag force on objects that are very small, move very slowly, and/or move through a more viscous fluid (such as oil or honey) turns out to be proportional to $|\vec{v}|$, and therefore is better modeled by

The drag on the objects that are very small, move very slowly, and/or are in a viscous medium (such as oil or honey)

$$|\vec{F}_D| \approx b |\vec{v}| \qquad (N5.10)$$

where b is some constant with SI units of N·s/m = kg/s.

N5.4   Linearly Accelerated Motion

Consider now an object moving along a straight line with nonzero acceleration. Newton's second law $\vec{F}_{net} = m\vec{a}$ implies that the net force on the object must point in the same direction as its observed acceleration. This means that analyzing the motion of an object becomes *much* simpler if we *orient the reference frame axes so that one axis coincides with the direction of the observed acceleration*. Then, the two components of the net force perpendicular to this axis are zero (just as in a constant-velocity problem), and analysis of the object's *motion* becomes a one-dimensional problem.

Some constrained-motion problems involving accelerating objects are *hybrid* problems in that we use information about the motion to determine things about the forces acting on the object, which in turn we use to determine things about the object's linear motion (such as its acceleration). Example N5.3 illustrates such a problem.

Example N5.3

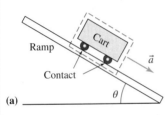

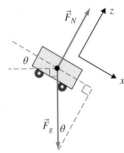

(b)

Figure N5.5

(a) A cart rolling down an incline. (b) A free-body diagram of this situation.

Problem: Consider a cart rolling on small frictionless wheels down a ramp inclined at an angle θ with respect to the horizontal. What is its acceleration? How does the magnitude of the normal force exerted on the cart by its interaction with the ramp compare to the magnitude of the object's weight?

Solution Figure N5.5 illustrates this situation. If there are no significant friction forces acting on the object, then the only forces acting on the cart are the normal force (which points perpendicular to the incline) and the cart's weight (which points vertically downward). Since the object is confined to moving on the surface of the incline, its acceleration must point parallel to the incline. Let us orient our reference frame so the x axis points down the incline and the z axis is perpendicular to the incline. If we express Newton's second law $m\vec{a} = \vec{F}_{net} = \vec{F}_g + \vec{F}_N$ in column-vector form, we get

$$\begin{bmatrix} ma_x \\ 0 \\ 0 \end{bmatrix} = \begin{bmatrix} +m|\vec{g}|\sin\theta \\ 0 \\ -m|\vec{g}|\cos\theta \end{bmatrix} + \begin{bmatrix} 0 \\ 0 \\ +|\vec{F}_N| \end{bmatrix} \qquad (N5.11)$$

This gives us two useful equations in the three unknowns m, a_x, and $|\vec{F}_N|$, so it does not look as if we have enough information to solve. However, we can cancel out the m in the equation corresponding to the first line of equation N5.11 and substitute $|\vec{F}_g|$ for $m|\vec{g}|$ in the third line to get

$$a_x = +|\vec{g}|\sin\theta \qquad \text{and} \qquad |\vec{F}_N| = mg\cos\theta = |\vec{F}_g|\cos\theta \qquad (N5.12)$$

We see that an object moving frictionlessly down an incline will have a constant acceleration whose magnitude is $|\vec{a}| = |\vec{g}|\sin\theta$, and the magnitude of the normal force is $\cos\theta$ times that of the object's weight. Since $\cos\theta < 1$ for nonzero incline angles, this means that $|\vec{F}_N| < |\vec{F}_g|$. (This last statement is true whether the object is accelerating or not.) So we are able to answer both questions definitively without knowing the object's mass m. Note that the units of both expressions in equation N5.12 make sense and that a_x is positive, which also makes sense considering how we defined our reference frame.

Exercise N5X.3

Suppose we were to analyze the situation described in example N5.3 using a reference frame where the *x* axis was horizontal. *Why* would this be more difficult?

N5.5 A Constrained-Motion Checklist

Constrained-motion problems are a subcategory of force-from-motion problems in which the object in question is constrained by its environment to move in a certain way (for example, with a constant velocity along a straight line, constant acceleration along a straight line, or constant speed in a circle), and we use this information to determine the forces acting on the object. Constrained-motion problems will be our focus through chapter N7; our focus in this particular chapter is on **linearly constrained motion** problems, in which the object must move along a straight line.

In section N4.4, we discussed a general approach for solving force-from-motion problems. This general framework works very well for constrained-motion problems if we make one adjustment. It is *especially* important in constrained-motion problems to understand and think about the constraint that the environment places on the object's *acceleration*. (This is partly because you want to be *sure* to align one of your coordinate axes with the direction of that acceleration.) Because of this, *drawing* an acceleration arrow on the situation diagram (if $\vec{a} \neq 0$) and *checking* that the arrow you draw is consistent with your free-body diagram are both essential steps.

So here is a general checklist for constrained-motion problems:

The importance of aligning a coordinate axis with the object's acceleration vector

Constrained-Motion Problem Checklist	**General Checklist**
☐ Draw an arrow for $\vec{a}$ in your main drawing and align one axis with $\vec{a}$.	☐ Define symbols.
☐ Show coordinate axes in all drawings.	☐ Draw a picture.
☐ Draw a *free-body diagram* (*in addition* to a general picture of the situation).	☐ Describe model and assumptions.
	☐ Use a master equation.
☐ Write down Newton's second law in *column-vector* form.	☐ List knowns and unknowns.
	☐ Do algebra symbolically.
☐ Make all signs *explicit* when possible.	☐ Track units.
	☐ Check result.

This checklist will be useful for most of the problems we will consider through chapter N7.

The commented examples on the next pages illustrate how we can use this checklist.

Example N5.4

Problem: A 14-kg box sits in the back of a 2800-kg pickup truck waiting at a stoplight. When the stoplight turns green, the driver of the truck drives forward with an acceleration of 5.0 m/s². If the coefficient of static friction between the box and the truck bed is 0.40, will the box be able to accelerate with the truck or will it slide backward relative to the truck bed?

Solution Here is a diagram of the situation, a free-body diagram for the box, and a list of knowns and unknowns. The free-body diagram assumes that the box *is* able to accelerate along with the truck.

Coordinate axes serve
both diagrams

Diagram of the situation

Acceleration arrow

Free-body diagram

List of knowns
and unknowns

$M = 2800$ kg
$m = 14$ kg
$\mu_s = 0.40$
$|\vec{a}| = 5.0$ m/s²
Is $|\vec{F}_{SF}| \leq |\vec{F}_{SF,\,max}|$?

Box (mass m)   Truck (mass M)

Contact

Statement about what
the object touches (so we
know what forces act)

Approximations and
assumptions stated

Check that acceleration
arrow is consistent with
free-body diagram

The box only touches the truck bed (which exerts an upward normal force and a *forward* static friction force that pulls the box along with the truck) and the air (which we will assume exerts negligible forces on the box). The box also interacts gravitationally with the earth. If the box accelerates along with the truck, the box's acceleration $\vec{a}$ points in the $+x$ direction. Note that the forces in the free-body diagram add up to a net force in the $+x$ direction, which is consistent with the box accelerating in that direction. Newton's second law in column-vector form then implies that

Master equation:
Newton's second law
in column vector form
(note explicit signs)

$$\begin{bmatrix} m|\vec{a}| \\ 0 \\ 0 \end{bmatrix} = \begin{bmatrix} 0 \\ 0 \\ -m|\vec{g}| \end{bmatrix} + \begin{bmatrix} 0 \\ 0 \\ |\vec{F}_N| \end{bmatrix} + \begin{bmatrix} +|\vec{F}_{SF}| \\ 0 \\ 0 \end{bmatrix} \qquad (\text{N5.13}a)$$

We can use the first line of this equation to compute the magnitude of the static friction force $\vec{F}_{SF}$ required to accelerate the box along with the truck and then see if this is greater or less than the maximum allowed static friction. The top and bottom components of equation N5.13a imply, respectively, that

$$m|\vec{a}| = |\vec{F}_{SF}| \quad \text{and} \quad |\vec{F}_N| = m|\vec{g}| \qquad (\text{N5.13}b)$$

Substituting the second of these equations into the right side of the equation for the maximum possible static friction force yields

Algebra with symbols

$$|\vec{F}_{SF,\,max}| = \mu_s|\vec{F}_N| = \mu_s m|\vec{g}| \qquad (\text{N5.14})$$

Now suppose we divide the first of the relations in equation N5.14 by this equation, to get

Calculation includes
and tracks units

$$\frac{|\vec{F}_{SF}|}{|\vec{F}_{SF,\,max}|} = \frac{m|\vec{a}|}{\mu_s m|\vec{g}|} = \frac{|\vec{a}|}{\mu_s|\vec{g}|} = \frac{5.0 \text{ m/s}^2}{0.40(9.8 \text{ m/s}^2)} = 1.28 \qquad (\text{N5.15})$$

Since the static friction force required to make the box accelerate along with the truck exceeds the maximum possible static friction force that can be exerted, the box will indeed slip backward as the truck accelerates forward. Note that the units work out and that this result is independent of the box's mass! The result is also

Final check of
plausibility

believable: an acceleration of 5.0 m/s² is pretty large ($\approx \frac{1}{2}|\vec{g}|$) and so might plausibly result in slipping.

Problem: A 55-kg person in an elevator traveling upward is standing on a spring scale that reads 420 N. What are the magnitude and direction of the elevator's acceleration? [*Note:* An ordinary scale does not directly register your weight, since weight is a force that acts directly on *you* and cannot be intercepted by the scale. Rather, the scale registers the magnitude of the upward normal force that its spring must exert to support you. This same contact interaction puts pressure on your feet that gives you a sense of what your "perceived weight" is. When you are at rest, this normal force is equal to your actual weight, but if you are accelerating vertically, the net vertical force on you will *not* be zero, and thus the normal force (and thus your perceived weight) will not be equal to your actual weight. This is why your perceived weight changes as you ride an elevator.]

Example N5.5

Solution A context diagram and a free-body diagram for this situation appear below.

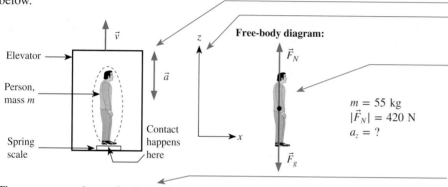

Diagram of the situation

Coordinate axes serve both diagrams

Free-body diagram

$m = 55$ kg
$|\vec{F}_N| = 420$ N
$a_z = ?$

List of knowns and unknowns

The person touches only the scale and the air. The scale exerts an upward normal force on the person, and we will ignore any forces resulting from contact with the air. The person also interacts gravitationally with the earth. Since the person is constrained to move vertically with the elevator, the person's acceleration (if there is any) must be vertical, though whether it is up or down is not yet clear. This is consistent with the free-body diagram: there are no horizontal forces, and the relative magnitudes of the vertical normal and gravitational forces will determine the person's (and thus by implication the elevator's) acceleration. Newton's second law in this case therefore reads

Statement about what the object touches (so we know what forces act)

Approximations and assumptions stated

Check that acceleration arrow is consistent with free-body diagram

$$\begin{bmatrix} 0 \\ 0 \\ ma_z \end{bmatrix} = \begin{bmatrix} 0 \\ 0 \\ -m|\vec{g}| \end{bmatrix} + \begin{bmatrix} 0 \\ 0 \\ +|\vec{F}_N| \end{bmatrix} \qquad (N5.16)$$

Master equation: Newton's second law in column vector form Note explicit signs

Only the bottom line (the z component) of this equation tells us anything useful, but it is sufficient for us to solve for our single unknown variable a_z:

$$a_z = \frac{|\vec{F}_N|}{m} - |\vec{g}| = \frac{(420 \text{ N})}{(55 \text{ kg})} \left(\frac{1 \text{ kg·m/s}^2}{1 \text{ N}} \right) - 9.8 \text{ m/s}^2 = -2.2 \text{ m/s}^2 \qquad (N5.17)$$

Algebra with symbols

Calculation includes and tracks units

Since a_z is negative, the elevator must be accelerating *downward* (perhaps as it slows when it approaches an upper-floor destination). This makes the person seem to weigh less, since the 420-N normal force is lower than the person's actual unchanging weight ($m|\vec{g}| \approx 540$ N). This coincides with our experience: when an elevator slows while going up, we do feel momentarily lighter. The acceleration also has the right units and has a reasonable magnitude ($<|\vec{g}|$). (Note that the full three-dimensional machinery of equation N5.16 is probably overkill here, but it is *essential* in more complicated problems.)

Final check of plausibility

TWO-MINUTE PROBLEMS

N5T.1 The magnitude of the normal force on a box sitting on an incline is equal to that of its weight. T or F?

N5T.2 A certain crate sits on a rough floor. You find that you have to apply a horizontal force of 200 N to get the crate moving. If you put some massive objects in the crate so that its mass is doubled, how much force does it take to get the crate moving now?
A. Still 200 N
B. 400 N
C. 800 N
D. It depends (specify)

N5T.3 Two boxes of the same mass sit on a rough floor. These boxes are made of the same kind of cardboard and are identical except that one is twice as large as the other. If it takes 200 N to start moving the smaller box, how much force does it take to start moving the larger one?
A. Still 200 N
B. 400 N
C. 800 N
D. It depends (specify)

N5T.4 The coefficient of static friction between Teflon and scrambled eggs is about 0.1. What is the smallest tilt angle from the horizontal that will cause the eggs to slide across the surface of a tilted Teflon-coated pan?
A. 0.002°
B. 5.7°
C. 15°
D. 33°
E. Other (specify)

N5T.5 If you want to stop a car without anti-lock brakes as quickly as possible on an icy road, you should
A. Jam on the brakes as hard as you can.
B. Push on the brakes as hard as you can without locking the car's wheels and thus making the car skid.
C. Pump the brakes.
D. Do something else (specify).

N5T.6 Putting wider tires on your car will clearly give you more traction. T or F?

N5T.7 Assume that the coefficient of static friction between your car's tires and a certain road surface is about 0.75. Your car can climb a 45° slope. T or F?

N5T.8 Imagine that an external force of 100 N must be applied to keep a bicycle and rider moving at a constant speed of 12 mi/h against opposing air drag. To double the bike's speed to 24 mi/h, we must increase the magnitude of the force exerted on the bike to
A. 141 N
B. 200 N
C. 400 N
D. It depends on the bike's shape and area
E. Other (specify)

N5T.9 Imagine that a certain engine can cause the road to exert a certain maximum forward force $|\vec{F}_{SF}|$ on a certain car. If we change the car's design to reduce its drag coefficient by a factor of 2, by what factor will the car's maximum speed increase (other things being equal)?
A. No increase
B. 1.41
C. 2
D. 4
E. Depends (specify)
F. Other (specify)

N5T.10 A truck is traveling down a steady slope such that for each 1 m the truck goes forward along the slope it goes down 0.04 m (we call this a 4% grade). Imagine that the truck's brakes fail. What is the approximate increase in the truck's speed after 30 s, assuming that the engine is not used and there is little drag or other friction? [$|\vec{g}| = 22$ (mi/h)/s.]
A. 0.9 mi/h
B. 11 mi/h
C. 26 mi/h
D. 53 mi/h
E. 660 mi/h
F. Other (specify)

N5T.11 We can express the kinetic friction law in the form $\vec{F}_{KF} = \mu_k \vec{F}_N$. T or F?

N5T.12 A car with four-wheel drive can climb a steeper slope than a car with two-wheel drive. T or F?

HOMEWORK PROBLEMS

Basic Skills

N5B.1 Suppose you must exert 200 N of horizontal force on a 30-kg crate to get it moving on a level floor. What is the value of μ_s for the surfaces involved here?

N5B.2 Suppose the coefficient of static friction μ_s between a 25-kg box and the floor is 0.55. How hard would you have to push on the box to get it moving?

N5B.3 Suppose you must exert a horizontal force of magnitude 80 N to push a 20-kg box at a constant speed of 3 m/s. What is the coefficient of kinetic friction μ_k between the box and the floor in this case?

N5B.4 Suppose that the coefficient of kinetic friction between a certain 15-kg box and the floor is 0.35. How hard must you push on the box to move it at a constant speed of 2 m/s across the floor?

N5B.5 Assuming that air drag is the main opposing force on a bike rider, that the area which the bike and rider present to the wind is about 0.75 m^2, and that C for such an irregular shape is about 1, what is the approximate forward force required to give the bicycle a constant speed of 10 m/s?

N5B.6 The drag force on a car moving at 65 mi/h is how many times bigger than that on a car moving at 45 mi/h?

N5B.7 Suppose a cart rolls without friction down a slope that makes an angle of 6° with respect to the horizontal. If the cart rolls for 12 s, how far does it go? (*Hint:* you can use the result of example N5.3.)

N5B.8 Suppose a glider slides without friction down a tilted air track. If it takes 3.0 s to slide the 1.5-m length of the track, at what angle was the track inclined? (*Hint:* You can use the result of example N5.3.)

Modeling

N5M.1 The coefficient of static friction between a certain car's tires and an asphalt road is about 0.60. Only the rear tires are powered. The magnitude of the car's acceleration can be at most what value? Explain your response carefully.

N5M.2 Why do the tires of a car grip the road better on level ground than when the car is going up or down an incline? Explain carefully. (You may find that a couple of force diagrams will help you make your case.)

N5M.3 A 2.0-kg box slides down a 25° incline at a constant velocity of 3.0 m/s. What are the magnitude and direction of the kinetic friction force acting on this box?

N5M.4 Suppose that the coefficient of static friction between the tires of a certain 2250-kg four-wheel-drive all-terrain vehicle and a typical gravel roadbed is 0.45. What is the maximum possible incline that the vehicle can climb?

N5M.5 A 12-kg box sits at rest on a 15° incline. If the coefficient of static friction is 0.3 between the box and the incline, what additional force would you have to exert directly down the incline to get the box to start sliding?

N5M.6 A certain car has a drag coefficient of 0.32. If you look at it from the front, the car's cross section looks roughly like a rectangle that is 1.5 m high by 2.1 m across. What is the engine's *minimum* horsepower if its top cruising speed is 45 m/s ($\approx$100 mi/h)?

N5M.7 A 20-μg nanobot you are testing is cruising in water in the +x direction at a speed of $|\vec{v}_0|$. Its motor suddenly dies at a time we will call $t = 0$, and you observe the nanobot's x-velocity thereafter to be $v_x(t) = |\vec{v}_0|e^{-qt}$, where $q = 0.5$ s^{-1}. Is the nanobot's

motion consistent with the idea that the drag force on it is given by $|\vec{F}_D| = b|\vec{v}|$? If so, determine the value of b. (*Hint:* The time derivative of e^{-qt} is $-qe^{-qt}$.)

N5M.8 A crane hauls a crate (mass 250 kg) upward at a constant acceleration of 2.2 m/s^2. What is the magnitude of the tension force exerted on the crate by the crane's cable?

N5M.9 A 65-kg person stands on a scale in an elevator moving downward. If the scale reads 720 N, what are the magnitude and direction of the elevator's acceleration?

N5M.10 A railroad flatcar is loaded with crates having a coefficient of static friction of 0.50 with respect to the car's floor. If a train is moving at 22 m/s ($\approx$ 48 mi/h), within how short a distance can the train be stopped without letting the crates slide?

N5M.11 A pickup truck carries cans of paint in the back bed. The coefficient of static friction between the cans and the bed of the truck is 0.54. There is no back gate to the truck. How long should the driver take to accelerate to a speed of 55 mi/h to avoid losing paint cans out of the truck's rear?

N5M.12 An 1100-kg car with a frontal cross-sectional area of 3.5 m^2 and a drag coefficient of 0.42 rolls down a slope that makes a constant angle of 8° with respect to the horizontal. After accelerating a while, the car will eventually reach a maximum constant speed. What is this speed? (Assume that the wheels rotate frictionlessly.)

N5M.13 An 1800-kg car with four-wheel drive travels up a 12° incline at a speed of 15 m/s. What can you infer about the coefficient of static friction between the tires of this car and the road? Which pieces of information provided are relevant, and which are not?

N5M.14 Anti-lock brakes keep a car's tires from skidding on a road surface. A certain 1500-kg car equipped with such brakes and initially traveling at 27 m/s ($\approx$ 59 mi/h) is able to come to rest within a time interval of 6.0 s.
(a) What is the minimum coefficient of static friction between the tires and the road in this case?
(b) How far does the car travel before it stops?

Derivations

N5D.1 The engine in a car that has a normal (two-wheel) drive turns *two* wheels against the road; the other wheels spin freely. Assume that the load is the same on each wheel.
(a) Explain why the forward force on such a car cannot exceed $\frac{1}{2}\mu_s|\vec{F}_N|$, where $|\vec{F}_N|$ is the total normal force that the road exerts on the car, no matter how powerful the engine might be.
(b) Explain why four-wheel drive on a car can exert *twice* as much forward force on the car.

N5D.2   We can use the following model to understand equation N5.9. Assume that an object with cross-sectional area A moving through air of density ρ has to accelerate all the air molecules that it touches from rest up to its speed. Show that if this were so, the drag force on the object would be

$$|\vec{F}_D| = \rho A |\vec{v}|^2 \qquad (N5.18)$$

(The actual drag force will generally be somewhat smaller, since some of the air will slip around the object without being fully accelerated to its speed.)

N5D.3   (This problem requires using some calculus involving exponentials and logarithms.) Imagine that an object of mass m is initially moving in the $+x$ direction through a viscous fluid at a speed $|\vec{v}_0|$. If the only force acting on this object is a viscous drag force whose magnitude is given by $|\vec{F}_D| = b|\vec{v}|$, prove that the object's x-velocity is given by

$$v_x(t) = |\vec{v}_0| e^{-qt} \qquad (N5.19)$$

and determine how the constant q depends on b and m. (*Hint:* This is actually a motion-from-force problem. Divide both sides of the x component of Newton's second law by v_x and take the indefinite integral of both sides.)

N5D.4   We have seen that the drag force between a fluid and an object moving through that fluid is proportional to $|v|^2$ if the object is big and/or fast and/or the fluid is not very viscous but is proportional to $|\vec{v}|$ if the object is small and/or slow and/or the fluid is very viscous. But how can you really tell what formula to use? It turns out that you can tell by calculating a unitless constant physicists call an object's **Reynolds number**. We define this number to be

$$\text{Re} \equiv \frac{|\vec{v}|L}{\nu_k} \qquad (N5.20)$$

where L expresses the object's size (it does not really matter much if this is the object's length, width, or depth), and ν is the fluid's **kinematic viscosity** ($\nu_k = 15 \times 10^{-6}$ m²/s for air, 1.0×10^{-6} m²/s for water, and 75×10^{-6} m²/s for honey, all at room temperature and standard pressure). The approximate rule is that if $\text{Re} \gtrsim 1000$, then $|\vec{F}_D| \propto |\vec{v}|^2$ and if $\text{Re} \lesssim 1$, then $|\vec{F}_D| \propto |\vec{v}|$. Between those values, neither formula works very well.
  (a) Consider a baseball ($L \approx 0.075$ m) moving at 30 m/s. Which formula should we use?
  (b) Consider a bacterium ($L \approx 1$ μm) moving at 3 μm/s in water. Which formula should we use?
  (c) Consider a crumpled piece of paper ($L = 0.1$ m) moving at 0.5 m/s through air. Which formula should we use?
  (d) Consider tiny droplets of water ($L = 0.1$ mm) moving at a few millimeters per second through air. Which formula should we use?
  (continues)

  (e) Consider a bumblebee moving a few centimeters per second through air. Will either formula work very well for this case?
  (f) You may find it surprising that the kinematic viscosity of air is larger than that for water. However, what we usually think colloquially of as "viscosity" is the **dynamic viscosity** μ, which more directly expresses, say, the force you would have to exert to drag a finger through the fluid. These quantities are related by the fluid density ρ: $\mu = \rho \nu_k$. What is the ratio of the dynamic viscosities of water and air?

N5D.5   At highway speeds, the drag force exerted by air on a moving car dominates over other forms of friction.
  (a) Show that the power the engine must produce to make up for the work done by the drag force on the car is proportional to $|\vec{v}|^3$, where $|\vec{v}|$ is the car's speed.
  (b) Argue that the fuel required to get to the car's destination at a constant speed is proportional to $|\vec{v}|^2$.
  (c) Compare the cost in fuel for traveling to your destination at 25 m/s (about 56 mi/h) to the cost at 30 m/s (about 66 mi/h).

Rich-Context

N5R.1   A bicyclist (whose mass is 54 kg and whose bike has a mass of 11 kg) coasting down an essentially endless 5° slope is observed to reach a maximum speed of 32 mi/h. At these kinds of speeds, air drag dominates over all other kinds of friction. *Estimate* the drag coefficient C for the bike and rider.

N5R.2   A certain car has a drag coefficient of 0.32. If you look at it from the front, the car's cross section looks roughly like a rectangle that is 1.5 m high by 2.1 m across. What is its engine's minimum horsepower if the car's top cruising speed is 50 m/s (112 mi/h)? (*Hint:* Note that power expresses the *energy* converted per unit time. Use the concept of work in addition to Newton's second law.)

N5R.3   You are driving a 12,000-kg truck at a constant speed of 65 mi/h down a 6% slope (that is, an incline that goes down 0.06 m for every 1 m that one goes along the incline). You suddenly see that a bridge is out 425 ft ahead, and you jam on the brakes. The coefficient of static friction between your tires and the wet asphalt road is 0.45, the coefficient of kinetic friction is 0.30, and the cross-sectional area of your truck is 6.6 m². Can you stop in time? (*Hints:* The magnitude of the drag force depends on speed, so it will not be constant as you slow down. Is the drag force significant? Make plausible estimates for quantities you are not given.)

ANSWERS TO EXERCISES

N5X.1 According to equation N5.6, $|\vec{F}_{SF}| = m|\vec{g}|\sin\theta$. Equations N5.6 and N5.7, on the other hand, also indicate that $|\vec{F}_{SF,\max}| = \mu_s|\vec{F}_N| = \mu_s m|\vec{g}|\cos\theta$. Therefore,

$$\frac{|\vec{F}_{SF}|}{|\vec{F}_{SF,\max}|} = \frac{\cancel{m}\cancel{|\vec{g}|}\sin\theta}{\mu_s \cancel{m}\cancel{|\vec{g}|}\cos\theta} = \frac{\tan\theta}{\mu_s} = 0.69 \qquad (N5.21)$$

N5X.2 $|\vec{F}_D| = \frac{1}{2}(0.3)(1.2\text{ kg/m}^3)(3.0\text{ m}^2)(25\text{ m/s})^2 = 340\text{ N}$. This is about the same as 76 lbs, so this is a large push.

N5X.3 In this case, Newton's second law would read

$$\begin{bmatrix} ma_x \\ 0 \\ ma_z \end{bmatrix} = \begin{bmatrix} F_{N,x} \\ F_{N,y} \\ F_{N,z} \end{bmatrix} + \begin{bmatrix} F_{g,x} \\ F_{g,y} \\ F_{g,z} \end{bmatrix} = \begin{bmatrix} |\vec{F}_N|\sin\theta \\ 0 \\ |\vec{F}_N|\cos\theta \end{bmatrix} + \begin{bmatrix} 0 \\ 0 \\ -m|\vec{g}| \end{bmatrix} \qquad (N5.22)$$

But neither a_x nor a_z is zero in this case, so it becomes much more difficult to solve this problem. It can be done, however: we can use what we know about the acceleration's direction to show that $a_x = |\vec{a}|\cos\theta$ and $a_z = -|\vec{a}|\sin\theta$, substitute these into the preceding equations, and then solve the two coupled equations for the two unknowns $|\vec{a}|$ and $|\vec{F}_N|$. Even so, this is much more complicated than the process illustrated in the example.

N6 Coupled Objects

Chapter Overview

Introduction

This chapter continues our exploration of what we can learn about forces by observing motion. In this chapter, we will consider pairs or sets of objects that are coupled together by internal interactions that constrain the objects to move with the same acceleration. We will see that Newton's third law can help us determine the magnitudes of these internal forces.

Section N6.1: Force Notation for Coupled Objects

Coupled objects are objects that are constrained by some connection so that their motions are linked in a well-defined manner. When we draw free-body diagrams for the objects, it helps greatly if we use a notation for forces that describes not only the type of force we are talking about (and thus the type of interaction involved), but also the two objects involved in that interaction. In this chapter, we will use notation of the form $\vec{F}_N^{A(B)}$, where the *subscript* defines the type of force (a normal contact force in this example) and the *superscripts* define the objects involved in the interaction (in this case, the force is acting on object A and arises from its interaction with object B).

A pair of forces are *third-law partners* if and only if

1. Each force acts on a different object.
2. The pair of forces represents the two ends of the *same* interaction.

This means that in our notation, any third-law partners will have reversed superscripts and the same subscript, for example, $\vec{F}_N^{A(B)}$ and $\vec{F}_N^{B(A)}$. Such partners are easy to spot on the free-body diagrams of coupled objects participating in a contact interaction: the partners are the forces applied to the surfaces in contact.

Section N6.2: Pushing Blocks

This section discusses in detail an example involving a block that is being pushed along a surface by another block. Since Newton's third law links the contact forces that the blocks exert on each other, we can use it in conjunction with Newton's second law to determine the magnitudes of the contact forces. We find that the pair of blocks behaves exactly as if it were a single object responding to the external forces exerted on the system.

Section N6.3: Strings, Real and Ideal

This section explores an example involving two blocks connected by a string that are being pulled upward by a known external force. The fully correct way to analyze this system involves applying Newton's second and third laws to the *three* objects involved (the string being the third object). This approach yields five equations that can be solved for five unknowns (four internal tension forces and the system's common acceleration). As before, we find that the entire set accelerates as if it were a particle responding to the external forces on the system.

Moreover, we find that the difference between the magnitudes of the forces exerted by each end of the internal string goes to zero as the string's mass goes to zero. The ideal string model assumes that the string is completely massless, inextensible, and flexible. In this ideal limit, tension forces exerted by the string's ends have equal magnitudes (we call this magnitude *the* **tension on the string**), and the two objects linked by the string accelerate at exactly the same rate. This model makes it much easier to solve problems involving objects linked by a cable, chain, rope, or string.

Section N6.4: Pulleys

An **ideal pulley** (that is, a frictionless, massless pulley) changes the direction of a string without affecting the tension forces it exerts. Real pulleys, of course, have nonzero mass and nonzero friction. Since rotating a real pulley thus requires a nonzero torque, the tension forces exerted on the pulley by the parts of the string entering and leaving the pulley must be different. However, this change in the string tension can be very small for a good pulley.

It is often helpful in pulley problems to use a separate coordinate system for each object with its *x* axis aligned with the object's direction of motion.

Section N6.5: Using the Constrained-Motion Checklist

When solving coupled-object problems, you can use the general force-from-motion checklist for solving coupled-object problems with the following adaptations:

1. Describe how the objects' motions are linked. On the main translation diagram in a pulley problem, draw a *separate* acceleration arrow for each object.
2. Draw a separate free-body diagram *for each object*, using separate coordinate axes in pulley problems.
3. Link any third-law partners in the free-body diagrams.
4. Apply Newton's second law in column-vector form to each object, and use Newton's third law to link the magnitudes of any third-law partners.

So a revised checklist for such problems looks like this:

Coupled-Object Problem Checklist

☐ In the main drawing, draw an arrow for the acceleration $\vec{a}$ *of each object* in the system.
☐ Provide single-letter labels for everything.
☐ State joint constraints on the objects' motions.
☐ Draw a free-body diagram *for each object*.
☐ Show *third-law partners* in these diagrams.
☐ Draw different coordinate axes *for each object* if necessary. Align one axis with each object's acceleration $\vec{a}$.
☐ Write down Newton's second law in column-vector form *for each object*.
☐ Make all signs *explicit* when possible.

General Checklist

☐ Define symbols.
☐ Draw a picture.
☐ Describe your model and assumptions.
☐ List knowns and unknowns.
☐ Use a master equation.
☐ Do algebra symbolically.
☐ Track units.
☐ Check your result.

Example N6.4 illustrates the use of this checklist.

Figure N6.1

A sketch of a person pushing two boxes up an incline and free-body diagrams for each of the boxes. The forces in both diagrams are labeled according to the notation convention established in this section. Note also that when multiple forces act in the same direction, we draw them head-to tail, as in the right-most drawing.

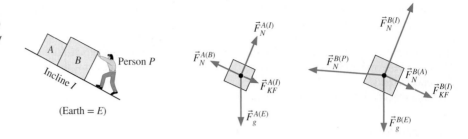

N6.1   Force Notation for Coupled Objects

What are coupled objects?

Our general task in chapters N4 through N8 is to expand our ability to apply Newton's laws to determine the forces acting on objects whose motion is constrained in various ways. In this chapter, we will focus on situations involving *coupled objects*. A pair of **coupled objects** consists of two objects that are constrained by some kind of connection so that their motions (which we will assume here to be linear) are linked in some well-defined manner. We can use this constraint to analyze the forces acting on the individual objects in the pair and describe their motion.

When we need to describe and analyze the forces acting on more than one object, we begin to run into problems with notation. For example, imagine a person pushing a box B up an incline, which in turn pushes box A in front of it (figure N6.1). How can we distinguish the symbol for the normal force exerted on A (by A's contact interaction with the incline) from the symbol for the normal force exerted on B (by B's interaction with the incline) from the symbol for the normal force exerted on B (by B's contact interaction with the pusher's hands) from the symbol for the normal force exerted on A (by A's contact interaction with the box B behind it)? We would get very confused if we assigned all these different forces the same symbol $\vec{F}_N$!

A notation convention for force symbols

Let us adopt the following notation convention for cases like this. We start with the basic symbol for the type of force involved, using the standard symbols described in section N2.2. We add to this a pair of superscripts, the first indicating the object on which this force is exerted and the second (in parentheses) indicating the other object involved in the interaction that gives rise to this force. For example, $\vec{F}_N^{A(B)}$ represents the normal force exerted on box A by its contact interaction with box B (which is pushing A up the incline), $\vec{F}_N^{B(P)}$ the normal force exerted on box B by the person P, and so on. While it is not the prettiest notation imaginable, this notation (which is summarized in figure N6.2) is simple, relatively easy both to typeset and to write by hand, and easy to interpret (as long as we have well-defined single letters for every object involved). Figure N6.1 shows free-body diagrams for the two boxes with all forces labeled using this notation.

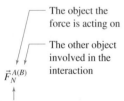

The object the force is acting on

The other object involved in the interaction

$\vec{F}_N^{A(B)}$

The type of force (which indicates the kind of interaction that is exerting the force)

Figure N6.2

A convention for force symbols used in coupled-object problems.

Recognizing third-law partners

Obviously, this notation is too cumbersome to use all the time. This notation is rarely helpful when we are analyzing the forces acting on a single object. In this chapter, though, it is more than helpful; this notation (or something equivalent) is essential for keeping things straight.

One advantage of this notation is that it makes it pretty easy to recognize third-law partners in a given situation. Newton's third law states that

When objects A and B interact, the force that the interaction exerts on A is equal in magnitude but opposite to the force it exerts on B.

This means that the two forces linked by this law (**third-law partners**) have the following characteristics:

1. The forces always act on *different* objects (when A and B interact, the interaction exerts one force on A and one on B).
2. They always reflect the *same* interaction.

This in turn implies that the symbols for third-law partners will have *subscripts* that are the *same* (since both forces must reflect the same interaction) and *superscripts* that are *reversed* (since the partner to the force exerted on A due to its interaction with B is the force on B due to its interaction with A). For example, if I exert a normal force $\vec{F}_N^{W(H)}$ on the wall W by pushing on it with my hand H, the contact interaction exerts a force $\vec{F}_N^{H(W)}$ back on my hand.

$$\vec{F}_N^{W(H)} = -\vec{F}_N^{H(W)} \tag{N6.1}$$

Wall   (Note the inter-
Hand   changed order)

Indicates the *same* interaction

by Newton's third law. The symbols for third-law partners will always have the characteristics noted in equation N6.1.

In the situation shown in figure N6.1, the only third-law partners among the forces shown are the normal forces $\vec{F}_N^{A(B)}$ and $\vec{F}_N^{B(A)}$ that the boxes exert on each other due to their contact interaction. Note that these symbols display the characteristics shown in equation N6.1.

Spotting third-law partners on free-body diagrams

Problem: Consider a book sitting at rest on a table. Draw free-body diagrams for both the book and the table, and determine which pairs of forces on these diagrams (if any) are third-law partners.

Solution Figure N6.3 shows the free-body diagrams for these objects. The only forces acting on the book are a downward gravitational force $\vec{F}_g^{B(E)}$ and an upward normal force $\vec{F}_N^{B(T)}$ due to its contact interaction with the table. The forces acting on the table are a downward gravitational force $\vec{F}_g^{T(E)}$, a set of upward normal forces $\vec{F}_N^{T(F)}$ exerted by the contact interaction between the floor F on the table legs, and the downward normal force $\vec{F}_N^{T(B)}$ exerted by the contact interaction between the book and the table.

The only third-law partners among these forces are the two forces that arise from the contact interaction between the book and the table. Note that this pair has all the characteristics described in this section and their symbols are consistent with the pattern shown in equation N6.1.

In contrast, note that the gravitational force on the book $\vec{F}_g^{B(E)}$ and the normal force $\vec{F}_N^{B(T)}$ are *not* third-law partners, even though these forces are opposite and have the same magnitude. These forces are equal and opposite because of Newton's *second* law: since the book is at rest, its acceleration is zero; so the net force on it is zero, and so $\vec{F}_g^{B(E)}$ and $\vec{F}_N^{B(T)}$ must be equal in magnitude and opposite in direction so that they cancel.

Note that even though the book's weight $\vec{F}_g^{B(E)}$ does not act directly on the table, the sum $\vec{F}_N^{T(F)}$ of the normal forces acting on the table legs is larger when the book sits on the table than it would be if the book were not there, because Newton's second law requires that this force cancel both the table's weight $\vec{F}_g^{T(E)}$ and the downward contact force $\vec{F}_N^{T(B)}$, which is equal in magnitude to the book's weight (as discussed in the previous paragraph).

Example N6.1

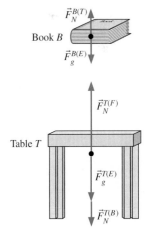

Figure N6.3
Free-body diagrams for the book and the table. Note that E = earth and F = floor.

Exercise N6X.1

Consider someone in an elevator moving at a constant speed. Draw free-body diagrams for both the person and the elevator (ignoring friction). Label the forces, using the convention described in this section, and determine which pairs of forces on these diagrams (if any) are third-law partners. Explain why there is more tension on the elevator's cable when the person is in the elevator than when it is empty, even though the gravitational force (weight) of the person does not act directly on the elevator (or the cable).

N6.2   Pushing Blocks

When two objects are connected in such a way that they become constrained to move together in a well-defined way, the constraint on their motion itself provides information we can use to learn something about the forces acting on those objects. Example N6.2 illustrates how we can use Newton's second and third laws together to extract information from this constraint.

Example N6.2

Problem: Consider two blocks pushed along a frictionless surface by someone who exerts a constant force with his or her hand (see figure N6.4a). Suppose block A has a mass of 4.0 kg, block B has a mass of 2.0 kg, and the hand exerts a force of 3.0 N (about 0.65 lb). What is the magnitude of the acceleration of the blocks? What is the magnitude of the force exerted by block B on block A?

Solution Figure N6.4a provides a sketch of the situation. The first step in the conceptual model is to specify a coordinate system and draw free-body diagrams of the two objects in question: these items are shown in figure N6.4b.

The contact interaction between blocks A and B exerts opposing normal forces on each block, which are labeled $\vec{F}_N^{A(B)}$ (read "the normal force on A due to its interaction with B") and $\vec{F}_N^{B(A)}$ (read "the normal force on B due to its interaction with A"). Newton's third law asserts that these forces have equal magnitudes, whether the blocks are accelerating or not.

In figure N6.4, $\vec{F}_N^{A(H)}$ is the force exerted on A by the interaction with the person's hand, $\vec{F}_N^{A(S)}$ and $\vec{F}_N^{B(S)}$ are the normal forces exerted on A and B by their contact interaction with the surface on which they slide, and $\vec{F}_g^{A(E)}$ and $\vec{F}_g^{B(E)}$ are the forces exerted on each by their gravitational interaction with the earth (E = earth). Note that figure N6.4a explicitly defines the symbols S and H to make these single-letter abbreviations clearer.

The next step in developing the model for this problem is to list the known quantities. We know that $|\vec{F}_N^{A(H)}| = 3.0$ N, that $m_A = 4.0$ kg, and that $m_B = 2.0$ kg. Knowing these masses would allow us to compute the magnitudes of the weight forces $\vec{F}_g^{A(E)}$ and $\vec{F}_g^{B(E)}$, should we so desire. The implicit constraints in the problem are (1) that the blocks move only in the x direction, which means that $\vec{a} = [a_x, 0, 0]$, and (2) that the two boxes have the *same* horizontal acceleration a_x (since they are moving as a unit). On the other hand, we do not know the value of a_x, nor the magnitudes of the forces $\vec{F}_N^{A(B)}$ and $\vec{F}_N^{B(A)}$ (the first of which we are asked to find), nor the magnitudes of $\vec{F}_N^{A(S)}$ and $\vec{F}_N^{B(S)}$ (which are probably of no concern), making a total of five unknowns.

The next model step in most constrained-motion problems is to apply Newton's second law in column-vector form. For block A we have

$$\begin{bmatrix} m_A a_x \\ 0 \\ 0 \end{bmatrix} = \begin{bmatrix} 0 \\ 0 \\ -m_A|\vec{g}| \end{bmatrix} + \begin{bmatrix} 0 \\ 0 \\ +|\vec{F}_N^{A(S)}| \end{bmatrix} + \begin{bmatrix} -|\vec{F}_N^{A(B)}| \\ 0 \\ 0 \end{bmatrix} + \begin{bmatrix} +|\vec{F}_N^{A(H)}| \\ 0 \\ 0 \end{bmatrix} \quad \text{(N6.2)}$$

The x and z components of this vector equation imply, respectively, that

$$m_A a_x = |\vec{F}_N^{A(H)}| - |\vec{F}_N^{A(B)}| \quad \text{and} \quad |\vec{F}_N^{A(S)}| = m_A|\vec{g}| \quad \text{(N6.3)}$$

Similarly, Newton's second law for block B implies that

$$\begin{bmatrix} m_B a_x \\ 0 \\ 0 \end{bmatrix} + \begin{bmatrix} 0 \\ 0 \\ -m_B|\vec{g}| \end{bmatrix} + \begin{bmatrix} 0 \\ 0 \\ +|\vec{F}_N^{B(S)}| \end{bmatrix} + \begin{bmatrix} +|\vec{F}_N^{B(A)}| \\ 0 \\ 0 \end{bmatrix} \quad \text{(N6.4)}$$

$$\Rightarrow \quad m_B a_x = |\vec{F}_N^{B(A)}| \quad \text{and} \quad |\vec{F}_N^{B(S)}| = m_B|\vec{g}| \quad \text{(N6.5)}$$

The z component equations for both blocks tell us that the vertical normal force on each block cancels its weight: there are no surprises here. Newton's third law also tells us that

$$|\vec{F}_N^{A(B)}| = |\vec{F}_N^{B(A)}| \quad \text{(N6.6)}$$

The two equalities in equation N6.3, the two equalities in equation N6.5 and equation N6.6 provide five equations for our five unknowns. Substituting equation N6.6 into the first of equations N6.5 yields

$$m_B a_x = +|\vec{F}_N^{A(B)}| \quad \text{(N6.7)}$$

If we add this equation to the first of equations N6.3, the unknown force $|\vec{F}_N^{A(B)}|$ cancels out, leaving us with

$$(m_A + m_B)a_x = |\vec{F}_N^{A(H)}| \quad \text{(N6.8a)}$$

Note that this tells us that the *system* consisting of the two objects accelerates as if it were a particle of mass $M = m_A + m_B$ subject to the net external force acting on the system ($\vec{F}_N^{A(H)}$ in this case). We already knew that this *should* be the case from chapter C4: the important thing here is to notice how Newton's third law ensures that it *does* happen. Solving this equation for a_x yields

$$a_x = \frac{|\vec{F}_N^{A(H)}|}{m_A + m_B} = \frac{3.0 \text{ N}}{4.0 \text{ kg} + 2.0 \text{ kg}}\left(\frac{1 \text{ kg} \cdot \text{m/s}^2}{1 \text{ N}}\right) = 0.50 \frac{\text{m}}{\text{s}^2} \quad \text{(N6.8b)}$$

Substituting this into equation N6.7 and using $|\vec{F}_N^{A(B)}| = |\vec{F}_N^{B(A)}|$, we find that

$$|\vec{F}_N^{A(B)}| = |\vec{F}_N^{B(A)}| = m_B a_x = (2.0 \text{ kg})(0.50 \text{ m/s}^2)\left(\frac{1 \text{ N}}{1 \text{ kg} \cdot \text{m/s}^2}\right) = 1.0 \text{ N} \quad \text{(N6.8c)}$$

Note that the net x-force on object A is $|\vec{F}_N^{A(H)}| - |\vec{F}_N^{A(B)}| = 3.0 \text{ N} - 1.0 \text{ N} = 2.0 \text{ N}$, which is just the force required to give this 4.0-kg object an x-acceleration of $a_x = 0.50 \text{ m/s}^2$. Note also that the net force on B is 1.0 N, just what is required to give it an x-acceleration of 0.50 m/s^2. Everything is self-consistent!

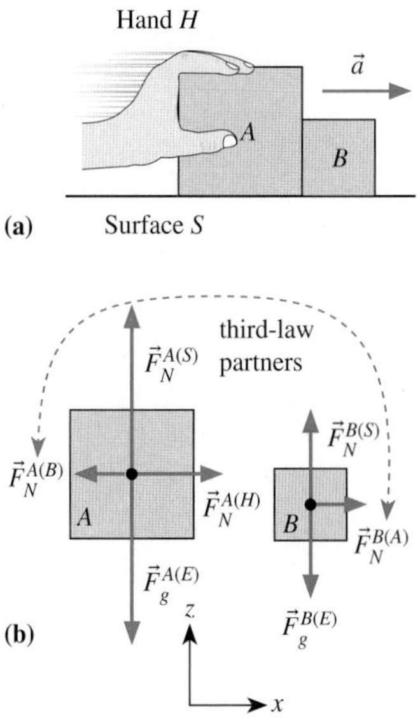

The point of example N6.2 is that when objects are coupled so that they are constrained to have the same acceleration, the forces associated with the interaction between the objects will adjust themselves to whatever common magnitude gives each object the *same* horizontal acceleration. (Such forces are invariably contact forces, which *can* adjust themselves in this way.)

(a) Surface S

(b)

Figure N6.4

(a) Two blocks being pushed along a frictionless surface. (b) Free-body diagrams of the blocks, also indicating the single third-law pair.

N6.3   Strings, Real and Ideal

Overview

Objects do not have to be in direct contact to exert forces on each other. Two objects connected by string or cord can exert forces on each other through the string that are similar to the forces they exert when in direct contact.

We can think of each end of a string as exerting a tension force on the object to which it is connected. While these two forces are generally *not* equal in direction, they are often at least approximately equal in magnitude (for reasons we will discuss shortly). Because of this, we sometimes refer to the common magnitude of these forces as being ***the* tension on** (or **of**) **the string**. It is not *always* the case that the forces exerted by the ends of a string have the same magnitude, and in such cases "the string's tension" is not well defined, but you should understand that when this phrase is used, it refers to the common magnitude of the forces exerted by each end of the string.

A specific example will help us understand these issues more clearly.

Example N6.3

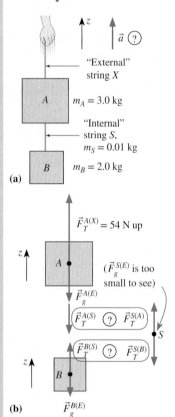

(a)

(b)

Figure N6.5

(a) A person pulls vertically on two blocks connected by a string. (b) Free-body diagrams for the two blocks and the string. Note that these diagrams also specify the knowns and unknowns (with question marks) in this problem.

Problem: Consider two blocks connected by a string (call this the *internal* string) and being pulled vertically upward by another string (the *external* string) attached to block A. Suppose that blocks A and B have masses of 3.0 kg and 2.0 kg, respectively, and suppose that the external string's tension has a magnitude of 54 N. How do the tension forces exerted by the ends of the internal string compare if the internal string's mass is 0.010 kg?

Solution Again, the usual first step is to draw a sketch (see figure N6.5a) and a set of free-body diagrams (figure N6.5b). I have defined symbols for the internal string S and the external string X, set up a reference frame (we only need a z axis for this problem), and indicated knowns and unknowns. The implicit constraints on the motion are that the objects only move vertically, and that all three objects have the same z-acceleration a_z.

To solve the problem, we apply Newton's second law in component form to each of the three objects involved (block A, block B, and the internal string S). In this case, only the z component of Newton's second law is interesting (the others all read $0 = 0$). So, rather than write a bunch of zeros, we will cut to the chase: the z component of Newton's second law for each of the three objects is, respectively,

$$m_A a_z = F^A_{net,z} = F^{A(E)}_{g,z} + F^{A(X)}_{T,z} + F^{A(S)}_{T,z} = -m_A|\vec{g}| + |\vec{F}^{A(X)}_T| - |\vec{F}^{A(S)}_T| \quad (N6.9a)$$

$$m_S a_z = F^S_{net,z} = F^{S(E)}_{g,z} + F^{S(A)}_{T,z} + F^{S(B)}_{T,z} = -m_S|\vec{g}| + |\vec{F}^{S(A)}_T| - |\vec{F}^{S(B)}_T| \quad (N6.9b)$$

$$m_B a_z = F^B_{net,z} = F^{B(E)}_{g,z} + F^{B(S)}_{T,z} = -m_B|\vec{g}| + |\vec{F}^{B(S)}_T| \quad (N6.9c)$$

In addition, Newton's third law tells us that connected pairs of forces in figure N6.5b have equal magnitudes:

$$|\vec{F}^{A(S)}_T| = |\vec{F}^{S(A)}_T| \quad (N6.9d)$$

$$|\vec{F}^{S(B)}_T| = |\vec{F}^{B(S)}_T| \quad (N6.9e)$$

Equations N6.9 represent five equations in the four unknown force magnitudes $|\vec{F}^{S(A)}_T|$, $|\vec{F}^{A(S)}_T|$, $|\vec{F}^{S(B)}_T|$, and $|\vec{F}^{B(S)}_T|$ and the one unknown acceleration a_z. We can therefore solve the problem. First, we can solve equation N6.9c for $|\vec{F}^{B(S)}_T|$. According to equation N6.9e, that force magnitude is equal to $|\vec{F}^{S(B)}_T|$, so we can substitute all this into equation N6.9b to eliminate $|\vec{F}_T^{S(B)}|$ and rearrange things a bit as follows:

$$m_S a_z = -m_S|\vec{g}| + |\vec{F}_T^{S(A)}| - (m_B a_z + m_B|\vec{g}|)$$

$$\Rightarrow |\vec{F}_T^{S(A)}| = m_S a_z + m_S|\vec{g}| + m_B a_z + m_B|\vec{g}| \qquad \text{(N6.10)}$$

Equation N6.9d tells us that $|\vec{F}_T^{S(A)}| = |\vec{F}_T^{A(S)}|$, so we can substitute equation N6.10 into equation N6.9a, eliminating $|\vec{F}_T^{A(S)}|$. Rearrangement then yields

$$(m_A + m_S + m_B)a_z = +|\vec{F}_T^{A(X)}| - (m_A + m_S + m_B)g \qquad \text{(N6.11)}$$

Exercise N6X.2

Verify equation N6.11.

This says that the system consisting of the two blocks and the internal string accelerates as if it were a single object of mass $M = m_A + m_S + m_B$ acted on by two external forces: (1) the tension force exerted by the external string and (2) the system's total weight. All references to the internal forces acting in the system have disappeared! This is yet another illustration of the general principle that a system of interacting objects responds to external forces as if it were a single object. The third law ensures this is true.

We can easily solve equation N6.11 for the unknown acceleration a_z:

$$a_z = \frac{|\vec{F}_T^{A(X)}|}{M} - |\vec{g}| = \frac{54 \; \cancel{N}}{5.01 \; \cancel{kg}} \left(\frac{1 \; \text{kg} \cdot \text{m/s}^2}{1 \; \cancel{N}} \right) - 9.8 \; \text{m/s}^2 = 0.98 \; \text{m/s}^2 \qquad \text{(N6.12)}$$

Once we know a_z, we can plug its value back into equations N6.9b and N6.9c to find the magnitudes of the tension forces exerted by the string. A simple rearrangement of equation N6.9c tells us that

$$|\vec{F}_T^{B(S)}| = m_B(|\vec{g}| + a_z) \qquad \text{(N6.13)}$$

Note that this implies that the string exerts a force on block B that is *larger* than the weight of block B. (It is necessary to have a nonzero *net* force on B to accelerate it upward.) You can verify that $|\vec{F}_T^{B(S)}| = 21.6$ N.

Equation N6.9b in combination with Newton's third law (equations N6.9d and N6.9e) implies that the difference in the magnitudes of the tension forces exerted by each end of the string is

$$m_S(|\vec{g}| + a_z) = |\vec{F}_T^{S(A)}| - |\vec{F}_T^{S(B)}| = |\vec{F}_T^{A(S)}| - |\vec{F}_T^{B(S)}| \qquad \text{(N6.14)}$$

This difference in forces is necessary to provide a nonzero net force to accelerate the string. The difference is 0.10 N when $m_S = 0.010$ kg.

Note that the difference in the tension forces between the two ends of the string is about 0.5% of the magnitude of the tension forces themselves. Therefore, with a string this light, the tension force magnitudes are almost imperceptibly different. Now, 0.010 kg = 10 g is a pretty large mass for a short string. A real 20-cm-long string might actually have a mass of 2 g or less.

The difference between the tension magnitudes at the string's ends decreases as its mass decreases

---

Exercise N6X.3

If the string's mass were 1 g, show that the system's acceleration would be 0.998 m/s², and find the difference in tension across the string as a fraction of the tension force exerted at either end.

Exercise N6X.3 makes it clear that as the string's mass approaches zero, we can begin to think of 21.6 N as being *the* tension of the string: the difference between the tensions at its ends becomes *very* small.

If the string were completely massless, then equation N6.14 tells us that the tension forces exerted by the string's ends would be exactly the same, and the tension on the string would be a precisely defined number. While no string is truly massless, strings connecting objects usually have masses so much smaller than the objects that they connect that the idea that the string is massless represents an excellent approximation to the real situation.

Treating strings as massless
makes problems easy!

If we adopt the fiction that the string is *massless*, then analyzing a situation like the one shown in figure N6.5 becomes much simpler. We can ignore the free-body diagram of the string altogether and simply assume that the forces exerted by the string's ends have the common magnitude $|\vec{F}_T^S|$. Instead of the *five* equations N6.9a through N6.9e, the problem is completely described by *two* equations in the unknowns a_z and $|\vec{F}_T^S|$ that express the z component of Newton's second law as it applies to the two blocks:

$$m_A a_z = F_{\text{net},z}^A = F_{g,z}^{A(E)} + F_{T,z}^{A(X)} + F_{T,z}^{A(S)} = -m_A|\vec{g}| + |\vec{F}_T^{A(X)}| - |\vec{F}_T^S| \qquad \text{(N6.15a)}$$

$$m_B a_z = F_{\text{net},z}^B = F_{g,z}^{B(E)} + F_{T,z}^{B(S)} = -m_B|\vec{g}| + |\vec{F}_T^S| \qquad \text{(N6.15b)}$$

Simply *adding* these equations eliminates the unknown tension $|\vec{F}_T^S|$, leaving

$$(m_A + m_B)a_z = F_{T,z}^{A(X)} - (m_A + m_B)|\vec{g}| \qquad \text{(N6.16)}$$

which can be quickly solved for a_z:

$$a_z = \frac{|\vec{F}_T^{A(X)}|}{m_A + m_B} - |\vec{g}| = \frac{54 \cancel{N}}{5.0 \cancel{kg}}\left(\frac{1\ \text{kg}\cdot\text{m/s}^2}{1\ \cancel{N}}\right) - 9.8\ \text{m/s}^2 = 1.0\ \text{m/s}^2 \qquad \text{(N6.17)}$$

Substituting this into equation N6.15b yields

$$|\vec{F}_T^S| = m_B(|\vec{g}| + a_z) = (2.0\ \text{kg})(10.8\ \text{m/s}^2) = 21.6\ \text{kg}\cdot\text{m/s}^2 = 21.6\ \text{N} \qquad \text{(N6.18)}$$

Note how using this massless string approximation allows us to analyze the situation shown in figure N6.5 in a few lines, whereas it took more than a page of work before. Moreover, we get essentially the same results!

The ideal string model

In physics, the technical term **ideal string** refers to a hypothetical massless, inextensible, and flexible cord binding two objects together. An ideal string is really a *model* that we can apply to any real string, chain, wire, or other cord connecting two objects as long as (1) the cord's mass is very much less than the masses of the objects it connects, (2) the cord is reasonably inextensible and flexible, and (3) we state that we are using the approximation. ***The* tension on a** (presumably ideal) **string** is defined to be the tension force exerted on the object at *either* end of the string as a result of its interaction with the string.

N6.4   Pulleys

The ideal pulley model

Consider the situation shown in figure N6.6. The pulley in this situation redirects the string so that the tension forces exerted by its ends need no longer be opposite in direction: the string pulls *rightward* on object A and *upward* on object B. An **ideal pulley** simply redirects the string without creating any kind of difference in tension between the string on one side of the pulley and the string on the other. Like an ideal string, an ideal pulley is a mental construct that we use as an approximate model for a real pulley.

A real pulley will be approximately ideal if (1) it has a very small mass (particularly out near the rim) and (2) it has nearly frictionless bearings. If the pulley has a significant mass, then making significant changes in the pulley's rotation rate requires applying a significant torque to the pulley, which in turn requires that there be a significant difference in the tension of the string on one side and the string on the other. To see this, imagine tugging on a rope that goes over a very massive pulley. Perhaps you can imagine that you will have to pull very hard on the rope to accelerate the massive pulley wheel, even if the other side of the rope is almost slack. Similarly, a difference in tension would be needed to rotate the pulley if its bearings had friction: this difference in tension supplies a net force and thus a net torque to the rim of the pulley that keeps it turning against the opposing torque due to friction.

Pulleys these days are sufficiently light and frictionless that the ideal pulley model is actually a fairly reasonable approximation in many situations. Like the ideal string model, the ideal pulley model makes problems involving pulleys *much* easier to do.

In working problems involving pulleys, you will often find it advantageous to use a *separate* set of coordinate axes for each object, as shown in figure N6.6. This is perfectly acceptable as long as you (1) keep straight which coordinate axes apply to which object and (2) are careful to correctly describe any quantities that link the objects using the appropriate set of axes for the object in question. For example, in the situation shown in figure N6.6, the axes have been chosen so that the acceleration of each object points entirely along the x axis, and a_x has the same positive value for both objects (since the string has a fixed length and thus any motion of object A along its x axis is exactly duplicated by object B along its x axis). Also note that the tension force the string exerts on object A pulls it in its $+x$ direction, while the tension force the string exerts on object B pulls it in its $-x$ direction.

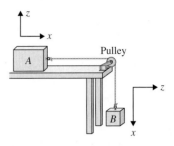

Figure N6.6
A situation involving two objects connected by a string going over a pulley.

Separate reference frames are often advantageous

N6.5 Using the Constrained-Motion Checklist

Coupled-object problems are almost exactly like the linearly constrained motion problems discussed in chapter N5, so we can pretty easily adapt the checklist we introduced in that chapter. The main *differences* are as follows. First, we need to do the following:

1. Draw a *separate* free-body diagram for each object in the system (and provide single-letter labels for all objects involved).
2. Indicate any *third-law partners* in these diagrams.
3. Describe in the *conceptual model* section any *joint* constraints on the objects' motion (for example, we might say, "Because the objects are connected, a_x is the same for both").
4. Apply Newton's second law in column-vector form to *each* object.

In addition (depending on the details of the problem), it *may* be necessary to do the following:

5. Draw individual coordinate axes for each object.
6. Draw separate acceleration arrows for each object.
7. Use Newton's third law to connect the magnitudes of third-law partners.

A modified version of the constrained-motion checklist that takes all this into account might look something like the following:

Coupled-Object Problem Checklist

☐ In the main drawing, draw an arrow for the acceleration $\vec{a}$ *of each object* in the system.
☐ Provide single-letter labels for everything.
☐ State joint constraints on the objects' motions.
☐ Draw a free-body diagram *for each object*.
☐ Show *third-law partners* in these diagrams.
☐ Draw different coordinate axes *for each object* if necessary. Align one axis with each object's acceleration $\vec{a}$.
☐ Write down Newton's second law in column-vector form *for each object*.
☐ Make all signs *explicit* when possible.

General Checklist

☐ Define symbols.
☐ Draw a picture.
☐ Describe your model and assumptions.
☐ List knowns and unknowns.
☐ Use a master equation.
☐ Do algebra symbolically.
☐ Track units.
☐ Check your result.

Example N6.4 illustrates how to use this adapted constrained-motion checklist to analyze a pulley problem like the situation shown in figure N6.6.

Example N6.4

Problem: In the situation shown in figure N6.6, assume that block *A* has a mass of 0.75 kg, block *B* has a mass of 0.25 kg, and the table is frictionless. What is the tension on the string connecting the blocks?

Coordinate axes for each object, one axis aligned with each object's acceleration

Diagram of the situation

Solution Here is a diagram of the situation.

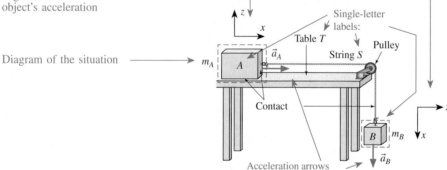

List of knowns and unknowns

$m_A = 0.75$ kg
$m_B = 0.25$ kg
$|\vec{a}_A| = |\vec{a}_B|$
no friction
$|\vec{F}_T^S| = ?$

Free-body diagrams of the objects appear below.

Free-body diagram for each object, with coordinate axes

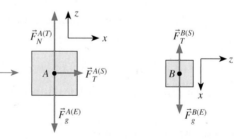

Statement about the joint constraints on the objects' motions

Check that acceleration arrow is consistent with free-body diagram

Statement about what the objects touch

Each object is constrained to move in its $+x$ direction (according to its own set of axes, as defined above), and the string also constrains the magnitudes of the objects' accelerations to have the same magnitude, so in this case $a_{Ax} = a_{Bx} \equiv a_x$. The forces shown in the free-body diagrams look consistent with accelerations in the $+x$ direction (as long as $|\vec{F}_g^{B(E)}| > |\vec{F}_T^{B(S)}|$). Block *A* touches the string and the table, which exert a normal force and a tension force on *A*, respectively (note we are excluding any kinetic friction force). Block B touches the string. Both blocks also touch the air, but we will ignore drag and buoyancy forces resulting from these interactions. Newton's second law thus implies the following two column-vector equations:

Approximations and assumptions stated

$$\begin{bmatrix} m_A a_x \\ 0 \\ 0 \end{bmatrix} = \begin{bmatrix} 0 \\ 0 \\ -m_A|\vec{g}| \end{bmatrix} + \begin{bmatrix} 0 \\ 0 \\ +|\vec{F}_N^{A(T)}| \end{bmatrix} + \begin{bmatrix} +|\vec{F}_T^{A(S)}| \\ 0 \\ 0 \end{bmatrix} \qquad (N6.19a)$$

$$\begin{bmatrix} m_B a_x \\ 0 \\ 0 \end{bmatrix} = \begin{bmatrix} +m_A|\vec{g}| \\ 0 \\ 0 \end{bmatrix} + \begin{bmatrix} -|\vec{F}_T^{B(S)}| \\ 0 \\ 0 \end{bmatrix} \qquad (N6.19b)$$

Master equations: Newton's second law in column-vector form for each object. Note the explicit signs.

Only three components of these equations are meaningful, but we have four unknowns ($a_x, |\vec{F}_N^{A(T)}|, |\vec{F}_T^{A(S)}|$, and $|\vec{F}_T^{B(S)}|$). However, if we assume the ideal string and pulley approximations, then $|\vec{F}_T^{A(S)}| = |\vec{F}_T^{B(S)}| \equiv |\vec{F}_T^S|$. This gives us enough information to solve the equations. The z component of equation N6.19a tells us that $m_A|\vec{g}| = |\vec{F}_N^{A(T)}|$, which gets rid of one unknown but is not really relevant. The x component of this equation tells us that

Counting knowns and unknowns

$$m_A a_x = |\vec{F}_T^S| \qquad (N6.20a)$$

Algebra with symbols

while the x component of equation N6.19b tells us that

$$m_B a_x = m_B|\vec{g}| - |\vec{F}_T^S| \qquad (N6.20b)$$

Adding these equations yields $(m_A + m_B)a_x = m_B|\vec{g}|$. Solving this for a_x and substituting the result into equation N6.20a yields

$$|\vec{F}_T^S| = m_A\left(\frac{m_B|\vec{g}|}{m_A + m_B}\right) = \frac{(0.25 \text{ kg})(0.75 \text{ kg})(9.8 \text{ m/s}^2)}{0.25 \text{ kg} + 0.75 \text{ kg}}\left(\frac{1 \text{ N}}{1 \text{ kg} \cdot \text{m/s}^2}\right)$$

$$= 1.84 \text{ N} \qquad (N6.21)$$

Calculation includes and tracks units

This has the right sign for a magnitude and the right units, and seems plausible, particularly as $m_B|\vec{g}| = 2.45 \text{ N} > 1.84 \text{ N} = |\vec{F}_T^S|$, as is required to produce a positive x-acceleration in block B.

Final check of plausibility

TWO-MINUTE PROBLEMS

N6T.1 A jet airplane flies at a constant velocity through the air. Its jet engines exert a constant force forward on the plane that exactly balances the force of air friction exerted backward on the plane. These forces are equal in magnitude and opposite in direction. Do we know this because of Newton's second law or Newton's third law?
A. Newton's second law
B. Newton's third law
C. Both laws
D. Neither (explain)

N6T.2 Which of the following pairs of forces are third-law partners? Answer T if the two forces described are third-law partners, F if they are not.
(a) A thrust force from its propeller pulls a plane forward; a drag force pushes it backward.
(b) A car exerts a forward force on a trailer; the trailer tugs backward on the car.
(c) A motorboat propeller pushes backward on the water; the water pushes forward on the propeller.
(d) Gravity pulls down on a person sitting in a chair; the chair pushes back up on the person.

N6T.3 A box B sits in the back of a truck T as the truck slows down for a stop (the box remains motionless relative to the truck). What is the appropriate symbol for the horizontal force that the contact interaction between the box and the truck exerts on the truck?
A. $\vec{F}_N^{B(T)}$
B. $\vec{F}_N^{T(B)}$
C. $\vec{F}_{SF}^{B(T)}$
D. $\vec{F}_{SF}^{T(B)}$
E. $\vec{F}_{KF}^{B(T)}$
F. Other (specify)

N6T.4 A child C pulls on a wagon W, using a string S; the wagon moves forward at a constant speed as a result. The third-law partner to the forward force exerted on the wagon is which of the following forces? (R = road.)
A. $\vec{F}_T^{S(W)}$
B. $\vec{F}_T^{W(S)}$
C. $\vec{F}_T^{W(C)}$
D. $\vec{F}_T^{C(W)}$
E. $\vec{F}_{KF}^{W(R)}$
F. Other (specify)

N6T.5 A small car pushes on a disabled truck, accelerating it slowly forward. Each exerts a force on the other as a result of their contact interaction. Which vehicle exerts the *greater* force on the other?
A. The car
B. The truck
C. Both forces have the same magnitude.

(more choices →)

D. The truck doesn't exert any force on the car.
E. One needs more information to answer.

N6T.6 A physicist and a chemist are playing tug-of-war. For a certain length of time during the game, the participants are essentially at rest. During this time, each person pulls on the rope (which can be treated as an ideal string) with a force of 350 N. What is the tension on the rope?
A. 700 N
B. 350 N
C. 175 N
D. Other (specify)

N6T.7 Object A ($m_A = 1.0$ kg) hangs at rest from an ideal string A connected to the ceiling. Object B ($m_B = 2.0$ kg) hangs at rest from an ideal string B connected to object A. The tension on string A is
A. Twice the tension on string B
B. 1.5 times the tension on string B
C. Equal to the tension on string B
D. $\frac{2}{3}$ the tension on string B
E. Other (specify)

N6T.8 A given rope will break if its tension exceeds 360 N. Suppose that each of two people can exert a pull of 200 N. Which strategy below can they use to break that rope?
A. They each take an end and pull.
B. They tie one end to the wall and both pull on the other.
C. They use either of the strategies above.
D. They cannot break the rope.

N6T.9 A spring scale typically indicates the magnitude of the tension force exerted on its bottom hook. What will the scale read in each of the cases shown in the diagram? (*Hint:* Draw a free-body diagram for the scale in each case.)
A. 4.9 N
B. 9.8 N
C. 19.6 N
D. Other (specify)

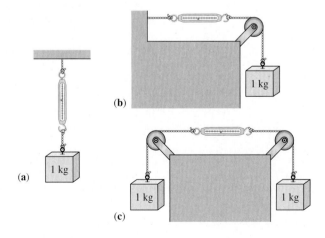
(a) (b) (c)

HOMEWORK PROBLEMS

Basic Skills

N6B.1   Which of the following force pairs are third-law partners? Explain your reasoning.
(a) The earth attracts a stone; the stone attracts the earth.
(b) A jet's engine thrusts it forward; drag pushes it back.
(c) You push on a box without moving it; the floor pushes back on the box.

N6B.2   Which of the following force pairs are third-law partners? Explain your reasoning.
(a) A mule pulls on a plow, moving it forward; the ground pulls backward on the plow.
(b) A team of dogs pulls a sled, moving it; the sled pulls backward on the dogs.
(c) Gravity tugs downward on a box sitting on the ground; the ground pushes up on the box.

N6B.3   In example N6.2, what would be the magnitude of the force that the blocks exerted on each other if blocks *A* and *B* had masses of 10 kg and 6.0 kg and we pushed on the blocks with a force of 4.0 N?

N6B.4   Find the acceleration of the system, the magnitude of the tension force on block *B*, and the difference in the magnitudes of the tension forces exerted by each end of the string in example N6.3 if the string has a mass of 100 g. Compare to the answers that we got before. Is this string even approximately ideal in your opinion?

N6B.5   For each of the situations described below, (1) draw a separate, isolated free-body diagram for each object involved, (2) assign an appropriate symbol to each force vector according to the conventions established in this chapter and in chapter N1, (3) indicate the approximate relative magnitudes of the forces by giving each arrow an appropriate length, and (4) link any third-law partners in the diagrams you draw.
(a) The moon orbits the earth. (Draw diagrams for both the moon and the earth.)
(b) You jump up off the floor. (Draw a diagram for both you and the earth during the interval of time while you are still in contact with the floor and are accelerating upward.)
(c) You are swinging a child around in a circle (the child's feet are off the ground).

N6B.6   For each of the situations described next, (1) draw a separate, isolated free-body diagram for each object involved, (2) assign an appropriate symbol to each force vector according to the conventions established in this chapter and in chapter N1, (3) indicate the approximate relative magnitudes of the forces by giving each arrow an appropriate length, and (4) link any third-law partners in the diagrams you draw.

(a) A little box sits on a bigger box. Both boxes are at rest. (Draw diagrams for both boxes.)
(b) A little box sits at rest on top of a bigger box sitting at rest on an incline. (Draw diagrams for both boxes.)
(c) A little box sits on top of a bigger box. The big box is sliding on a rough but level floor and as a result is slowing down. (Draw diagrams for both boxes.)

N6B.7   For each of the situations described below, (1) draw a separate, isolated free-body diagram for each object involved, (2) assign an appropriate symbol to each force vector according to the conventions established in this chapter and in chapter N1, (3) indicate the approximate relative magnitudes of the forces by giving each arrow an appropriate length, and (4) link any third-law partners in the diagrams you draw.
(a) A tractor pulls a plow at a constant velocity in a field. (Draw diagrams for both the tractor and the plow.)
(b) A person hangs from a helicopter by a rope as the helicopter begins to accelerate upward. (Draw diagrams for the person, the helicopter, and the rope.)
(c) A small car pushes a large disabled truck so both accelerate gently forward. (Draw diagrams for both the car and the truck. Ignore air resistance.)

Modeling

N6M.1   Two teams of people are involved in a tug-of-war. Since the forces exerted by each team on the other are equal and opposite by the third law, how is it possible for either team to win? Explain carefully, using appropriate free-body diagrams, how one team *can* win.

(Credit: Robert Kneschke/Shutterstock)

N6M.2   A block with a substantial mass *M* is suspended from the ceiling by a light string *A*. An identical string *B* hangs from the bottom of the block. If you jerk suddenly on string *B*, it will break, but if you pull steadily on string *B*, string *A* will break. Using a force diagram of the block, carefully explain why.

N6M.3 A 12,000-kg tugboat pushes on a 420,000-kg barge in still water. If the tug and barge accelerate at a rate of 0.20 m/s^2, what magnitude of force must the water exert on the tugboat's propellers? What is the magnitude of the force that the tug exerts on the barge? Ignore friction.

N6M.4 A 32-kg child puts a 15-kg box into a 12-kg wagon. The child then pulls horizontally on the wagon with a force of 65 N. If the box does not move relative to the wagon, what is the static friction force on the box?

N6M.5 A helicopter with mass M lifts a crate with mass m that is suspended from its fuselage.
(a) Assume the crate is suspended using steel cables of negligible mass. If the helicopter accelerates upward with acceleration $\vec{a}$ (and the force that the downward-flowing air exerts on the crate is negligible), (1) what is the magnitude of the force that the helicopter rotors must exert on the surrounding air, and (2) what is the magnitude of the force the cables exert on the point where they are connected to the fuselage?
(b) Answer the same questions if thick chains whose total mass is m_C are used instead of the cables.
(c) What is the difference between the total forces exerted by the top and bottom ends of the chains in the situation described in part (b)?
(d) Provide numerical answers for all parts above if $M = 2500$ kg, $m = 1500$ kg, $|\vec{a}| = 2.0$ m/s^2, and $m_C = 150$ kg.

N6M.6 A 65-kg crate slides on a rough plane inclined upward at an angle of 28°. The crate is hauled up the plane by a lightweight rope that runs parallel to the incline and then over a pulley at the top of the incline. A worker standing below the pulley pulls vertically downward on the rope. If the 55-kg worker hangs his or her entire weight from the rope, it is barely sufficient to move the crate up the incline at a constant velocity of 1.2 m/s.
(a) What is the magnitude of the kinetic friction force on the box?
(b) What is the coefficient of kinetic friction between the box and the incline?

N6M.7 An 85-kg crate sits in a 280-kg boat. What is the acceptable range of forces that a boat's propeller can exert on the water if we want to ensure that the crate remains at rest relative to the boat as the boat accelerates? (Estimate any information you are not provided.)

N6M.8 An 82-kg worker clings to a lightweight rope going over a lightweight, low-friction pulley. The other end of the rope is connected to a 67-kg barrel of bricks. If the worker is initially at rest 15 m above the ground, how fast will he or she be moving when he or she hits the ground?

N6M.9 Consider the situation shown in figure N6.6, except assume that the table is inclined at an angle of 22°, with the pulley at the top of the incline. Does block A slide up or down the incline? Assume that the table is frictionless, $m_A = 0.75$ kg, $m_B = 0.25$ kg, and that the masses of the pulley and string are negligible.

Figure N6.7
A drawing of the situation discussed in problem N6M.10.

N6M.10 Imagine that a person is seated in a chair that is suspended by a rope that goes over a pulley. The person holds the other end of the rope in his or her hands, as shown in figure N6.7. Assume that the combined mass of the person and chair is M.
(a) What is the magnitude of the downward force the person must exert on the rope to raise the chair at a constant speed? Express your answer in terms of M and $|\vec{g}|$. (*Hint:* The answer is not $M|\vec{g}|$!)
(b) What is the magnitude of the required force if the person is accelerating upward with $|\vec{a}| = 0.10|\vec{g}|$?

N6M.11 Consider the situation shown below: a cart of mass m_A on an incline tilted at angle θ is connected by a rope to a hanging weight of mass m_B. Assume that m_B is large enough that the cart accelerates up the incline.

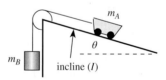

Assume also that the cart's wheels are frictionless and that the pulley and rope are ideal.
(a) Make a guess: will the tension on the rope be greater, less than, or equal to the weight of mass B, or does the answer depend on the values of m_A, m_B, and/or θ? Don't calculate anything yet: just reason qualitatively.
(b) Draw what might be appropriate coordinate systems on a copy of the drawing above, and explain why you drew them as you did.
(c) Then, actually calculate $|\vec{F}_T|$ in terms of m_A, m_B, $|\vec{g}|$, and θ using free-body diagrams based on these coordinate systems. Did you guess well in part (a)?

Rich-Context

N6R.1 A mule is asked to pull a plow. The mule resists, explaining that "If I tug on the plow, Newton's third law asserts that the plow will tug on me with an equal and opposite force. Since these forces will cancel each other out, it is obvious that we're not going anywhere. Therefore, there is no point in trying." Carefully (but politely) explain to the mule the error in its reasoning, and using appropriate free-body diagrams, explain why it is possible for the mule to accelerate the plow.

N6R.2 A glider with mass M slides frictionlessly on an air track inclined at an angle of θ. The glider is connected to a light-weight string which passes over a low-friction pulley at the top of the air track. The other end of the string is connected to a spring scale (whose mass is m_S), which is connected to a weight with mass m.
(a) If you hold the glider at rest, what will the scale read?
(b) If you release the glider, what equation determines whether the glider accelerates up or down the incline?
(c) What will the scale read as the system accelerates?
(d) Provide numerical answers to all the questions above if $M = 0.62$ kg, $\theta = 12°$, $m_S = 0.05$ kg, and $m = 0.17$ kg.

N6R.3 Imagine that Chris, whose mass is 45 kg, has fallen into a crevasse in a glacier. Chris's partner Pat, whose mass is 65 kg, has thrown a rope to Chris, and intends to lift Chris out by holding

on to the other end of the rope while sliding down the glacier's icy slope, which is tilted at an angle of 35° with respect to the horizontal. Assume that Pat has used some camping equipment to cobble together a pretty good pulley to change the rope's direction at the lip of the crevasse. Also assume that Chris hangs freely when suspended below the lip of the crevasse. Is it possible for Pat to pull Chris out, considering the likely coefficients of friction involved? Pat has a 30-kg backpack. Could this help?

Advanced

N6A.1 Imagine a train consisting of N frictionless cars following a locomotive that is accelerating the whole train forward with acceleration of magnitude $|\vec{a}|$. Assume that the first car behind the locomotive has mass M. If the tension in the coupling at the rear of *each* car is 10% smaller than the tension in the coupling in the front of the car, what is the mass of each successive car as a fraction of M? What is the total mass of the cars in terms of M? [*Hint:* You might find it helpful to know that $1 + x + x^2 + \cdots + x^n = (1 - x^{n+1})/(1 - x)$.]

ANSWERS TO EXERCISES

N6X.1 Free-body diagrams for the elevator and the person look like this:

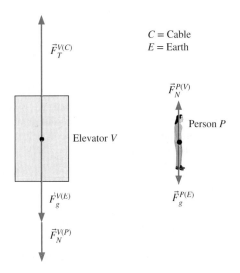

The only third-law partners that appear in these diagrams are the normal forces $\vec{F}_N^{V(P)}$ and $\vec{F}_N^{P(V)}$ that the person and elevator floor exert on each other due to their mutual contact interaction. Note that $|\vec{F}_N^{P(V)}|$ must be equal to the person's weight if the person is not accelerating vertically. Similarly, the magnitude $|\vec{F}_T^{V(C)}|$ of the force the cable exerts on the elevator must be equal to the sum of $|\vec{F}_g^{V(E)}|$ and $|\vec{F}_N^{V(P)}|$. The latter force would not be there if there

were no person in the elevator, so the tension on the cable must increase by this amount (which happens to be equal to the person's weight) when the person is in the elevator (and there is no vertical acceleration). So, even though the person's weight acts on the person, not the elevator, the effect of this weight is communicated to the elevator (and ultimately to the cable) by the contact interaction between the person and the floor (in a manner constrained by Newton's second and third laws).

N6X.2 Substituting equation N6.10 into N6.9*d* and the result into N6.9*a*, we get

$$m_A a_z = -m_A|\vec{g}| + |\vec{F}_T^{A(X)}| - (m_S a_z + m_S|\vec{g}| + m_B a_z + m_B|\vec{g}|)$$

(N6.22)

Adding $m_S a_z + m_B a_z$ to both sides yields equation N6.11.

N6X.3 Equation N6.12 now reads

$$a_z = \frac{54 \ \cancel{N}}{5.001 \ \text{kg}}\left(\frac{1 \ \text{kg} \cdot \text{m/s}^2}{1 \ \cancel{N}}\right) - 9.8 \ \text{m/s}^2 = 0.998 \ \text{m/s}^2 \quad \text{(N6.23)}$$

Equation N6.13 now reads

$$|\vec{F}_T^S| = (2.0 \ \text{kg})(9.8 \ \text{m/s}^2 + 0.998 \ \text{m/s}^2) = 21.6 \ \text{N} \quad \text{(N6.24)}$$

remembering that 1 kg·m/s² = 1 N. Note that this is essentially the same as before, and equation N6.14 reads

$$|\vec{F}_T^{A(S)}| - |\vec{F}_T^{B(S)}| = (0.001 \ \text{kg})(10.8 \ \text{m/s}^2) = 0.011 \ \text{N} \quad \text{(N6.25)}$$

Note that (0.011 N)/(21.6 N) ≈ 0.0005, or 0.05%.

N7

CORE

Circularly Constrained Motion

Chapter Overview

Introduction

We continue our examination of how we can determine forces from motion by examining the case in which an object is constrained to move in a circle, extending our work on uniform circular motion in chapter N1 to cover cases in which the object moves in a circle with non-constant speed. This chapter lays important foundations for chapters N8, N10, and N11.

Section N7.1: Uniform Circular Motion

In this section, we use a motion diagram to argue more carefully and rigorously than in chapter N1 that a particle that moves in a circle of radius R with constant speed $|\vec{v}|$ has an acceleration that points directly toward the center and has magnitude $|\vec{a}| = |\vec{v}|^2/R$. The section also notes that in such cases, we can write the particle's speed as

$$|\vec{v}| = \frac{2\pi R}{T} \qquad \text{for circular motion at constant speed} \qquad (N7.5)$$

- **Purpose:** This equation expresses the speed $|\vec{v}|$ of an object moving around a circle in terms of the circle's radius R and the time T it takes the object to go around the circle.
- **Limitations:** This expression is true only if $|\vec{v}|$ is *constant*.

Section N7.2: Unit Vectors

We will find it convenient in the future to express the direction of an object's acceleration using **unit vectors**. A unit vector (which we can also call a **directional**) is a symbol representing a pure direction. In this section, we will see that we can consider an arbitrary unit vector to be a vector with magnitude 1 (no units!). We can construct a unit vector $\hat{u}$ that points in the same direction as a given vector $\vec{u}$ as follows:

$$\hat{u} \equiv \frac{\vec{u}}{|\vec{u}|} \qquad (N7.8)$$

We define the unit vector $\hat{r}$ to be the direction of an object's position vector $\vec{r}$ from some origin. This unit vector is thus equivalent to the direction "directly away from the origin" at the particle's location. This unit vector is useful in a variety of contexts. In this chapter, we can use it to express the acceleration of an object moving with constant speed in a circle as follows:

$$\vec{a} = -\frac{|\vec{v}|^2}{R}\hat{r} \qquad\qquad (N7.11)$$

- **Purpose:** This equation describes the acceleration $\vec{a}$ of a particle (or an object's center of mass) that is moving in a circle of radius R with constant speed $|\vec{v}|$, where $\hat{r}$ is a unit vector meaning "directly away from the circle's center." (Since the acceleration in this case points *toward* the circle's center, it points in the $-\hat{r}$ direction, which is the point of the minus sign.)
- **Limitations:** The particle (or object) must both move in a *circle* and have a *constant* speed.

Section N7.3: Nonuniform Circular Motion

In this section, we extend the argument presented in section N7.1 to cover the situation in which an object moves in a circle with a speed that is not necessarily constant. The result is

$$\vec{a} = \frac{d|\vec{v}|}{dt}\hat{v} - \frac{|\vec{v}|^2}{R}\hat{r} \qquad\qquad (N7.18)$$

- **Purpose:** This equation describes the acceleration $\vec{a}$ of a particle (or an object's center of mass) that is moving in a circle of radius R with speed $|\vec{v}|$, where $\hat{r}$ is a unit vector meaning "directly away from the circle's center," and $\hat{v}$ is a unit vector pointing parallel to the particle's velocity.
- **Limitations:** The particle must move in a *circle*.

In circular motion, the velocity unit vector $\hat{v}$ and the position unit vector $\hat{r}$ are always perpendicular, so equation N7.18 indicates that the particle's or object's acceleration is constructed of the sum of two perpendicular component vectors in this case. Also, note that the first component vector points in the $+\hat{v}$ direction if the particle's speed is increasing ($d|\vec{v}|/dt > 0$) and in the $-\hat{v}$ direction if its speed is decreasing ($d|\vec{v}|/dt < 0$).

Section N7.4: Banking

This section discusses why an airplane must bank to make a turn, and it uses Newton's second law and equation N7.11 to calculate the banking angle. It also discusses why roads are often banked to make it safer for cars to navigate tight turns at high speeds.

Section N7.5: Examples

This section discusses how we can use the constrained-motion checklists developed in chapters N4 and N5 to solve circular motion problems. For example, in solving circular motion problems, it is useful to orient frame axes so the circle lies in the plane defined by a suitable pair of coordinate axes, and orient one of the axes to point in either the $+\hat{r}$ or the $-\hat{r}$ direction. It is also useful to note that in most banking problems it looks as if you don't have enough information to solve the problem, but you can usually manipulate things so that unknown quantities cancel out. The chapter closes with examples that illustrate the use of the checklist.

N7.1   Uniform Circular Motion

We introduced the concept of *uniform circular motion* in chapter N1, where we saw that an object moving at a constant speed in a circle was accelerating toward the center of the circle and that the magnitude of this acceleration was plausibly equal to $|\vec{v}|^2/R$, where $|\vec{v}|$ is the object's speed and R is the radius of its circular trajectory. Our task in this section is to put these results on a firmer mathematical foundation, partly so we can more easily go on to study *nonuniform* circular motion.

Consider a particle (or the center of mass of an object) moving in a circle of radius R at a constant speed $|\vec{v}|$. Figure N7.1 shows a motion diagram for the particle as it travels past points 1, 2, and 3, and it displays how we can construct the change-in-velocity vector $\Delta\vec{v}$ from the average velocity vectors $\vec{v}_{12}$ and $\vec{v}_{23}$.

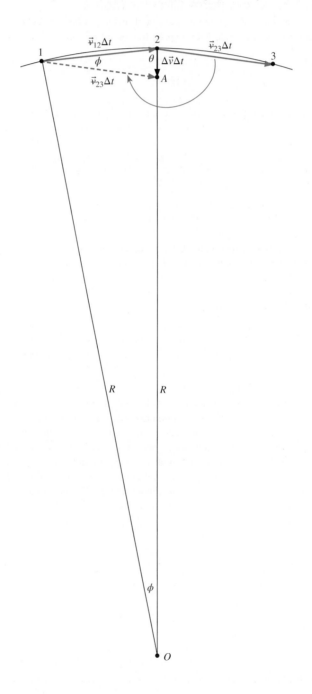

Figure N7.1

A motion diagram for an object moving at a constant speed in a circular path. Note that triangles $A12$ and $2O1$ are similar triangles.

We can use this diagram to prove that the particle's acceleration as it passes point 2 points directly toward the circle's center and has a magnitude of $|\vec{a}| = |\vec{v}|^2/R$ as follows.

Since the particle's speed is constant by hypothesis, the vectors $\vec{v}_{12}\Delta t$ and $\vec{v}_{23}\Delta t$ in the diagram have the same length. This means that triangles $A12$ and $2O1$ are similar: both are isosceles triangles with a common base angle θ. This means that the lengths of their sides must be proportional:

$$\frac{|\Delta\vec{v}\Delta t|}{|\vec{v}_{12}\Delta t|} = \frac{|\vec{v}_{12}\Delta t|}{R} \qquad (N7.1)$$

Now, $|q\vec{u}| = q|\vec{u}|$ for any ordinary (positive) number q and vector $\vec{u}$. Therefore, if we divide both sides of equation N7.1 by Δt, we get

$$\frac{|\Delta\vec{v}|}{|\vec{v}_{12}|}\frac{1}{\Delta t} = \frac{|\vec{v}_{12}|}{R} \quad\Rightarrow\quad \left|\frac{\Delta\vec{v}}{\Delta t}\right| = \frac{|\vec{v}_{12}|^2}{R} \qquad (N7.2)$$

Now, the object's speed $|\vec{v}|$ in its circular path is the arclength between points 1 and 2 divided by Δt, whereas $|\vec{v}_{12}|$ is the length of the straight line between those points divided by Δt. In the limit that $\Delta t \to 0$, points 1, 2, and 3 become infinitesimally separated, and the arclength and the straight-line distance between 1 and 2 become indistinguishable. Therefore,

$$\lim_{\Delta t \to 0}\left|\frac{\Delta\vec{v}}{\Delta t}\right| = \lim_{\Delta t \to 0}\frac{|\vec{v}_{12}|^2}{R} = \frac{|\vec{v}|^2}{R} \qquad (N7.3)$$

But the particle's acceleration $\vec{a}$ is defined to be $\lim_{\Delta t \to 0} (\Delta\vec{v}/\Delta t)$, so

$$\frac{|\vec{v}|^2}{R} = \lim_{\Delta t \to 0}\left|\frac{\Delta\vec{v}}{\Delta t}\right| = \left|\lim_{\Delta t \to 0}\frac{\Delta\vec{v}}{\Delta t}\right| \equiv |\vec{a}| \qquad (N7.4)$$

We can also see from the diagram that the $\Delta\vec{v}\Delta t$ vector points directly toward the center of the circle, so the $\vec{a}$ vector (which is simply a scalar multiple of $\Delta\vec{v}/\Delta t$, even in the limit) must point toward the center. Q.E.D.

There is nothing special about point 2: the argument would be the same at *any* point along the particle's circular trajectory. Therefore, this result describes the particle's acceleration at *any* point along its path.

In uniform circular motion problems, we are often given the time T that it takes to go around the circle once instead of the object's speed. Since the object goes a distance of $2\pi R$ in this time, if its speed is constant, it is given by

$$|\vec{v}| = \frac{2\pi R}{T} \qquad \text{for circular motion at constant speed} \qquad (N7.5)$$

- **Purpose:** This equation expresses the speed $|\vec{v}|$ of an object moving around a circle in terms of the circle's radius R and the time T it takes the object to go around the circle.
- **Limitations:** This expression is true only if $|\vec{v}|$ is *constant*.

A proof of the expression for an object's acceleration when it moves in uniform circular motion

N7.2   Unit Vectors

It is awkward to keep explaining in words that the direction of this vector is "toward the center of the circle." We can more compactly express this information by using what we call a *unit vector*. A **unit vector** is a shorthand description of a pure direction (which is why I prefer to call it a **directional**). For example, you should read the unit vector $\hat{x}$ as a shorthand for "in the $+x$ direction." Therefore, the equation $\vec{v} = -(2.0 \text{ m/s})\hat{x}$ describes a velocity vector with a magnitude of 2.0 m/s pointing in the $-x$ direction.

We can consider a directional to be a vector …

Now, in column-vector notation, we write a velocity vector whose magnitude is 2.0 m/s and which points in the $-x$ direction as

$$\begin{bmatrix} v_x \\ v_y \\ v_z \end{bmatrix} = \begin{bmatrix} -2.0 \text{ m/s} \\ 0 \\ 0 \end{bmatrix} = (-2.0 \tfrac{\text{m}}{\text{s}})\begin{bmatrix} 1 \\ 0 \\ 0 \end{bmatrix} \tag{N7.6}$$

We see that the column vector with the components [1, 0, 0] plays exactly the same role in the expression above that the unit vector $\hat{x}$ plays in the expression $\vec{v} = -(2.0 \text{ m/s})\hat{x}$. This suggests that we can think of a unit vector as being a special case of a vector (as the name suggests).

Indeed, let $\hat{u}$ represent an *arbitrary* direction, and say that we can write another vector $\vec{w}$ in the form $\vec{w} = \pm|\vec{w}|\hat{u}$. If we treat $\hat{u}$ as a vector and take the equation $\vec{w} = |\vec{w}|\hat{u}$ at face value, then taking its magnitude yields

… with a magnitude of 1 (and no units)

$$|\vec{w}| = |\pm|\vec{w}|\hat{u}| = |\vec{w}||\hat{u}| \quad \Rightarrow \quad |\hat{u}| = 1 \ (!) \tag{N7.7}$$

because $|\vec{w}|$ is a positive ordinary number and the magnitude of a positive ordinary number times a vector is just the magnitude of the vector times the ordinary number. So, if we *do* consider the unit vector $\hat{u}$ to be a vector, then it *must* have magnitude 1 (with no units!). This is *the* essential characteristic of a unit vector and the reason that we call it a *unit* vector. While the term *unit vector* describes the mathematical character of $\hat{u}$ and is conventional, I personally prefer the term *directional* because it more clearly describes the *role* it plays in indicating a vector's direction. I urge you to develop the habit of thinking "direction" whenever you see a hat on a vector symbol or see the term "unit vector."

We can construct a unit vector (directional) $\hat{u}$ that indicates the direction of an arbitrary vector $\vec{u}$ as follows:

How to construct the unit vector corresponding to a given vector's direction

$$\hat{u} \equiv \frac{\vec{u}}{|\vec{u}|} \tag{N7.8}$$

Since any positive scalar multiple of a vector has the same direction as the vector, $\hat{u}$ as defined above has the same direction as $\vec{u}$. Moreover, if we take the magnitude of both sides of the equation above, we find that (as required)

$$|\hat{u}| = \left|\frac{\vec{u}}{|\vec{u}|}\right| = \frac{1}{|\vec{u}|}|\vec{u}| = 1 \tag{N7.9}$$

The $\hat{r}$ unit vector

Now that we have discussed the general concept of a unit vector, let's get back to the issue of describing the direction of an object's acceleration when it is in uniform circular motion. A useful unit vector in a variety of contexts is

$$\hat{r} \equiv \frac{\vec{r}}{|\vec{r}|} \tag{N7.10}$$

where $\vec{r}$ is the position vector of a certain point relative to some specified origin. The unit vector $\hat{r}$ evaluated at a given point therefore indicates the direction "directly away from the origin" at that point. If we define the circle's center to be the origin in situations involving circular motion, we can write the acceleration of an object in uniform circular motion as follows:

$$\vec{a} = -\frac{|\vec{v}|^2}{R}\hat{r} \tag{N7.11}$$

- **Purpose:** This equation describes the acceleration $\vec{a}$ of a particle (or an object's center of mass) that is moving in a circle of radius R with constant speed $|\vec{v}|$, where $\hat{r}$ is a unit vector (directional) meaning "directly away from the circle's center." (Since the acceleration in this case points *toward* the circle's center, it points in the $-\hat{r}$ direction, which is the point of the minus sign.)
- **Limitations:** The particle (or object) must both move in a *circle* and have a *constant* speed.

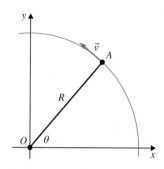

Figure N7.2
What are the components of the unit vector $\hat{r}$ for a particle passing through point A?

Exercise N7X.1

Suppose a particle is passing through point A, shown in figure N7.2, as it moves in a circular trajectory of radius R. Write $\hat{r}$ as a column vector, expressing its components in terms of θ.

N7.3   Nonuniform Circular Motion

Now consider the case in which a particle (or an object's center of mass) moves along a circular path with non-constant speed. Figure N7.3 shows a motion diagram for a particle that is speeding up as it moves to the right. Note that we can consider the object's total displacement $\vec{v}_{23}\Delta t$ between times t_2 and t_3 to be the sum of the displacement $\vec{v}_{23a}\Delta t$ that it *would* have moved if its speed had remained the same, plus an increment $\vec{v}_{23b}\Delta t$ indicating how much farther it made it along the circle due to its increased speed. The change in velocity vector $\Delta\vec{v}\Delta t$ that we would construct at point 2 is thus

How to find the acceleration vector $\vec{a}$ for nonuniform circular motion

$$\Delta\vec{v}\Delta t = \vec{v}_{23}\Delta t - \vec{v}_{12}\Delta t = \vec{v}_{23b}\Delta t + (\vec{v}_{23a}\Delta t - \vec{v}_{12}\Delta t) \qquad (N7.12)$$

Dividing both sides by Δt^2 and taking the limit as $\Delta t \to 0$, we get

$$\vec{a} \equiv \lim_{\Delta t \to 0}\frac{\Delta\vec{v}}{\Delta t} = \lim_{\Delta t \to 0}\frac{\vec{v}_{23b}}{\Delta t} + \lim_{\Delta t \to 0}\frac{\vec{v}_{23a}\Delta t - \vec{v}_{12}\Delta t}{\Delta t^2} \qquad (N7.13)$$

But the right-most quantity above is the same as the vector $\Delta\vec{v}/\Delta t$ we evaluated in section N7.1, which we found to have magnitude $|\vec{v}|^2/R$ and point toward the circle's center in the limit that $\Delta t \to 0$. So equation N7.13 becomes

$$\vec{a} = \lim_{\Delta t \to 0}\frac{\vec{v}_{23b}}{\Delta t} - \frac{|\vec{v}|^2}{R}\hat{r} \qquad (N7.14)$$

Figure N7.3
A motion diagram of a particle that follows a circular trajectory but speeds up. If the particle's speed were constant, it would make it to point a by time t_3, but because its speed has increased, it actually makes it to point 3 by that time.

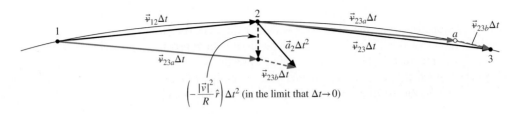

Figure N7.3
(A repeat of the figure on the previous page, for easy reference.)

To finish determining this acceleration vector, we need to evaluate the limit of $\vec{v}_{23b}/\Delta t$ as $\Delta t \to 0$. Notice that even for the generously large value of Δt shown in figure N7.3, the vectors $\vec{v}_{23a}\Delta t$, $\vec{v}_{23b}\Delta t$, and $\vec{v}_{23}\Delta t$ point pretty much along the same line. This approximation becomes even better as Δt becomes small. This means that

$$|\vec{v}_{23b}| \approx |\vec{v}_{23}| - |\vec{v}_{23a}| = |\vec{v}_{23}| - |\vec{v}_{12}| \qquad \text{(N7.15)}$$

since $|\vec{v}_{23a}| = |\vec{v}_{12}|$ by definition. Now each of these average velocities most closely represents the instantaneous velocity halfway through its respective interval, so $\Delta|\vec{v}| \equiv |\vec{v}_{23}| - |\vec{v}_{12}|$ very closely represents the change in the object's speed during a time interval of duration Δt centered on time t_2. All these approximations become increasingly exact as $\Delta t \to 0$. So

$$\lim_{\Delta t \to 0} \frac{|\vec{v}_{23b}|}{\Delta t} = \lim_{\Delta t \to 0} \frac{\Delta|\vec{v}|}{\Delta t} \equiv \frac{d|\vec{v}|}{dt} \qquad \text{(N7.16)}$$

where $d|\vec{v}|/dt$ is now the time derivative of the object's *speed*. Note also that as $\Delta t \to 0$, the direction of $\vec{v}_{23b}$ becomes increasingly horizontal, which is the same as the direction of the particle's velocity $\vec{v}$ at time t_2. So in that limit,

$$\lim_{\Delta t \to 0} \frac{\vec{v}_{23b}}{\Delta t} = \frac{d|\vec{v}|}{dt}\hat{v} \qquad \text{(N7.17)}$$

where $\hat{v} \equiv \vec{v}/|\vec{v}|$ is a unit vector in the direction of the particle's velocity at t_2.

So the complete expression for the acceleration of a particle (or object's center of mass) moving in a circular trajectory with a nonuniform speed is

The acceleration of a particle (or an object's CM) that moves in a circle with varying speed

$$\vec{a} \equiv \frac{d|\vec{v}|}{dt}\hat{v} - \frac{|\vec{v}|^2}{R}\hat{r} \qquad \text{(N7.18)}$$

- **Purpose:** This equation describes the acceleration $\vec{a}$ of a particle (or an object's center of mass) that is moving in a circle of radius R with (a possibly varying) speed $|\vec{v}|$, where $\hat{r}$ is a unit vector meaning "directly away from the circle's center," and $\hat{v}$ is a unit vector pointing parallel to the particle's velocity.
- **Limitations:** The particle must move in a *circle*.

Note that this expression reduces to equation N7.11 when $|\vec{v}|$ is constant, because $d|\vec{v}|/dt = 0$ in that case. Note also that if the particle is speeding up, $d|\vec{v}|/dt$ is positive, so the acceleration vector "leans forward" in toward the direction of the particle's velocity. However, if the particle is slowing down, $d|\vec{v}|/dt$ is negative and the acceleration vector "leans backward." Finally, note that since $\hat{v}$ is always perpendicular to $\hat{r}$ for circular motion, the magnitude of an object's acceleration in nonuniform circular motion is

$$|\vec{a}| = \sqrt{\left(\frac{d|\vec{v}|}{dt}\right)^2 + \frac{|\vec{v}|^4}{R^2}} \qquad\qquad (N7.19)$$

Exercise N7X.2

A car is traveling around a circular bend in the road, which has a radius of 450 m. At a certain instant, the car is due north of the circle's center, is traveling due west at a speed of 22 m/s, and is slowing down at a rate of 1.5 m/s². What are the magnitude and direction of the car's acceleration at this instant?

N7.4   Banking

In the rest of this chapter, we will consider examples in which we use either equation N7.11 or equation N7.18 to infer things about the forces acting on objects in circular motion.

One interesting application relates to the phenomenon of *banking*. You perhaps know that when a pilot turns a plane, she or he has to bank the plane into the turn—that is, lower one wing and raise the other. Why?

Figure N7.4a shows a rear-view free-body diagram of the plane. Two forces act vertically on the plane: its weight pulls it downward, and the lift from its wings pushes it upward. When the plane is in straight and level flight, these forces are directly opposite to each other and cancel.

We have seen, though, that a plane flying in a circular path has an acceleration toward the circle's center, which is *leftward* in figure N7.4b. Newton's second law implies that the net force on the plane must also point in that direction. Where does this leftward net force come from? It can't come from the plane's engines, since the thrust force that they exert is always forward.

The easiest way to exert a leftward force on a plane is to tilt the wings at an angle θ so that the lift force they exert (which acts perpendicular to the wings) has both an upward and a leftward component (see figure N7.4c). If the plane maintains its altitude as it turns, the upward part of the lift force must still balance the downward weight of the plane, leaving the leftward part as the net force on the plane. This net force is what causes the plane to accelerate away from its natural straight-line path into the circular path.

Why a plane banks when it turns

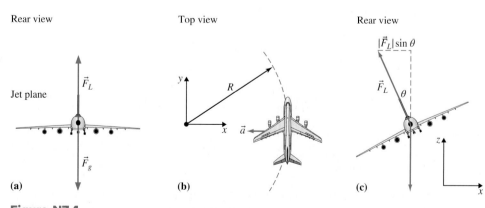

Figure N7.4
(a) A free-body diagram (rear view) of a jet plane flying straight and level. (b) A top view of a plane in a circular turn to the left. (c) A free-body diagram of the plane turning left.

Example N7.1

Problem: A jet flies at a constant speed of 260 mi/h in a holding pattern that is a horizontal circle of radius 5.0 mi. What is the jet's banking angle?

Solution Here are sketches of the plane's motion and a free-body diagram:

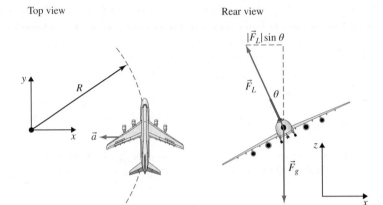

If the plane is moving in a *horizontal* circle, its acceleration in the z direction is zero: $a_z = 0$. On the other hand, because of its circular motion the plane must have a *nonzero* acceleration toward the circle's center: at the instant shown in the figure, $a_x = -|\vec{v}|^2/R$ (negative because the acceleration is to the left, that is, in the $-x$ direction as we have defined our coordinates) and $a_y = a_z = 0$. Newton's second law then reads

$$\begin{bmatrix} -m|\vec{v}|^2/R \\ 0 \\ 0 \end{bmatrix} = m\vec{a} = \vec{F}_g + \vec{F}_L = \begin{bmatrix} 0 \\ 0 \\ -m|\vec{g}| \end{bmatrix} + \begin{bmatrix} -|\vec{F}_L|\sin\theta \\ 0 \\ |\vec{F}_L|\cos\theta \end{bmatrix} \qquad \text{(N7.20)}$$

If we multiply both sides of the x component of equation N7.20 by -1 and add $m|\vec{g}|$ to both sides of the z component of equation N7.20, we get

$$\frac{m|\vec{v}|^2}{R} = |\vec{F}_L|\sin\theta \qquad m|\vec{g}| = |\vec{F}_L|\cos\theta \qquad \text{(N7.21)}$$

This problem looks hopeless at first: while we know $|\vec{v}|$ and R, we do not know the plane's mass m, the magnitude of the lift force $\vec{F}_L$, or the angle θ, meaning we have more unknowns than equations! But it turns out that we can solve the problem anyway. If we divide the first equation by the second, the unknowns m and $|\vec{F}_L|$ divide out, and we are left with

$$+\frac{|\vec{v}|^2}{R|\vec{g}|} = \frac{\sin\theta}{\cos\theta} = \tan\theta \qquad \Rightarrow \qquad \theta = \tan^{-1}\left(\frac{|\vec{v}|^2}{R|\vec{g}|}\right) \qquad \text{(N7.22a)}$$

$$\theta = \tan^{-1}\left[\frac{(260 \text{ mi}/\text{h})^2}{(5.0 \text{ mi})[22(\text{mi}/\text{h})/\text{s}]}\left(\frac{1 \text{ h}}{3600 \text{ s}}\right)\right] = 9.7° \qquad \text{(N7.22b)}$$

Exercise N7X.3

You may notice that the pilot when beginning a sharp turn will power up the engines somewhat if he or she wants to keep the plane from losing altitude. Why would the pilot do this?

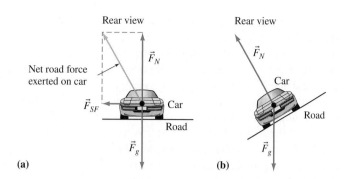

Figure N7.5
(a) A free-body diagram of a car rounding a curve to the left on a level roadbed. Static friction has to provide the leftward force required to keep the car following the curve. (b) A free-body diagram of a car on an ideally banked roadbed. Here the normal force alone provides the necessary sideward force.

A car rounding a corner does *not* need to bank into the curve. When you turn the steering wheel to the left, the road exerts a static friction force on the tires to the left, which provides the leftward net force necessary to accelerate the car away from its natural straight-line motion (see figure N7.5a). The car does not need to lean to the left to do this. (In fact, if you consider the torques that have to be exerted on the car, you will find that the car actually leans somewhat to the *right* when turning to the left. See problem N7M.10.)

An engineer designing a road will often call for a tight curve to be banked at an angle. This is often easiest to see on freeway overpasses that curve in one direction or another. (It is even more obvious in racetracks and the tracks of roller coasters, which are often banked at very large angles.) If the banking angle θ is chosen just right for the radius of the curve and the typical speed of a car on that curve, then the tilted *normal* force acting on the car provides the required leftward acceleration (see figure N7.5b). This is advantageous because the car does not then have to *depend* on static friction to keep it traveling with the curve. Under bad weather conditions, the coefficient of static friction between the roadbed and the tires may become so low that the static friction force cannot keep the car following the curve. The normal force, on the other hand, is not affected by bad conditions: the car cannot move *into* a wet or icy roadbed any more easily than into a dry roadbed! Banking the curve thus cuts down on accidents in bad weather. (The advantages in the case of roller coasters are even more obvious!)

A person riding a bicycle, on the other hand, *does* have to lean into the curve as an airplane does. The reasons are superficially similar to the reason a plane has to bank, but a careful analysis requires techniques that we will develop in chapter N8 (see example N8.4).

A car doesn't need to bank

Roads are sometimes banked for safety's sake

Bicyclists also have to lean into the curve

N7.5   Examples

Examples N7.2 through N7.5 illustrate how we can adapt the problem-solving checklists developed in chapters N4 and N5 to solve various kinds of problems involving circular motion.

Circular motion problems, like problems involving linear acceleration, are generally easier if you orient the reference frame axes correctly. Usually, the best orientation is such that one axis points parallel to the line connecting the object to the center of its circular path. In the case of *uniform* circular motion, the object's acceleration will thus nicely lie along this axis direction, and the components of the net force in other axis directions will be zero. In the case of *nonuniform* circular motion, the two axis directions in the plane of the circle will then correspond in a simple way to the $\hat{r}$ and $\hat{v}$ unit vectors of equation N7.18. Note that since these unit vectors change as the object moves, the reference frame you set up will only

How to choose an appropriate reference frame

Banking problems *seem* to have
too many unknowns

be useful at one instant. This is usually good enough to solve the problem, since
other instants will be analogous.

Most banking angle problems, like the airplane problem discussed in example N7.1, may look at first as though they have too many unknowns to solve. Press ahead anyway and see if some unknowns divide out.

In most other respects, circular motion problems are like any other constrained-motion problem: we take what we know about the object's motion (that it moves in a circle) and use that to determine unknown forces, banking angles, the time to complete the circular path, and the like.

Here, for the sake of reference, is the constrained-motion checklist from chapter N5 (adjusted a bit for circular motion):

Circular-Motion Problem Checklist
- ☐ Draw an arrow for $\vec{a}$ in your main drawing and point one axis toward or away from the circle's center.
- ☐ Show coordinate axes in all drawings.
- ☐ Draw a *free-body diagram* (*in addition* to a general picture of the situation).
- ☐ Write down Newton's second law in *column-vector* form.
- ☐ Make all signs *explicit* when possible.

General Checklist
- ☐ Define symbols.
- ☐ Draw a picture.
- ☐ Describe the model and assumptions.
- ☐ Use a master equation.
- ☐ List knowns and unknowns.
- ☐ Do algebra symbolically.
- ☐ Track units.
- ☐ Check result.

Check the example solutions that follow yourself to ensure that each expresses *all* of the items on the checklist.

Example N7.2

Problem: A 1500-kg car travels at a constant speed of 22 m/s over the top of a hill whose cross section near the top is approximately a circle whose radius is 150 m. What is the magnitude of the normal force of the car just as it passes the top of the hill, and how does this compare to the car's weight?

Solution Here are the diagrams we need:

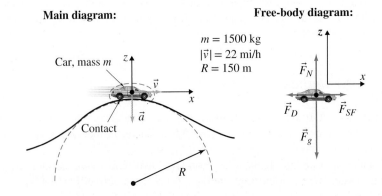

In the free-body diagram for the car, $\vec{F}_{SF}$ is from the tires' interaction with the road, and $\vec{F}_D$ is from the car's interaction with the air. The constraint on the car's motion is that we are told it moves in a vertical circle (which we have defined to be the xz plane) at a *constant speed*. This means that the car's acceleration is *downward* (in the $-z$ direction in our coordinate system) with a magnitude of $|\vec{a}| = |\vec{v}|^2/R$. This in turn means that $\vec{F}_{SF}$ must be directed *forward* (so that it cancels $\vec{F}_D$), and $\vec{F}_N$ must be smaller than $\vec{F}_g$ (as drawn) to yield a downward net force. The free-body diagram is then consistent with the observed acceleration. We are assuming that the hill's cross section really *is* circular (it probably isn't quite), and that drag is the only significant opposing force. If this is true, then Newton's second law in this case tells us that

$$\begin{bmatrix} 0 \\ 0 \\ -m|\vec{v}|^2/R \end{bmatrix} = \begin{bmatrix} 0 \\ 0 \\ -m|\vec{g}| \end{bmatrix} + \begin{bmatrix} 0 \\ 0 \\ +|\vec{F}_N| \end{bmatrix} + \begin{bmatrix} +|\vec{F}_{SF}| \\ 0 \\ 0 \end{bmatrix} + \begin{bmatrix} -|\vec{F}_D| \\ 0 \\ 0 \end{bmatrix} \quad \text{(N7.23)}$$

The x component of this equation simply tells us that $|\vec{F}_{SF}| = |\vec{F}_D|$, which we knew intuitively. Solving the z component for $|\vec{F}_N|$, we get

$$|\vec{F}_N| = m|\vec{g}| - \frac{m|\vec{v}|^2}{R} = \left[(1500 \text{ kg})\left(9.8 \frac{\text{m}}{\text{s}^2}\right) - (1500 \text{ kg})\frac{(22 \text{ m/s})^2}{150 \text{ m}}\right]\left(\frac{1 \text{ N}}{1 \text{ kg·m/s}^2}\right)$$

$$= 14{,}700 \text{ N} - 4800 \text{ N} = 9900 \text{ N} \quad \text{(N7.24)}$$

We can read from this calculation that $m|\vec{g}| = 14{,}700$ N, and that $|\vec{F}_N|/|\vec{F}_g| = (9900 \text{ N})/(14{,}700 \text{ N}) = 0.67$, so the normal force is two-thirds of the weight. The units check out, and the magnitudes look reasonable here ($|\vec{F}_N| < |\vec{F}_g|$ is a good sign). The sign of $|\vec{F}_N|$ is also positive, which is appropriate.

Problem: A roller-coaster car of mass m passes the top point of a vertical loop in the track traveling at a speed $|\vec{v}|$. The loop has radius R. Ignore friction. **(a)** What is the direction of the car's acceleration at this point? **(b)** What is the magnitude of the normal force that the track exerts on the car at this point?

Example N7.3

Solution Here are the diagrams we need for this situation:

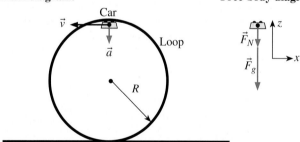

(a) Note that if we ignore friction, the only forces that can act on the car are its weight $\vec{F}_g$ and the normal force $\vec{F}_N$ that the track exerts on the car. The latter force must be perpendicular to the track and *away* from the track, so it must point *downward* if the car touches the track at all. Therefore, the net force on the car is downward, and so (by Newton's second law) the acceleration must be downward. This means that (at the instant in question anyway) the acceleration is toward the circle's center, meaning that $d|\vec{v}|/dt = 0$.

(b) The magnitude of Newton's second law in this case implies that

$$|\vec{F}_N| + |\vec{F}_g| = m|\vec{a}| = m\frac{|\vec{v}|^2}{R} \Rightarrow |\vec{F}_N| = m\frac{|\vec{v}|^2}{R} - |\vec{F}_g| = m\left(\frac{|\vec{v}|^2}{R} - |\vec{g}|\right) \quad \text{(N7.25)}$$

The units are right, but note that $|\vec{v}|$ must be large enough so that $|\vec{v}|^2/R > |\vec{g}|$ for $|\vec{F}_N|$ to be positive. A zero value for $|\vec{F}_N|$ would mean that the car is no longer touching the track, and $|\vec{F}_N| < 0$ (which is impossible), would mean that an assumption we have made is breaking down. In this case, the assumption would be that the car is moving in a circle: if its speed $|\vec{v}|$ is too small, the car will simply fall away from the track before reaching the top. Therefore, for the riders' safety, we want to make sure that $|\vec{v}|$ is sufficiently large at the loop's top to keep the car firmly engaged with the track ($|\vec{F}_N|$ large).

Example N7.4

Problem: What is the banking angle needed to keep a 1500-kg car following a circular bend in the road of radius 330 m when it is going 25 m/s ($\approx$58 mi/h) without requiring that *any* static friction force act on its tires?

Solution Rear- and top-view diagrams of the situation look like this:

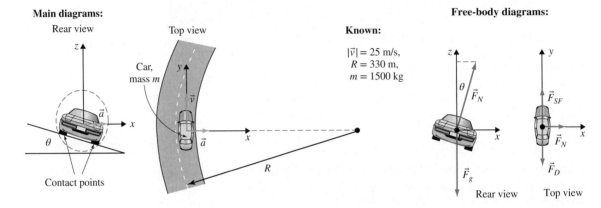

Main diagrams:

Rear view   Top view

Car, mass m

Contact points

Known:

$|\vec{v}| = 25$ m/s,
$R = 330$ m,
$m = 1500$ kg

Free-body diagrams:

Rear view   Top view

The constraint here is that the car is moving in a horizontal circle at a constant speed. Its acceleration therefore points directly inward, which (as we have defined the frame axes) is in the $+x$ direction at the instant shown. This is consistent with the free-body diagrams, which show the net force to point in that direction (due to the tilting of the normal force). We are assuming that the bend is level as well as circular, and that the car's speed is constant. If this is so, then there is no component of acceleration in either the y direction or the z direction. I am also assuming that the force opposing the car's forward motion is mostly drag. Newton's second law in this case implies that

$$\begin{bmatrix} +m|\vec{v}|^2/R \\ 0 \\ 0 \end{bmatrix} = \begin{bmatrix} 0 \\ 0 \\ -m|\vec{g}| \end{bmatrix} + \begin{bmatrix} +|\vec{F}_N|\sin\theta \\ 0 \\ +|\vec{F}_N|\cos\theta \end{bmatrix} + \begin{bmatrix} 0 \\ +|\vec{F}_{SF}| \\ 0 \end{bmatrix} + \begin{bmatrix} 0 \\ -|\vec{F}_D| \\ 0 \end{bmatrix} \quad \text{(N7.26)}$$

The y component of this equation tells us that $|\vec{F}_D| = |\vec{F}_{SF}|$, which is not relevant to what we want to find. The z component, on the other hand, tells us that $m|\vec{g}| = |\vec{F}_N|\cos\theta$, whereas the x component tells us that $m|\vec{v}|^2/R = |\vec{F}_N|\sin\theta$. If we divide the second equation by the first, we get

$$\frac{m|\vec{v}|^2}{mR|\vec{g}|} = \frac{|\vec{F}_N|\sin\theta}{|\vec{F}_N|\cos\theta} \quad \Rightarrow \quad \tan\theta = \frac{|\vec{v}|^2}{R|\vec{g}|} = \frac{(25 \text{ m/s})^2}{(330 \text{ m})(9.8 \text{ m/s}^2)} = 0.19 \quad \text{(N7.27)}$$

So $\theta = \tan^{-1}(0.19) = 11°$, so this is the ideal banking angle for the road. Note how the units work out so that $\tan\theta$ is correctly unitless, and θ is positive (consistent with the pictures) and its magnitude is reasonable ($\theta > 45°$, for example, would be unreasonable). Note also that the result does not depend on m but does depend on $|\vec{v}|$. A car not going at exactly the right speed will need some x component of static friction to hold it on its path.

Problem: A satellite in a circular orbit is a freely falling object that happens to be moving at just the right speed so that the gravitational force acting on it provides exactly the right acceleration to hold it in its circular path. How fast would a satellite have to travel to be in a circular orbit just above the earth's surface? How long would it take the object to go around the earth at this speed? Ignore the earth's atmosphere (!).

Example N7.5

Solution Here are the diagrams for this situation:

Main diagram:

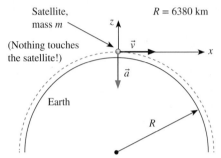

Free-body diagram:

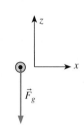

If there is no atmosphere, the only force that can possibly act on the satellite is the force of gravity. A satellite in a circular orbit is (by definition) moving in a circular path. We will assume that its speed $|\vec{v}|$ is constant. If this is so, then the satellite's acceleration will have a magnitude of $|\vec{a}| = |\vec{v}|^2/R$ and will point toward the center of the earth. This is consistent with the net force drawn on the diagram, since it must also point toward the center of the earth. At the representative instant shown, then, both the net force and the satellite's acceleration have components only in the $-z$ direction (according to our coordinate definitions). At that instant, Newton's second law says that

$$
\begin{bmatrix} 0 \\ 0 \\ -m|\vec{v}|^2/R \end{bmatrix} = \begin{bmatrix} 0 \\ 0 \\ -m|\vec{g}| \end{bmatrix}
\tag{N7.28}
$$

Solving the z component of this equation for $|\vec{v}|^2$ and then $|\vec{v}|$, we get

$$
|\vec{v}|^2 = R|\vec{g}| \Rightarrow |\vec{v}| = \sqrt{R|\vec{g}|} = \sqrt{(6{,}380{,}000 \text{ m})(9.8 \text{ m/s}^2)} = 7900 \text{ m/s}
\tag{N7.29}
$$

This is pretty fast! (Note that the satellite's mass is irrelevant.) Since the distance around the earth is $2\pi R$, the time T required for a complete orbit is

$$
T = \frac{2\pi R}{|\vec{v}|} = \frac{2\pi(6380 \text{ km})}{7.9 \text{ km/s}} \left(\frac{1 \text{ min}}{60 \text{ s}} \right) = 85 \text{ min} \; (=1.4 \text{ h})
\tag{N7.30}
$$

Note how the units all work out in both cases, and $|\vec{v}|$ comes out positive (as a magnitude should). The speed is pretty fast, but this is correct (one of the many reasons that space travel is difficult and expensive).

The problem with orbiting "just above" the earth's surface (in addition to the hazard posed to airplanes and mountains) is that drag is *not* going to be negligible at all at 7.9 km/s. (Indeed, vehicles re-entering the atmosphere at speeds comparable to this glow red-hot from friction!) Satellites usually orbit more than 100 km above the earth's surface, above virtually all of the atmosphere.

TWO-MINUTE PROBLEMS

N7T.1   When a speeding roller-coaster car is at the bottom of a loop, the magnitude of the normal force exerted on the car's wheels due to its interaction with the track is
A. Greater than the weight of the car and its passengers.
B. Equal to the weight of the car and its passengers.
C. Less than the weight of the car and its passengers.

N7T.2   A child grips tightly the outer edge of a playground merry-go-round as other kids push on it to give it a dizzying rotational velocity. When the other kids let go, the horizontal component of the net force on the child points most nearly
A. Inward toward the center of the merry-go-round.
B. Outward away from the center of the merry-go-round.
C. In the direction of rotation.
D. Nowhere: the net force's horizontal component is zero.
E. In some other direction (specify).

N7T.3   A car is traveling counterclockwise along a circular bend in the road whose effective radius is 100 m. At a certain instant of time, the car is traveling due *north*, has a speed of 10 m/s, and is in the process of *increasing* that speed at a rate of 1 m/s². The direction of the car's acceleration at that instant is most nearly
A. North
B. Northeast
C. East
D. Northwest
E. West
F. Southwest
T. Zero

N7T.4   A plane in a certain circular holding pattern banks at an angle of 8° when flying at a speed of 150 mi/h. If a second plane flies in the same circle at 300 mi/h, what is its banking angle?
A. A bit less than 16°
B. Exactly 16°
C. A bit more than 16°
D. A bit less than 32°
E. Exactly 32°
F. A bit more than 32°
T. The answer depends on the planes' masses.

N7T.5   Car 1 with mass m rounds an unbanked curve of radius R traveling at a constant speed $|\vec{v}|$. Car 2 with mass $2m$ rounds a curve of radius $2R$ traveling at a constant speed $2|\vec{v}|$. How does the magnitude $|\vec{F}_2|$ of the sideward static friction force acting on car 2 compare with the magnitude $|\vec{F}_1|$ of the sideward static friction force on car 1?
A. $|\vec{F}_1| = 4|\vec{F}_2|$
B. $|\vec{F}_1| = 2|\vec{F}_2|$
C. $|\vec{F}_1| = |\vec{F}_2|$
D. $|\vec{F}_1| = \frac{1}{2}|\vec{F}_2|$
E. $|\vec{F}_1| = \frac{1}{4}|\vec{F}_2|$
F. Other (specify)

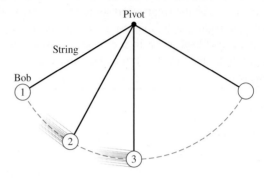

Figure N7.6
(For problems N7T.6 through N7T.8.) A pendulum bob swings from the end of a string. At point 1, the bob is at the extreme point of the swing and thus is instantaneously at rest. At point 3, the bob is directly below its suspension point and has its maximum speed.

N7T.6   (See figure N7.6.) Which one of the arrows below most closely indicates the direction of the bob's acceleration when it is at point 1?

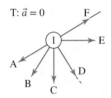

N7T.7   (See figure N7.6.) Which one of the arrows below most closely indicates the direction of the bob's acceleration when it is at point 2?

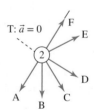

N7T.8   (See figure N7.6.) Which one of the arrows below most closely indicates the direction of the bob's acceleration when it is at point 3?

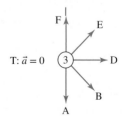

HOMEWORK PROBLEMS

Basic Skills

N7B.1   A car traveling at a constant speed of 50 mi/h travels around a curve. An accelerometer in the car measures its sideward acceleration to be $0.1|\vec{g}|$. What is the effective radius of the curve? [*Hint:* $|\vec{g}| = 22$ (mi/h)/s.]

N7B.2   A plane is traveling in a circular path 32 km in diameter. It is banking at an angle of 12°, which indicates that its sideward acceleration is 2.1 m/s². What is its speed?

N7B.3   An airplane banks at an angle of 11° while flying in a level circle at 320 mi/h. What is its path's radius?

N7B.4   A car is traveling counterclockwise along a circular bend in the road with an effective radius of 200 m. At a certain instant, the car's speed is 20 m/s, but it is slowing down at a rate of 1.0 m/s². What is the magnitude of the car's acceleration?

N7B.5   The position $\vec{r}$ of an object relative to the origin has components [3.0 m, 5.0 m, −2.0 m]. What are the components of the unit vector $\hat{r}$ that points in the same direction?

N7B.6   The position $\vec{r}_{B(A)}$ of object B relative to object A has components [7 m, −12 m, 0 m]. What are the components of the unit vector $\hat{r}_{B(A)}$ that points in the same direction?

N7B.7   Answer the questions posed in problems N7T.6 through N7T.8, and carefully explain your reasoning.

Modeling

N7M.1   A stunt driver drives a car over the top of a hill having a cross section that can be approximated by a circle of radius 250 m. What is the greatest speed the car can reach before it leaves the road at the top of the hill?

N7M.2   A car travels at a constant speed of 23 m/s through a small valley whose cross section is like a circle of radius 310 m. What is the magnitude of the normal force on the car, expressed as a multiple of the car's weight?

N7M.3   A child places a lunch box on the rim of a playground merry-go-round that has a radius of 2.0 m (≈6 ft). If the merry-go-round goes once around every 6.0 s, what is the speed of the box? What must be the coefficient of static friction between the box and the merry-go-round if the box is to stay on?

N7M.4   The acceleration of gravity near the surface of the moon has about one-sixth the magnitude it does on earth. The radius of the moon is 1740 km. How long would a satellite orbiting just above the moon's surface take to go once around the moon? (The moon has no atmosphere!)

N7M.5   Suppose you are designing a circular curve in a highway that must have a radius of 330 ft and will carry traffic moving at 60 mi/h.
   (a) If the roadway is *not* banked, what would the necessary coefficient of static friction between the tires and the asphalt road have to be to enable the car to follow the turn? (*Hint:* 1 m = 3.3 ft, 1 m/s = 2.24 mi/h.)
   (b) Is such a coefficient reasonable?
   (c) At what angle should the roadbed be banked?
   (d) Explain why this banking angle is necessary for safety.

N7M.6   A 150-lb student rides a Ferris wheel that rotates at a constant rate. At the highest point, the seat exerts a normal force of magnitude 110 lb on the student. What would the magnitude of this normal force be at the lowest point? (*Hint:* An object with mass 1 kg weighs 2.2 lb.)

N7M.7   A ball of mass m is tied to one end of a string of length L, the other end of which is fixed to the ceiling. The ball is then set in motion at a constant speed in a horizontal circle of radius $R < L$ around the axis that the string would make if the ball were to hang at rest (the string thus makes a constant angle θ with the vertical direction). Determine how long it takes the ball to go around the circle once in terms of m, $|\vec{g}|$, L, and θ. (Comment: An object moving as described is called a *conical pendulum*.)

N7M.8   As you are riding in a 1650-kg car, you approach a hairpin curve in the road whose radius is 50 m. The roadbed is banked inward at an angle of 10°.
   (a) Suppose the road is very icy, so that the coefficient of static friction is essentially zero. What is the maximum speed at which you can go around the curve?
   (b) Now suppose the road is dry and that the static friction coefficient between the tires and the asphalt road is 0.6. What is the maximum speed at which you can safely go around the curve?

N7M.9   At a certain instant of time, a 1200-kg car traveling along a curve 250 m in radius is moving at a speed of 10 m/s (22 mi/h) but is slowing down at a rate of 2 m/s². Ignoring air friction, what is the total static friction force on the car as a fraction of its weight at that instant?

N7M.10   Consider a car turning a corner to the left. Using the methods of chapter N4, show that in order to balance the torques on the car that seek to rotate the car around an axis going through its center of mass along the direction of its motion, the roadbed has to exert a greater normal force on the right wheels of the car than on the left wheels. Since this normal force will be transmitted to the car's body through the springs of the car's suspension, this means the suspension springs on the right will have to compress more than those on the left, and thus the car will lean to the right, as I asserted in section N7.4.

N7M.11 An unpowered roller-coaster car starts at rest at the top of a hill of height H, rolls down the hill, and then goes around a vertical loop of radius R. Determine the minimum value for H required if the car is to stay on the track at the top of the loop. (*Hints:* At the top of the loop, the car is upside down. If it is in contact with the track, though, the contact interaction will exert a normal force on the car perpendicular to the track and *away* from the track, since the normal force is a compression force. You may find it helpful to use conservation of energy here.)

Derivations

N7D.1 This problem explores a purely mathematical method for deriving equation N7.11. A particle moving in a circle of fixed radius R around an origin point O has a position vector that can be written $\vec{r}(t) = R\hat{r}$, where $\hat{r}$ is the direction of $\vec{r}$, which changes with time.
 (a) In a coordinate system oriented so the particle's path lies in the xy plane, show that at any given instant

$$\hat{r} = \begin{bmatrix} \cos\theta \\ \sin\theta \\ 0 \end{bmatrix} \qquad (N7.31)$$

 where the angle θ is the angle that the object's position vector makes with the x axis at that instant. Note that θ varies with time as the object moves around the circle.
 (b) By taking the time derivative of both sides of $\vec{r}(t) = R\hat{r}$ in column-vector form and using the chain rule (see appendix NA), show that

$$\vec{v}(t) = R\frac{d\theta}{dt}\begin{bmatrix} -\sin\theta \\ \cos\theta \\ 0 \end{bmatrix} \qquad (N7.32)$$

 (c) Argue from the expression above that

$$|\vec{v}(t)| = R\left|\frac{d\theta}{dt}\right| \qquad (N7.33)$$

 and that we can consider the column vector in brackets in equation N7.32 to be the unit vector $\hat{v}$.
 (d) Note that equation N7.33 implies that if the particle's speed is constant, then $|d\theta/dt|$ must be a constant. *Assuming* this, take the time derivative of both sides of equation N7.32 to show that

$$\vec{a}(t) = -\frac{|\vec{v}|^2}{R}\hat{r} \qquad (N7.34)$$

N7D.2 Equations N7.31 through N7.33 in problem N7D.1 apply even when the particle's speed in its circular trajectory is *not* constant. Read that problem carefully, then use the product rule to calculate the time derivative of equation N7.32, and argue that your result is consistent with

$$\vec{a}(t) = \frac{d|\vec{v}|}{dt}\hat{v} - \frac{v^2}{R}\hat{r} \qquad (N7.35)$$

when $d\theta/dt$ is *not* necessarily constant. (*Hint:* You will have to separately discuss the cases in which $d\theta/dt > 0$ and $d\theta/dt < 0$.

Show that in both cases, the term involving $d\theta/dt$ reduces to the first term on the right side of equation N7.35.)

N7D.3 In the limit that $R \rightarrow \infty$, a circular trajectory becomes a straight line. Argue carefully that in this limit, equation N7.18 becomes what you'd expect for linear motion with non-constant speed.

N7D.4 As we will see in unit E, a particle with charge q moving with velocity $\vec{v}$ in a uniform magnetic field $\vec{B}$ experiences a magnetic force $\vec{F}_m$ that is always perpendicular to its motion and has a magnitude $|\vec{F}_m| = |q||\vec{v}||\vec{B}|$. Assume this is the only force acting on the particle.
 (a) If $|\vec{F}_m| = |q||\vec{v}||\vec{B}|$, what must the SI units of $\vec{B}$ be?
 (b) Argue that if the magnetic force is really always perpendicular to the particle's velocity, then the particle's speed must remain constant.
 (c) This magnetic force will actually cause the particle to move in a circle of radius r. Find r as a function of the particle's charge q, its constant momentum magnitude $|\vec{p}|$, and the magnetic field's magnitude $|\vec{B}|$.

N7D.5 An object swings from the end of a string of length L in a vertical plane as a simple pendulum. The maximum angle that the string makes with the vertical is θ_0.
 (a) Use conservation-of-energy concepts to argue that

$$|\vec{v}|^2 = 2|\vec{g}|L(\cos\theta - \cos\theta_0) \qquad (N7.36)$$

 (b) Show that the angle ϕ that the object's acceleration vector makes with its velocity vector is such that

$$\tan\phi = \frac{2(\cos\theta - \cos\theta_0)}{-\sin\theta} \qquad (N7.37)$$

 where θ is the angle that the string makes with the vertical. (*Hints:* Take the derivative of equation N7.36 with respect to t. Also, note that θ is conventionally negative when the string is to the left of vertical. Assume that the bob is swinging to the right.)
 (c) Evaluate ϕ when $\theta = \pm\theta_0$ and when $\theta = 0$. Do your results make sense?
 (d) Evaluate ϕ when $\theta_0 = -45°$ and $\theta = -30°$.

Rich-Context

N7R.1 You are the technical consultant for a car-chase sequence in an action movie. In a certain part of the scene, the director wants a car to round a certain curve while braking from 66 mi/h to rest before it travels 290 ft along the curve. You measure the radius of the curve to be 590 ft. Is this scene possible? Defend your response and suggest an alternative scene if it is not possible.

N7R.2 "Rotor" is a ride found in many amusement parks that consists of a hollow cylindrical room (roughly 8 ft in radius) that rotates around a central vertical axis. Riders enter the room and stand against the canvas-covered wall. The room begins to rotate, and when a certain speed is reached, the floor of the room drops away, revealing a deep pit. The riders do not fall, though: they are supported by a static friction force exerted by the person's contact interaction with the wall.

(a) Suppose you are designing such a ride. What do you think would be a safe value to assume for the coefficient of static friction μ_s between peoples' clothes and the canvas wall. (Think about the liability suits that will result if you are wrong.)

(b) Estimate the rate at which the room should rotate (in revolutions per minute) to safely pin the riders to the wall. (You may have to make some estimates.)

People enjoying Rotor (see problem N7R.2).
(Credit: Joern Sackermann/Alamy Stock Photo)

ANSWERS TO EXERCISES

N7X.1   We can write the particle's position at point A as a column vector as follows:

$$\vec{r} = \begin{bmatrix} R\cos\theta \\ R\sin\theta \\ 0 \end{bmatrix} \qquad (N7.38)$$

Therefore, the unit vector $\hat{r}$ is

$$\hat{r} = \frac{\vec{r}}{r} = \frac{1}{R}\begin{bmatrix} R\cos\theta \\ R\sin\theta \\ 0 \end{bmatrix} = \begin{bmatrix} \cos\theta \\ \sin\theta \\ 0 \end{bmatrix} \qquad (N7.39)$$

N7X.2   The radial part of this car's acceleration has a magnitude of $|\vec{v}|^2/R = (22\text{ m/s})^2/(450\text{ m}) = 1.1\text{ m/s}^2$. We are told that the car is slowing down at the rate of 1.5 m/s^2, so the component of the acceleration in the direction of motion is $d|\vec{v}|/dt = -1.5\text{ m/s}^2$. The magnitude of the total acceleration (according to the Pythagorean theorem) is

$$|\vec{a}| = \sqrt{(1.1\text{ m/s}^2)^2 + (-1.5\text{ m/s}^2)^2} = 1.85\text{ m/s}^2 \qquad (N7.40)$$

Note that $\vec{a}$ points somewhat backward here. The angle ϕ that $\vec{a}$ makes with the backward direction is

$$\phi = \tan^{-1}\left|\frac{|\vec{v}|^2/R}{d|\vec{v}|/dt}\right| = \tan^{-1}\left|\frac{1.1\text{ m/s}^2}{1.5\text{ m/s}^2}\right| = 36° \qquad (N7.41)$$

(south of east).

N7X.3   Notice that in figure N7.4a, the whole lift force goes to supporting the plane: $m|\vec{g}| = |\vec{F}_L|$. In figure N7.4c, though, we see that the plane is supported by only the vertical component of the lift force $m|\vec{g}| = |\vec{F}_L|\cos\theta$. This means that the lift force must increase in magnitude as the plane banks to keep its vertical component equal in magnitude to the plane's weight. The easiest way to make the wings exert greater lift is to increase the speed of the plane through the air.

N8 Noninertial Frames

Chapter Overview

Introduction

Throughout this subdivision, we have been using Newton's second law and an object's observed motion to infer the forces acting on that object. However, this approach can backfire if we are observing the object's motion in a noninertial reference frame, because it can lead us to infer forces that do not physically exist. This chapter discusses this problem in depth.

Section N8.1: Fictitious Forces

In certain situations, such as in an airplane accelerating for takeoff, we seem to experience forces linked to the plane's motion that do *not* arise from physical interactions with other objects. The purpose of this chapter is to explain why such forces are in fact *inventions of our imagination*, and that we can adequately explain what we observe *without* inventing such forces if we use an appropriate reference frame.

Section N8.2: The Galilean Transformation

The first step in the process is to describe mathematically how observations in one reference frame S are linked to observations in another frame S'. If we orient these frames' axes so they point in the same direction in space, then we can show that

$$\vec{v}\,'(t) = \vec{v}(t) - \vec{\beta}(t) \qquad (N8.2)$$

- **Purpose:** This equation allows us to compute an object's velocity $\vec{v}\,'$ as measured in the S' frame if we know its velocity $\vec{v}$ in the S frame and the velocity $\vec{\beta}$ of the S' frame relative to the S frame.
- **Limitations:** All speeds must be much smaller than that of light.
- **Notes:** We call this the **Galilean velocity transformation equation**.

$$\vec{a}\,'(t) = \vec{a}(t) - \vec{A}(t) \qquad (N8.7)$$

- **Purpose:** This equation allows us to compute an object's acceleration $\vec{a}\,'$ in a frame S' given the same object's acceleration $\vec{a}$ in frame S and the acceleration $\vec{A}$ of frame S' relative to frame S ($\vec{A} \equiv d\vec{\beta}/dt$).
- **Limitations:** The speeds of the objects and frames involved must be much less than the speed of light.

Section N8.3: Inertial Reference Frames

Newton's first law states that an isolated object has zero acceleration. An **inertial reference frame** is a frame in which Newton's first law holds. In **noninertial reference frames**, where the first law is violated, *none* of Newton's laws apply.

We can test whether a frame is inertial by placing an object at rest relative to that frame, isolating it from external effects, and seeing whether it remains at rest. This is tricky in a gravitational field (since an object cannot be isolated from the effects of gravity), but we can imagine an object (such as a puck floating on an air table) that is at least isolated from external effects in the horizontal plane.

Equation N8.7 implies that if such **first-law detectors** establish that frame S is inertial, then another frame S' will also be inertial if and only if its acceleration $\vec{A}$ relative to S is zero. A frame attached to the ground is nearly inertial, so any frame that accelerates significantly relative to it will be noninertial.

Section N8.4: Linearly Accelerating Frames

In a frame attached to a plane accelerating for takeoff, a force appears to press objects backward. However, one comes to this conclusion only by inappropriately applying Newton's second law in the noninertial plane frame. If we analyze the situation in the ground frame, we see that we can easily explain the observed effects if we understand that the objects are simply trying to remain at rest as the plane accelerates forward.

Section N8.5: Circularly Accelerating Frames

Similarly, the "centrifugal" force you seem to feel in a turning car arises from your attempt to use Newton's second law to explain why objects appear to be thrown to the right when your car turns left. In the ground frame, though, we see that the objects are really attempting to travel in a straight line while the car frame veers left. Again in the *ground* frame, we can explain what we observe without invoking forces unconnected to physical interactions. Such forces are *not real*.

Section N8.6: Using Fictitious Forces

Having said that, we know that in some situations, analysis using a noninertial frame is much easier than using an inertial frame. In such cases, equation N8.7 implies that

$$m\vec{a}' = -m\vec{A} + \vec{F}_1 + \vec{F}_2 + \cdots \qquad \text{(N8.12)}$$

- **Purpose:** We can use this version of Newton's second law in a noninertial frame *if* we know that frame's acceleration $\vec{A}$ relative to an inertial frame. Here m is the mass of an object, $\vec{a}'$ is its acceleration in the noninertial frame, and $\vec{F}_1$, $\vec{F}_2$, ... are real forces acting on the object.
- **Limitations:** The relative speed of the two frames and the speed of the object relative to each must be much smaller than the speed of light.

We call the term $-m\vec{A}$ a **frame-correction force** (or *inertial force*). This force acts as if it were an additional gravitational force applied to the object. Note that this "force" does not physically exist: the *term* exists in this equation only to correct for our using a noninertial frame. Use this equation only if (1) an analysis in an inertial frame is very difficult and (2) you clearly indicate you are using a NONINERTIAL frame in your analysis.

Section N8.7: Freely Falling Frames and Gravity

Equation N8.12 implies that since all objects fall with the same acceleration $\vec{g}$ in an external gravitational field, the frame-correction and gravitational forces in a freely falling frame cancel out. This is the Newtonian explanation for why we can ignore external gravitational fields in a frame floating in space (as claimed in chapter C4).

In Einstein's theory of general relativity, however, freely falling frames are *really* inertial, and frames at rest on the surface of the earth are not! The section and selected homework problems discuss this issue in greater detail.

N8.1   Fictitious Forces

Examples of *fictitious forces* in daily life

According to the Newtonian model, *all* physical forces express the interaction of two objects, and all are either long-range forces or contact forces. The latter are easy to recognize since you can easily see when two objects are touching each other, and the only long-range forces that operate over macroscopic distances are gravitational and electromagnetic forces.

Yet in certain situations, forces appear to act that do *not* fit these categories. For example, imagine yourself in a jet accelerating for takeoff. When the pilot opens up the throttle, a magical force seems to appear that pushes you backward in your seat. The more extreme the plane's acceleration, the stronger this apparent force seems to be. When the plane lands and the pilot applies the brakes and reverses the thrust on the engines, the reverse happens: a mysterious force appears that tugs you forward. If you are in an automobile and the driver suddenly applies brakes sharply, you feel the same kind of force pulling you forward.

As a car turns a corner, a magical force seems to press you against the side of the car away from the turn. Again, this force seems to be associated with the car's acceleration, because we know that a car going around a curve is accelerating toward the center of the curve. As the car's acceleration increases (that is, the tighter the curve or the higher the car's speed), the magnitude of this force appears to increase as well.

These alleged forces do not fit in any of the categories described in chapter N2. They are obviously not contact forces (nothing that you are touching presses you back into the plane seat during takeoff). Such forces feel like gravitational forces, but they cannot *be* gravitational (an extremely large and massive object does not magically appear behind the plane when the pilot opens the throttle). Nor are they electrostatic or magnetic (you do not become electrically charged or magnetized when your car turns a sharp corner). These alleged forces do not in fact seem to be a consequence of the presence of *any* external object.

Our task: to explain the effects without the forces

Even though these forces "feel" quite real, *I claim that these alleged forces are in fact inventions of your imagination.* Ladies and gentlemen of the jury, I will show you beyond a shadow of a doubt that it is possible within the context of the Newtonian model to explain *all* the described *effects* of these forces *without* assuming that these forces really exist. Since the argument for the existence of these alleged forces is based *entirely* on the circumstantial evidence of their effects, an alternative explanation of those effects makes the argument for the existence of these **fictitious forces** disappear. Let me begin my case.

N8.2   The Galilean Transformation

Fictitious forces are associated with *accelerating* reference frames

The first step in understanding this situation is to recognize that these forces only seem to arise when you are riding in something (such as a plane or car) that is *accelerating* relative to the earth's surface. While being pushed back into your chair is something you might expect during an airplane takeoff, you would be very surprised (even terrified) if a mysterious force were to spontaneously push you back into your chair while you were sitting at home reading a book!

When you are riding in a plane, car, or even a playground merry-go-round, you automatically and unconsciously use your surroundings (the cabin of the plane, the frame of the car, or the structure of the merry-go-round) as your

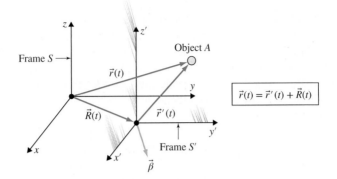

Figure N8.1
This diagram shows two reference frames S and S'. Frame S' moves relative to S at a velocity $\vec{\beta}$ as shown. The positions of object A as measured in the two reference frames are related by the vector equation in the box.

frame of reference, and you judge your motion and the motion of objects around you in terms of that frame. When your jet or car is cruising at a constant speed, no magical forces appear: only when the jet or car changes its velocity do strange things seem to happen. *The presence of these forces has something to do with observing motion from within an accelerating reference frame.*

The next step is to find a means of mathematically connecting observations made in one reference frame with observations made in another. Figure N8.1 shows two abstract reference frames S and S' (the latter is read "S prime"). Frame S' is moving with respect to S at some (possibly time-dependent) relative velocity $\vec{\beta}$ (the Greek letter beta here refers to the "boost" in velocity required to take you from frame S to frame S'). Let's also take advantage of our ability to choose the orientation of reference frames to set them up so that their axes point in the same directions (this makes the analysis easier). Assume we measure the position of object A as a function of time in both reference frames: let $\vec{r}(t)$ be its position vector as measured in frame S at time t, and let $\vec{r}\,'(t)$ be its position as measured in S' at the same instant, and let the position of S' relative to S at that instant be $\vec{R}(t)$. As shown in figure N8.1, the definition of vector addition means that the mathematical relationship between $\vec{r}(t)$, $\vec{r}\,'(t)$, and $\vec{R}(t)$ at any time (no matter how S' is moving) is given by

$$\vec{r}(t) = \vec{r}\,'(t) + \vec{R}(t) \qquad (N8.1)$$

The transformation equation for positions

In words, this equation tells us that the object's position in frame S is the vector sum of its position in frame S' and the position $\vec{R}(t)$ of the origin of frame S' in frame S.

If we solve this equation for $\vec{r}\,'(t)$, take the time derivative of both sides, and note that $d\vec{R}/dt$ is the same as the velocity $\vec{\beta}$ of S' relative to S, we get

$$\vec{v}\,'(t) = \vec{v}(t) - \vec{\beta}(t) \qquad (N8.2)$$

The (Galilean) transformation equation for velocities

- **Purpose:** This equation allows us to compute an object's velocity $\vec{v}\,'$ as measured in the S' frame if we know its velocity $\vec{v}$ in the S frame and the velocity $\vec{\beta}$ of the S' frame relative to the S frame.
- **Limitations:** All speeds must be much smaller than that of light.
- **Notes:** We call this the **Galilean velocity transformation equation**.

Does this make sense? Let's consider some examples.

Example N8.1

Problem: Suppose you are on a train traveling in the $+x$ direction at a speed of 25 m/s with respect to the ground. If you throw a baseball in the $+x$ direction at a speed of $+12$ m/s relative to the train, what is the ball's x-velocity relative to the ground? *Intuitively*, you might say that the ball's velocity with respect to the ground should be the *sum* of the train's velocity with respect to the ground and the ball's velocity with respect to the train, that is, 37 m/s. Is this correct?

Solution To apply equation N8.2 correctly, we must do two things: (1) determine what the reference frames are and which ones we will take to be S and which S', and (2) determine which of the stated velocities correspond to which of the symbolic quantities $\vec{v}$, $\vec{v}\,'$, and $\vec{\beta}$.

The first of these steps is actually a fairly arbitrary decision. As long as we keep everything straight, we would get the same ultimate answer no matter which frame (ground or train) we took to be frame S. But note that $\vec{\beta}$ is defined to specify the velocity of frame S' *relative* to S. In the problem description, the train's velocity is specified relative to the ground, so it is *convenient* to let S' be the train frame and S be the ground frame, because then we know immediately from the problem description that $\vec{\beta} = (+25 \text{ m/s})\hat{x}$.

We are told that the ball moves at a speed of 12 m/s in the x direction with respect to the train, so this velocity must be $\vec{v}\,'$ since the train is the "primed" frame. We are trying to find the velocity of the ball with respect to the ground, so $\vec{v}$ is the unknown quantity. Solving equation N8.2 for $\vec{v}$ yields

$$\vec{v} = \vec{v}\,' + \vec{\beta} = \begin{bmatrix} +12 \text{ m/s} \\ 0 \\ 0 \end{bmatrix} + \begin{bmatrix} +25 \text{ m/s} \\ 0 \\ 0 \end{bmatrix} = \begin{bmatrix} +37 \text{ m/s} \\ 0 \\ 0 \end{bmatrix} \qquad \text{(N8.3)}$$

So the ball's velocity with respect to the ground is $(+37 \text{ m/s})\hat{x}$ as expected.

Example N8.2

Problem: Suppose you are in a train traveling in the $+x$ direction at a speed of 35 m/s relative to the ground, and you observe a car traveling in the same direction at a speed of 29 m/s relative to the ground. What is the car's speed relative to you?

Solution Again, we are *given* that the train's velocity with respect to the ground is 35 m/s in the x direction. If we take the train to be frame S' and the ground to be frame S, then this means that $\vec{\beta} = (+35 \text{ m/s})\hat{x}$. We are given that the car's velocity in the ground frame is $\vec{v} = (+29 \text{ m/s})\hat{x}$. The car's velocity $\vec{v}\,'$ in the train frame is the unknown quantity. Equation N8.2 then directly implies that

$$\vec{v}\,' = \vec{v} - \vec{\beta} = \begin{bmatrix} 29 \text{ m/s} \\ 0 \\ 0 \end{bmatrix} - \begin{bmatrix} 35 \text{ m/s} \\ 0 \\ 0 \end{bmatrix} = \begin{bmatrix} -6 \text{ m/s} \\ 0 \\ 0 \end{bmatrix} \qquad \text{(N8.4)}$$

This means that the car will appear to drift backward at a speed of 6 m/s relative to the train. If you visualize the situation, perhaps you will agree that this must be right.

Problem: An airplane is flying with a velocity of 75 m/s due north relative to the air. If the wind is blowing 12 m/s west relative to the ground, what is the plane's velocity (magnitude and direction) relative to the ground?

Solution Take the ground to be frame S and the air to be frame S'. According to the description of the situation, the air is moving relative to the ground at 12 m/s west. If both our frames are oriented in the usual way relative to the earth's surface, this means that $\vec{\beta} = [-12 \text{ m/s}, 0, 0]$. The plane's velocity with respect to the air is $\vec{v}' = [0, +75 \text{ m/s}, 0]$. We want to find the plane's velocity with respect to the ground, which is $\vec{v}$. Solving equation N8.2 for $\vec{v}$ yields

$$\vec{v} = \vec{v}' + \vec{\beta} = \begin{bmatrix} 0 \\ +75 \text{ m/s} \\ 0 \end{bmatrix} + \begin{bmatrix} -12 \text{ m/s} \\ 0 \\ 0 \end{bmatrix} = \begin{bmatrix} -12 \text{ m/s} \\ +75 \text{ m/s} \\ 0 \end{bmatrix} \qquad (N8.5)$$

The magnitude of this velocity is

$$|\vec{v}| = \sqrt{v_x^2 + v_y^2 + v_z^2} = \sqrt{(-12 \text{ m/s})^2 + (75 \text{ m/s})^2 + 0} = 76 \text{ m/s} \qquad (N8.6a)$$

The angle that this velocity makes with the y (north) axis is

$$\theta = \tan^{-1}\left(\frac{12 \text{ m/s}}{75 \text{ m/s}}\right) = 9° \qquad (N8.6b)$$

Since v_x is negative, the plane is flying at 76 m/s, 9° *west* of north.

Now, if we take the time derivative of both sides of equation N8.2, we get

$$\vec{a}'(t) = \vec{a}(t) - \vec{A}(t) \qquad (N8.7)$$

- **Purpose:** This equation allows us to compute an object's acceleration $\vec{a}'$ in a frame S' if we know the same object's acceleration $\vec{a}$ in frame S and the acceleration $\vec{A}$ of frame S' relative to frame S ($\vec{A} \equiv d\vec{\beta}/dt$).
- **Limitations:** The speeds of the objects and frames involved must be much less than the speed of light.

Equation N8.7 implies that if frame S' moves at a constant velocity with respect to S (so that $\vec{A} = 0$), the object's acceleration is the same in both frames:

$$\vec{a}'(t) = \vec{a}(t) \quad \text{if } S' \text{ moves at a constant velocity relative to } S \qquad (N8.8)$$

We will build our case regarding fictitious forces on an analysis of equations N8.7 and N8.8.

Exercise N8X.1

Suppose that in an elevator whose vertical acceleration is 3.0 m/s² upward, you drop a ball that falls with a downward acceleration of $|\vec{g}| = 9.8 \text{ m/s}^2$ in the frame of the earth. What is the ball's acceleration in the elevator's frame?

N8.3   Inertial Reference Frames

Suppose that an object is completely isolated from external interactions. This means that $\vec{F}_{net} = 0$ and by Newton's second law, the object should have *zero* acceleration. Newton's first law says this even more directly: *an isolated object moves at a constant velocity.* Now, either an object is isolated from external interactions, or it is not. You do not have to use a reference frame to determine whether an object is massive or whether it is electrically charged or whether it touches something else. Observers in all reference frames should therefore agree as to whether a given object is isolated.

An isolated object doesn't obey Newton's laws in a noninertial reference frame

Suppose that a universally accepted isolated object is observed to move at a constant velocity (zero acceleration) in some frame S. Equation N8.7 implies that all *other* frames can be divided into two categories. In those frames that move at a constant velocity with respect to S (so that their acceleration $\vec{A}$ relative to S is zero), the isolated object will *still* be measured to have zero acceleration and thus be observed to obey Newton's first and second laws. We call such frames **inertial reference frames** (because Newton's first law is sometimes called the *law of inertia*). In frames that are accelerating relative to frame S, the isolated object will be measured to have a nonzero acceleration $\vec{a}' = -\vec{A}$, meaning that Newton's first and second laws *fail* to work in such reference frames. We call such frames **noninertial reference frames**.

Let me emphasize again that *Newton's first and second laws do not apply in noninertial reference frames:* such frames are "bad" for analyzing motion in the Newtonian model. If we want to apply Newton's laws to analyze an object's motion, we must do the analysis in an *inertial* reference frame.

Constructing an idealized first-law detector

We can use a **first-law detector** to distinguish an inertial frame from a noninertial frame *without* determining the frame's motion with respect to something else. Figure N8.2 shows such a detector. Electrically actuated "fingers" hold an electrically uncharged and nonmagnetic ball in the center of a spherical container from which the air has been removed (figure N8.2a). When the fingers are retracted, the ball is (at least momentarily) at rest and completely isolated from contact and electromagnetic interactions (figure N8.2b). An isolated object at rest will *remain* at rest if the frame is inertial. If the frame to which the container is attached is noninertial, though, the ball will accelerate relative to that frame and thus the container (see equation N8.7). This can easily be detected, because if the ball drifts away from rest in any direction, it will eventually hit the container wall, where touch-sensitive sensors can register the violation of the first law (and trigger the "fingers" to reset the ball).

Actually, the detector as just described will only distinguish inertial from noninertial frames in deep space (far away from any gravitating objects) because there is no other way to isolate the internal ball from gravity. We might adapt our detector for operation in a gravitational field by changing the spherical container to a cylindrical one and allowing the ball to drop from rest. Since the ball's acceleration should be perfectly vertical by definition, any sideward deviation that would cause the ball to strike a container wall will indicate that the frame of the container is noninertial. Also, since all objects fall with the same acceleration, the ball *should* take a well-defined amount of time to fall the length of the cylinder. If it arrives at the bottom early or late, the frame is not inertial.

A floating-puck first-law detector

A more practical way to make a first-law detector for use in a gravitational field is to replace the ball with a puck that floats on a cushion of air above a flat, level air table. This puck is isolated from external interactions that might affect its *horizontal* motion, so if we place it at rest at the center of the table, it should *remain* at rest. If we observe the puck to accelerate away from the center, the frame in which the table sits is not inertial. This detector will not register violations of

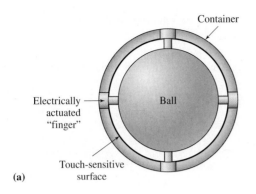

Container

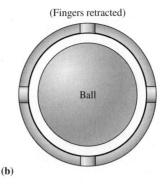

(Fingers retracted)

Electrically actuated "finger"

Ball

Ball

Touch-sensitive surface

(a)

(b)

Figure N8.2

(a) A cross-sectional view of a floating-ball first-law detector. Electrically actuated "fingers" hold the ball initially at rest in the spherical container. (b) After the fingers are retracted, the ball should continue to float at rest in the container if the container frame is inertial.

the first law due to vertical accelerations, but it is easier to imagine and construct than a dropped-ball detector.

Imagine now attaching first-law detectors in a number of places throughout a reference frame. A frame is *inertial* only if *no* detector in the frame registers any violation of Newton's first law; otherwise, the frame is *noninertial*.

Formal definitions of inertial *and* noninertial *frames*

If we apply this definition to a frame attached to the surface of the earth, we find that such a frame is *not* perfectly inertial, because the earth is rotating. But the earth rotates fairly slowly, so the deviations from perfection are *usually* negligible. Unless otherwise stated, we will assume that a reference frame attached to the earth's surface is sufficiently inertial for our purposes.

N8.4 Linearly Accelerating Frames

Consider again the example of an airplane accelerating for takeoff. Imagine that you hold a floating-puck first-law detector in your lap. Then the pilot turns on the engines, and the plane begins to accelerate down the runway. You feel a force pushing you back into your seat. You will also see the puck accelerate toward the rear of the plane. In the plane frame, then, it certainly looks as if some magical force is acting on you and the puck (see figure N8.3a on the next page), pulling everything backward. Yet the puck is horizontally *isolated*, and therefore there is no way for an external force to act on you and the puck. How can we resolve this paradox?

What you experience in a linearly accelerating frame

The problem is that we are inappropriately trying to make Newton's second law work in an accelerating frame. Let's look at the puck's motion from the perspective of the ground frame. When the pilot turns up the engines, the plane begins to accelerate. The puck is horizontally isolated, so if it was initially at rest relative to the ground, it will remain at rest. On the other hand, as the plane accelerates forward relative to the ground, the air table is carried with the plane forward with respect to the ground and thus with respect to the puck. This makes the puck *appear* to accelerate backward with respect to the table. In the inertial frame of the ground, no "magical forces" are needed to explain the puck's behavior (see figure N8.3b).

We can also see this from equation N8.7. Let the ground frame be frame S and the plane frame be frame S'. Since the puck has zero net physical force on it, its acceleration will be zero in the inertial ground frame S ($\vec{a} = 0$). If the plane has a nonzero acceleration with respect to the ground frame ($\vec{A} \neq 0$), then the puck's acceleration measured in the plane frame is

An explanation using equation N8.7

$$\vec{a}' = \vec{a} - \vec{A} = 0 - \vec{A} = -\vec{A} \qquad \text{(N8.9)}$$

Figure N8.3

(a) In a plane accelerating for takeoff, a puck floating on an air table accelerates to the rear of the plane relative to the table. (b) When viewed in the ground frame, the puck remains at rest. It only accelerates relative to the table because the *table* is accelerating forward with you and the plane. (I have displaced successive images of the situation to the right only for clarity's sake. The plane is accelerating straight forward, not also to the right.)

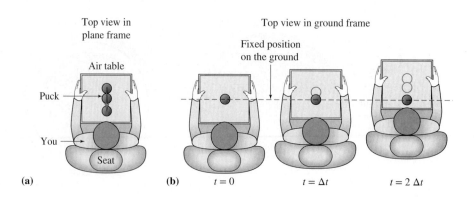

Resolving intuitive objections to this explanation

(This applies to *any* object not accelerating in the *S* frame.) This tells us that the puck's acceleration as measured in the plane is equal in magnitude and opposite in direction to the plane's acceleration relative to the ground. This is what we would expect if the puck had no acceleration relative to the ground.

The point is that we do not need a magical rearward force to explain the puck's behavior: we can completely explain that behavior in the ground frame *without* appealing to such a force. This alleged force therefore vanishes if we analyze the situation by using an *inertial* reference frame: *it is not real.*

"Wait," you say. "If this force is not real, what causes the puck to drift backward?" Think! The air table is being accelerated *forward* along with the plane, as observed in the ground frame. If the puck were to remain at rest relative to the table, it would *also* have to accelerate forward relative to the ground frame. But there is no force acting on the puck that can do this. Therefore, it is unable to keep up with the table as the latter accelerates forward.

"But," you cry, "we can *feel* this backward force. It *must* be real!" Not so, I say. What you *feel* is the airplane seat pressing you forward. In an *inertial* reference frame, if your seat were to press you forward but you don't accelerate forward, some other force would have to be pressing you backward. But in the noninertial plane frame, this logic does not apply, because Newton's laws do not apply. In the inertial ground frame, the seat presses you forward not because something else is pressing you backward, but because the seat *must* press you forward to accelerate you along with the plane (figure N8.4).

Part of the reason you seem to *feel* this force is that parts of your body must flex and stretch to accelerate other parts of your body along with the plane. For example, if you hold your hand in front of you, it feels as if something were pushing it toward you. But what you really feel is that you must flex the muscles in your arm to push your hand forward so it accelerates along with you and the plane. You feel this muscular action and intuitively interpret it as a response to something pushing your hand toward you.

In short, when you feel the seat pressing you forward, you intuitively *invent* a force pressing you backward, because otherwise your lack of acceleration in

Figure N8.4

(a) In an accelerating plane, you seem to be at rest. You feel the chair pressing forward on you, so you *assume* that a force must be pressing back on you to make the net force on you zero. (b) But the net force is *not* really zero: the chair must push forward on you to accelerate you along with the plane. In the ground frame, there is no need for a backward force.

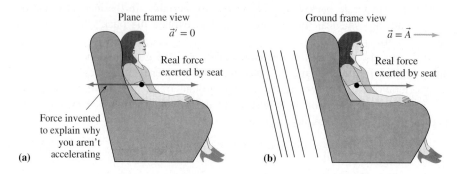

the plane frame and the way your body feels would be inexplicable. In doing this, though, you are illegally applying Newton's second law in an accelerated reference frame (an offense punishable throughout the universe by nonsensical answers!). In the inertial reference frame of the ground, such a force (1) cannot be made consistent with the idea that *forces express interactions* and (2) is entirely *unnecessary* to explain what we observe if we use an inertial frame. Therefore, from the point of view of the Newtonian model, this force is *not real*.

Exercise N8X.2

If you are in a car that suddenly slows down, you seem to feel a magical force acting to draw you forward. What is *really* going on? In a serious accident, people without seat belts can go through the windshield. If they are not being thrown forward by this magical force, how is this possible?

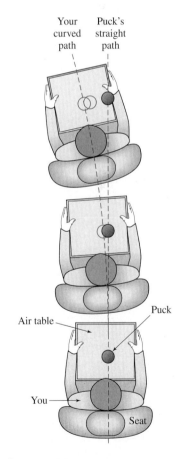

Your curved path   Puck's straight path

Air table

Puck

You

Seat

Figure N8.5

A puck floating on an air table seems to accelerate to the right as you ride in a car turning left because the car and you accelerate to the left under the puck while it moves straight forward.

N8.5 Circularly Accelerating Frames

Now imagine being in a car that is turning a corner or riding on a playground merry-go-round. The outward force that you seem to feel under such circumstances is such a vivid and common part of daily experience that it even has a colloquial English name: **centrifugal force**. This force is so deeply embedded in our pre-Newtonian intuition that even some popular science writers and precollege textbook authors discuss this force as if it were real.

We can analyze what you experience in the car turning a corner by using the same approach we used in section N8.4. When viewed in the inertial ground frame, as the car turns left (say), a floating puck continues to move forward in the direction of the car's original velocity, as shown in figure N8.5. In the car's frame, the air table is accelerating toward the left, so the puck seems to accelerate to the right relative to the table. In the car's frame, it *looks* as if the puck is pushed outward (to the right) by some centrifugal force. What is *really* happening, though, is that the table (along with the car) is accelerating left "underneath" the puck.

Your body, like the puck, would move forward in a straight line if your seat belt or the car's side did not exert a leftward force on you to accelerate you along with the car. In the car's frame, you do not *seem* to be accelerating, so you might use Newton's second law to infer that the leftward force exerted by the belt or car wall is due to a rightward force pulling you outward. But you are not *allowed* to use Newton's second law in the accelerating frame of the car, so this inference is *false*.

Really, we have two choices in situations like these. We could in principle *insist* that Newton's laws apply in *all* reference frames. This would require us to invent forces in some reference frames that are not due to contact or any other interaction with an external object. Moreover, these forces seem to exist at some times and not at others, and exist in one reference frame (for example, the plane) but not exist in another (for example, the ground). It seems to me that forces that either exist or don't exist depending on one's arbitrary choice of reference frame have an extremely poor claim on being "real."

It is better, simpler, and entirely sufficient to simply assert that Newton's laws *don't apply* in noninertial reference frames and insist that we use only *inertial* reference frames when analyzing motion by using Newton's laws. We have no physical evidence for these alleged forces other than the effects they seem to produce. We can entirely explain these effects *without* the use of such forces by using an inertial reference frame. These alleged forces are, therefore, not real in any useful sense of the word: they are *fictitious*.

Closing argument

Ladies and gentlemen of the jury, I have shown beyond a shadow of a doubt that motion can be easily interpreted and analyzed in an inertial reference frame without invoking magical forces that others claim to exist. Are these forces real? Only one verdict is logical, sensible, and right. Do your duty. I rest my case.

N8.6   Using Fictitious Forces

Now that you *completely understand* that the mysterious forces that appear to arise in a noninertial reference frame are totally fictitious and without any basis in physical reality, I will reluctantly admit that in a few cases, a given situation is much easier to analyze in a noninertial reference frame and therefore it is useful to *pretend* that these fictitious forces exist (as long as one completely understands what one is doing).

Imagine that we have an inertial frame S and a noninertial frame S' that is accelerating relative to S with acceleration $\vec{A}$. Suppose that real physical forces $\vec{F}_1$, $\vec{F}_2$, ... act on a certain object of mass m. Newton's second law applies in the inertial frame S, so in that frame, we have

$$m\vec{a} = \vec{F}_{\text{net}} = \vec{F}_1 + \vec{F}_2 + \cdots \qquad (N8.10)$$

Equation N8.7 states that the object's acceleration in the noninertial S' frame is $\vec{a}' = \vec{a} - \vec{A}$, so $\vec{a} = \vec{a}' + \vec{A}$. Substituting this into equation N8.10 yields

$$m\vec{a}' + m\vec{A} = \vec{F}_1 + \vec{F}_2 + \cdots \qquad (N8.11)$$

The $m\vec{A}$ term in this equation should *rightly* be considered to be a correction to the acceleration side of Newton's second law to enable us to use the acceleration $\vec{a}'$ measured in the noninertial frame. But if we subtract $m\vec{A}$ from both sides, we can put this term on the force side of the equation and *pretend* that it is just another force acting on the object:

A version of Newton's second law that we can use in a noninertial frame

$$m\vec{a}' = -m\vec{A} + \vec{F}_1 + \vec{F}_2 + \cdots \qquad (N8.12)$$

- **Purpose:** We can use this version of Newton's second law in a noninertial frame *if* we know that frame's acceleration $\vec{A}$ relative to an inertial frame. Here, m is the mass of an object, $\vec{a}'$ is its acceleration in the noninertial frame, and $\vec{F}_1$, $\vec{F}_2$, ... are real forces acting on the object.
- **Limitations:** The relative speed of the two frames and the speed of the object relative to each must be much smaller than the speed of light.

Let's call this "force" $-m\vec{A}$ a **frame-correction force** to emphasize that we are supplying this "force" *only* to compensate for measuring the object's acceleration in a noninertial reference frame. Such a "force" is also often called an *inertial force*. This "force" has the same mathematical form as the gravitational force $\vec{F}_g = m\vec{g}$: both are directly proportional to the object's mass. Indeed, in the noninertial frame, an object behaves *exactly as if it were experiencing an additional gravitational force* whose field vector is $\vec{g}_{\text{FC}} = -\vec{A}$.

We use the frame-correction "force" in situations where an analysis in an inertial frame is awkward. Example N8.4 illustrates such a situation.

Problem: A bicyclist goes around a curve of radius R at a speed $|\vec{v}|$. Why does the bicyclist have to lean into the curve? What is the cyclist's lean angle?

Solution We *could* analyze this situation in the inertial frame of the ground, but the analysis is complicated by the fact that the center of mass of the bike and rider is accelerating and both the bike and rider are also rotating in that frame. In the NONINERTIAL frame of the bike, however, this situation reduces to a straightforward statics problem. Figure N8.6 shows a torque diagram of the situation as viewed in this frame. Note that since the bike's acceleration $\vec{A}$ relative to the ground points in the $-x'$ direction and has magnitude $|\vec{A}| = |\vec{v}|^2/R$, the frame-correction force $\vec{F}_{FC}$ in the bike's frame points in the $+x'$ direction and has magnitude $m|\vec{v}|^2/R$, where m is the combined mass of the bike and rider.

In this NONINERTIAL frame, the bike is at rest and not rotating, so both the net force and the net torque on the system must be zero. Let's choose the point of contact between the tire and road to be the origin O for computing torques. Both $\vec{F}_N$ and $\vec{F}_{SF}$ exert zero torque around O (since both are applied *at* point O). Both $\vec{F}_{FC}$ and $\vec{F}_g$ behave as if they are applied to the center of mass of the bike and rider, which we will say has position $\vec{r}_{CM}$ relative to O (see the diagram on the right side of figure N8.6). Using the right-hand rule, we see that $\vec{F}_g$ exerts a torque directly toward us in figure N8.6, while $\vec{F}_{FC}$ exerts a torque in the opposite direction. These torques will add to zero (and thus the net torque on the bike will be zero) only if their magnitudes are equal.

We can most easily compute these torque magnitudes by using the fact that $|\vec{r} \times \vec{F}| = |\vec{r}_\perp||\vec{F}|$, where $|\vec{r}_\perp|$ is the magnitude of the part of $\vec{r}$ that is perpendicular to $\vec{F}$. We can see on the diagram that this component has a magnitude of $|\vec{r}_{CM}|\sin\theta$ in the case of $\vec{F}_g$ and $|\vec{r}_{CM}|\cos\theta$ in the case of $\vec{F}_{FC}$, so to balance the torques, we must have

$$(|\vec{r}_{CM}|\sin\theta)\,|\vec{F}_g| = (|\vec{r}_{CM}|\cos\theta)\,|\vec{F}_{FC}| \tag{N8.13}$$

Canceling $|\vec{r}_{CM}|$ from both sides and using $|\vec{F}_g| = m|\vec{g}|$ and $|\vec{F}_{FC}| = m|\vec{v}|^2/R$, we get $m|\vec{g}|\sin\theta = (m|\vec{v}|^2/R)\cos\theta$, so dividing both sides by $m|\vec{g}|\cos\theta$ yields

$$\tan\theta = \frac{\sin\theta}{\cos\theta} = \frac{|\vec{v}|^2}{R|\vec{g}|} \quad \Rightarrow \quad \theta = \tan^{-1}\left(\frac{|\vec{v}|^2}{R|\vec{g}|}\right) \tag{N8.14}$$

Note that the angle here is the same as we computed for a banking airplane in example N7.1. We see that the rider must lean into the curve so that the net torque on the bike is zero in the bike's NONINERTIAL frame. (Another way we could express this is to say that in the bike frame, the rider feels an *effective* total gravitational force of $\vec{F}_g + \vec{F}_{FC}$ that points at an angle θ outward from directly down, and that the rider must position his or her center of mass directly "above" the tire's contact point with the ground in this effective gravitational field to keep from falling over.)

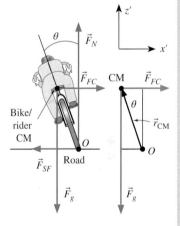

Rear view in the NONINERTIAL bike frame

Figure N8.6
A torque diagram for a turning bicyclist, as we would construct it in the NONINERTIAL frame of the bike. The right-hand diagram shows the two forces that exert nonzero torques around point O.

Exercise N8X.3

Redo example N8.4, using the system's center of mass as the origin O for computing torques, and show that you *still* have to lean at the angle given by equation N8.14 to balance torques. [*Hint:* Use equation N8.12 to argue that $|\vec{F}_N| = m|\vec{g}|$ and $|\vec{F}_{SF}| = |\vec{F}_{FC}|$.]

When to use frame-correction forces

I strongly recommend that you use equation N8.12 only in situations in which an analysis in an inertial frame would be horrendously complex. Even then, indicate very clearly in your solution that you are *deliberately* using a noninertial frame (as I did in example N8.4 by capitalizing NONINERTIAL), and denote the frame-correction "force" using the symbol $\vec{F}_{FC}$ to make it clear that you understand its nature.

N8.7   Freely Falling Frames and Gravity

A Newtonian explanation of why we can treat a falling frame as if it were inertial

In chapter C4, we discussed the fact that when we use a freely falling reference frame, we can *ignore* the external gravitational fields in which the frame falls. We are now in a position to understand *why*. Consider an inertial frame S in a gravitational field $\vec{g}$ and consider a frame S' that is freely falling in that field. (*Freely falling* in this context means that no significant forces *other* than gravity act on the frame.) Since all objects fall with the same acceleration $\vec{g}$ in a gravitational field, the acceleration of frame S' relative to S will be $\vec{A} = \vec{g}$.

Now suppose we analyze the motion of an object of mass m that is subject to the force of gravity $\vec{F}_g = m\vec{g}$, as well as additional forces $\vec{F}_1$, $\vec{F}_2$, ... According to equation N8.12, the object's motion as observed in the falling frame S' will obey the equation

$$m\vec{a}' = -m\vec{A} + \vec{F}_g + \vec{F}_1 + \vec{F}_2 + \cdots = -m\vec{g} + m\vec{g} = \vec{F}_1 + \vec{F}_2 + \cdots$$

$$\Rightarrow \quad m\vec{a}' = \vec{F}_1 + \vec{F}_2 + \cdots \tag{N8.15}$$

where in the second step I used the fact that $\vec{A} = \vec{g}$. Since the frame-correction force cancels the gravitational force, the final equation has the same form as Newton's second law *ignoring* that gravitational force: an object in a freely falling frame thus behaves as if Newton's second law is valid, *ignoring the gravitational field in which the frame falls* (see figure N8.7). This is why we could treat freely falling or "floating" frames as if they were inertial. From the perspective of Newtonian mechanics, however, such frames are *not* inertial frames, and the effective erasure of the consequences of real external gravitational interactions is a symptom of that.

Einstein's alternative explanation: *the equivalence principle*

However, in 1907, Albert Einstein presented an alternative take on this phenomenon. He pointed out that if one is placed in a freely falling enclosed room, equation N8.15 implies that no *mechanical* experiment that one could perform entirely within the room could tell whether that room is freely falling in a uniform gravitational field or is a genuine inertial frame floating in deep space far from any gravitating objects. Even the first-law detectors that we discussed in section N8.3 would certify that such a frame was inertial. To express his point in analogy to a famous cliché about ducks, we might say, "If it looks like an inertial frame and physically behaves as an inertial frame, it *is* an inertial frame." Einstein boldly hypothesized that a freely falling frame in a uniform gravitational field is *completely equivalent* to an idealized inertial frame floating in deep space. There is *no* experiment that one can perform will distinguish the two. By extension, since a frame at rest on the surface of the earth is accelerating *upward* relative to any freely falling frame, then (to the extent that we can consider the earth's gravitational field to be uniform) a frame at rest on the earth's surface is physically indistinguishable from an *accelerating* frame in deep space.

Einstein made a variety of testable predictions based on this hypothesis (some of which are discussed in the homework problems). These predictions have been experimentally tested and have all been found to be true.

Figure N8.7
Astronauts Shannon Walker and Soichi Nuguchi test Newton's first law (by using a bag of candy and some tomatoes, respectively) in the freely falling frame of the International Space Station. The earth's gravity at the altitude of the station (400 km) is about 90% as strong as on the earth's surface, but is completely erased in the freely falling station's frame. (Credit: NASA)

This hypothesis, called the **equivalence principle**, is the foundation of Einstein's theory of general relativity. In that theory, what we normally think of as gravity is a *fictitious* force (!) that appears in a frame at rest on the surface of the earth only because such a frame is *accelerating* upward relative to genuinely inertial (that is, freely falling) frames in the earth's vicinity. (The theory's main task, therefore, is to explain why *inertial* frames near the earth are *accelerating* rather than at rest relative to the earth's surface.)

TWO-MINUTE PROBLEMS

N8T.1 In a Western movie, a person shoots an arrow backward from a fleeing horse. If the velocity of the horse relative to the ground is 13 m/s west and the arrow's velocity relative to the horse is 38 m/s east, what is the arrow's velocity with respect to the ground?
A. 41 m/s east
B. 41 m/s west
C. 25 m/s east
D. 25 m/s west

N8T.2 A blimp has a velocity of 8.2 m/s due west relative to the air. There is a wind blowing at 3.5 m/s due north. The speed of the balloon relative to the ground is
A. 11.7 m/s
B. 8.9 m/s
C. 7.4 m/s
D. 4.7 m/s
E. Some other speed (specify)

N8T.3 An elevator accelerates downward at a rate of 6.2 m/s². A ball dropped from rest by a passenger will have what *downward* acceleration relative to the elevator?
A. 3.6 m/s²
B. 6.2 m/s²
C. 9.8 m/s²
D. 16.0 m/s²
E. Some other acceleration (specify)

N8T.4 When you go over the crest of a hill in a roller coaster, a force appears to lift you up out of your seat. This is a fictitious force. True (T) or false (F)?

N8T.5 If your car is hit from behind, you are suddenly pressed back into the seat. The normal force that the seat exerts on you is a fictitious force. T or F?

N8T.6 You are pressed downward toward the floor as an elevator begins to move upward. The force pressing you down is a fictitious force. T or F?

N8T.7 A beetle (black dot on the top view shown below) sits on a rapidly rotating turntable. The table rotates faster and faster, and eventually the beetle loses its grip. What is its subsequent trajectory relative to the ground (assuming it cannot fly)?

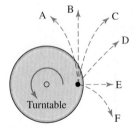

N8T.8 A cork floats in an inverted jar sitting on a cart, as shown in the left diagram below. If we suddenly accelerate the cart to the right (as shown in the right diagram) what will the cork do: (A) lean backward, (B) remain vertically floating directly above the base, (C) lean forward, (D) sink, or (E) explode? (*Hint:* Think of the jar as an accelerating reference frame. What is the direction of the effective force of gravity in such a frame?)

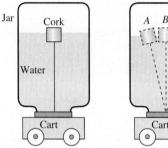

N8T.9 When is an elevator a reasonably good inertial reference frame?
A. Never, under any circumstances.
B. Always, since it is attached to the surface of the earth.
C. It is good except when it is changing speed.
D. Other conditions (describe).

N8T.10 According to Newton, is a freely falling elevator an accurate inertial reference frame?
A. No, such a frame is accelerating toward the earth.
B. It is not a *real* inertial frame, but we can treat it as one if we ignore external gravitational forces acting on objects inside the frame.
C. Yes.
D. It depends (explain).

N8T.11 According to Einstein, is a frame at rest on the earth's surface a reasonably accurate inertial frame (ignoring the earth's rotation)?
A. No, such a frame is accelerating upward.
B. It is not a *real* inertial frame, but we can treat it as one if we ignore external gravitational forces acting on objects inside the frame.
C. Yes.
D. It depends (explain).

N8T.12 Centrifugal forces are
A. Real.
B. Fictitious.
C. Real if we are in a rotating frame.

HOMEWORK PROBLEMS

Basic Skills

N8B.1   A person shoots a tranquilizer dart at a fleeing bear. If the dart's speed is 33 m/s relative to the ground and the bear is fleeing at 9 m/s relative to the ground, what is the dart's speed relative to the bear?

N8B.2   In a certain baseball game, a well-hit line drive moves almost horizontally eastward at 125 mi/h. The center fielder runs toward the ball at a speed of 15 mi/h. What is the ball's speed relative to the fielder?

N8B.3   An airplane flies due north at a speed of 145 km/h relative to the ground. If there is a wind blowing east at 15 km/h, what is the plane's speed relative to the air?

N8B.4   Two cars approach an intersection, one traveling north at 18 m/s and the other traveling west at 14 m/s. What is the cars' relative speed?

N8B.5   A boat originally traveling at a speed of 5.2 m/s due east hits a sandbar and comes to rest in 1.3 s. What is the approximate average acceleration (magnitude and direction) of a piece of slippery ice sitting on a table in the ship's galley
(a) relative to the ground and
(b) relative to the ship? Explain your answers.

N8B.6   Two cars in a race start from rest. Both cars accelerate forward, car *A* at a rate of 12 (mi/h)/s and car *B* at a rate of 9 (mi/h)/s. What is the acceleration (magnitude and direction) of car *B* relative to car *A*?

N8B.7   Suppose you toss a baseball to a friend while you are both in the back of a bus moving due west at a constant speed of 15 m/s. While the baseball is in the air, its acceleration with respect to the ground is 9.8 m/s^2 downward (since only the force of gravity acts on it once it leaves your hand). What is the ball's acceleration relative to the bus? Explain.

N8B.8   A car originally moving at 25 mi/h hits a brick wall and comes to rest in 0.12 s. As the car comes to rest, about how fast will an unbelted passenger accelerate
(a) relative to the ground and
(b) relative to the windshield? Explain your response. (Ignore friction.)

Modeling

N8M.1   A boat moves directly toward a dock 2.0 km due west across a river flowing 1.0 m/s due south. How fast must the boat travel relative to the water to make it across in 20 min?

N8M.2   An airplane flies due north at 250 mi/h relative to the ground in air that is moving east at 25 mi/h. At what angle is the plane pointed relative to north? What is its speed relative to the air?

N8M.3   An airplane accelerates for takeoff, reaching a speed of 150 mi/h in 30 s. A 65-kg person feels pressed back into his or her seat during this time.
(a) What is the person's acceleration relative to the ground?
(b) What is the magnitude of the forward force that the seat must exert on the person during this time?
(c) What is the person's acceleration in the plane frame?
(d) If you were to naively apply Newton's second law in this frame, what is the magnitude of the fictitious force that you infer must be pushing the person backward?

N8M.4   A 35-kg child is holding on for dear life on a playground merry-go-round that has a radius of 2.5 m and is turning at a rate of one turn every 5.0 s.
(a) What is the child's acceleration relative to the ground?
(b) What is the magnitude of the force acting on the child through the child's arms?
(c) What is the child's acceleration relative to the place on the merry-go-round where the child is standing?
(d) If you were to naively apply Newton's second law to this situation, what is the magnitude of the apparent centrifugal force acting on the child?

N8M.5   An elevator is accelerating upward at 1.8 m/s^2. Someone is standing on a bathroom scale in the elevator, and it reads 180 lb. Use equation N8.12 to calculate the person's true weight.

N8M.6   A test tube is placed in a centrifuge. When the centrifuge rotates at full speed, the test tube's bottom is about 38 cm from the axis of rotation. The material near the bottom of the test tube experiences an apparent effective gravitational force 18 times larger than normal. How many times does the centrifuge rotate per second?

N8M.7   Imagine that the ball in a floating-ball detector can travel only 1.0 cm before hitting the side of its container. Imagine also that we have figured out a way to cancel exactly the true effects of gravity. If we place our detector on the earth's equator, how long will it take for the ball (after it is released from rest) to reach the side, and in which direction will it drift? Compare this to the time it would take in a frame that accelerates from rest to walking speed (about 1 m/s) in 1 min.

N8M.8   Long-term space missions require some form of artificial gravity to prevent astronauts losing bone mass. One method would be to use a tether to connect a spent booster to the astronauts' capsule and set the system rotating to

Figure N8.8
A centrifuge for training astronauts. (Credit: NASA)

provide artificial gravity in the non-inertial frame of the capsule. Tests suggest that rotation rates less than 2 rpm are needed to avoid the astronauts experiencing inner-ear problems as they move about. About how long must the tether be, if you need an apparent gravitational field in the capsule of at least half of the earth's field? (Assume that the capsule and booster have roughly the same mass.)

N8M.9 Astronauts train for the high accelerations they experience during launch by using a centrifuge that consists of a little room suspended at the end of a long boom that rotates in a horizontal plane (see figure N8.8). If the *total* apparent effective gravitational field strength in the room (including the *real* gravitational force) is supposed to have a magnitude of $5\,|\vec{g}|$, and the distance between the room's center and the rotation axis is 10 m, about how long should it take the boom to rotate once?

Derivations

N8D.1   An object's inertial mass m_I expresses its resistance to being accelerated by a given force: $\vec{F}_{net} = m_I \vec{a}$ according to Newton's second law. An object's gravitational mass m_G expresses the strength of the gravitational force the object feels in a given gravitational field $\vec{g}$: $\vec{F}_g = m_G \vec{g}$. These are completely distinct ideas, and even though Newton recognized that one could explain much by assuming that $m_G = m_I$, Newtonian mechanics offers no explanation about why these quantities should be equal or even related. However, one of the consequences of the equivalence principle is that an object's inertial mass must be *exactly* equal to its gravitational mass, no matter what the object is made of. Carefully explain why this must be so. (*Note:* Recent experiments show that these quantities are experimentally equal to roughly 13 significant digits.)

N8D.2   One of the consequences of the principle of equivalence is that light should bend in a gravitational field. To see this, imagine a laboratory in deep space that is being accelerated by a rocket engine so that its acceleration relative to inertial frames is upward (relative to the laboratory's floor) and has a magnitude of A. Imagine that a beam of light enters a hole in one side of the laboratory traveling in an initially horizontal direction relative to the floor. If the laboratory frame were inertial, the light would travel in a straight line, and thus would hit the laboratory's opposite wall at the same height as the hole. However, the laboratory is accelerating upward, so during the time it takes the light to move across the laboratory, the laboratory's floor will move upward relative to the beam, which will make it appear to curve toward the floor in the laboratory frame. This will ultimately cause the beam to hit the far wall a distance d below the expected position.
(a) If the laboratory's width is L, what is d in terms of A and the speed of light c?
(b) The principle of equivalence implies that a laboratory at rest on the surface of the earth is physically equivalent to a laboratory in deep space whose acceleration relative to inertial frames is $|\vec{A}| = |\vec{g}|$. Making a reasonable estimate for the size of the laboratory, make an estimate of the distance through which a light beam will be deflected by the earth's gravitational field. Would this be easy to detect? (*Note:* In 1919, astronomers observing an eclipse measured the deflection of starlight passing near the obscured sun and verified the prediction of general relativity that a gravitational field would bend light.)

N8D.3   A freely falling frame does not behave *precisely* as an inertial frame because $\vec{g}$ is not precisely constant near a gravitating object.[†] For example, $\vec{g}$ near the earth always points directly toward the earth's center, which is in different directions at different points. Consider a boxcar that is 10 m long and that is freely falling toward the earth (assume the boxcar is oriented horizontally). Two marbles float initially at rest at opposite ends of the boxcar. Estimate the marbles' acceleration relative to each other. (*Hint:* $\sin\theta \approx \tan\theta \approx \theta$ if θ is measured in radians and $\theta \ll 1$.)

[†]*Note:* A *uniform* gravitational field would be *completely erased* in a freely falling frame. According to general relativity, a uniform gravitational field is therefore as fictitious as a centrifugal force. However, the *real* gravitational fields created by real physical objects are always nonuniform, and therefore they cannot be completely erased even by going to a freely falling frame. In general relativity, what we normally consider "gravity" is a fictitious force that arises from using a noninertial reference frame; for general relativity, what is *real* about gravity is the "tidal effects" of the gravitational field that remain and are observable even in a freely falling frame. Einstein was able to develop a logically coherent theory of general relativity partly because the equivalence principle eventually allowed him to focus on what was *real* about gravitation instead of being distracted by the more obvious but fictitious part.

N8D.4 A jet flies a distance L and back, first against and then with a wind. The wind moves with a speed $|\vec{u}|$. One might think that since the wind will slow the jet one way and speed it equally the other way, the effects would cancel and the trip time would be $2L/|\vec{v}|$, where $|\vec{v}|$ is the jet's speed relative to the air. Show that the round-trip time is *actually* $(2L/|\vec{v}|)(1 - |\vec{u}|^2/|\vec{v}|^2)^{-1}$.

Rich-Context

N8R.1 You are kidnapped and put blindfolded in an elevator at the ground floor of a building. As the elevator starts, you notice that your weight seems to increase by 10% for 3 s, remain normal for 24 s, then decrease by 10% for 3 s. You are then taken out and put into a locked room. What is the approximate floor number your room is on?

N8R.2 In the 1997 movie *Speed II*, a cruise ship crashes through a dock and moves some distance into a seaside town before coming to rest. The movie shows people on the ship being strongly affected by the ship's deceleration, crashing through forward windows and the like. The intrepid helmsman helpfully calls out the speed in knots as the ship slows down: "Six knots! Five knots!" and so on. The camera shows the digital speedometer, and it is clear that the ship's speed is decreasing at a rate of between 0.2 and 0.3 knots per second. What is wrong with this picture? (*Hints:* I mean, of course, besides the cheesy acting and the appalling screenwriting. 1 knot $\equiv$ 1 nautical mile/hour, where 1 nautical mile = 1852 m. You can find the scene on YouTube: search for "speed 2 cruise ship crash.")

N8R.3 In the 1968 movie *2001: A Space Odyssey,* a space shuttle is shown docking with a space station. The station consists of two rings, each connected by four struts to a hub. The entire station rotates at a rate of about 1 turn in 40 s, presumably to provide artificial gravity in the rotating reference frame of the ring. An exterior shot shows that the space station's ring has a radius of about 8 times the shuttle's length. An interior view of the shuttle shows that the passenger cabin is about 25 m long, and exterior shots suggest that the shuttle's total length is perhaps twice that. Did the film's science advisors get things about right (considering the uncertainties of the estimates here)? (You can see the entire scene on YouTube: search for "2001: Space Odyssey space station.")

ANSWERS TO EXERCISES

N8X.1 Let the elevator and the ground be the S' and S frames, respectively. Given $A_z = +3.0$ m/s^2 and that the ball's acceleration in S is $a_z = -9.8$ m/s^2, we want to find its acceleration a'_z in S'. The z component of equation N8.7 tells us that $a'_z = a_z - A_z = -12.8$ m/s^2.

N8X.2 The "magical force" that seems to draw a person forward is simply the tendency of the person's body to continue moving at a constant velocity while the car accelerates backward. If the person is not held to the car by a seat belt, the person will continue to move through the windshield as the car comes to rest. Thus, the person is not "thrown through the windshield" so much as the windshield is thrown backward by the collision past the person's body as the latter moves naturally forward.

N8X.3 In this case, the gravitational and frame-correction forces exert zero torque around the system's center of mass, so we need to calculate the torque contributed by the normal force and the static friction force. These forces contribute torques in opposite directions, so their torques will cancel if they have the same magnitude.

The net force on the system in the bike frame must be zero, so we must have

$$0 = \begin{bmatrix} 0 \\ 0 \\ +|\vec{F}_N| \end{bmatrix} + \begin{bmatrix} -|\vec{F}_{SF}| \\ 0 \\ 0 \end{bmatrix} + \begin{bmatrix} 0 \\ 0 \\ -m|\vec{g}| \end{bmatrix} + \begin{bmatrix} +m|\vec{v}|^2/R \\ 0 \\ 0 \end{bmatrix} \quad \text{(N8.16)}$$

implying that $|\vec{F}_N| = m|\vec{g}|$ and $|\vec{F}_{SF}| = m|\vec{v}|^2/R$. In this case, the position vector $\vec{r}$ of the point where both these forces are applied relative to the origin is the inverse of $\vec{r}_{CM}$ in figure N8.6, so the magnitudes of its components perpendicular to $\vec{F}_N$ and $\vec{F}_{SF}$ are $|\vec{r}|\sin\theta$ and $|\vec{r}|\cos\theta$, respectively. Therefore, setting the magnitudes of the torques exerted by these forces equal, we find that

$$|\vec{F}_N||\vec{r}|\sin\theta = |\vec{F}_{SF}||\vec{r}|\cos\theta$$

$$\tan\theta = \frac{|\vec{F}_{SF}|}{|\vec{F}_N|} = \frac{\cancel{m}|\vec{v}|^2/R}{\cancel{m}|\vec{g}|} = \frac{|\vec{v}|^2}{R|\vec{g}|} \quad \text{(N8.17)}$$

This is what we got before.

Projectile Motion

Chapter Overview

Introduction

This chapter launches a four-chapter subdivision exploring practical situations where we can use our knowledge about forces acting on an object to determine the object's motion. The three situations we will examine in depth are projectile motion (this chapter), oscillatory motion (chapter N10), and orbital motion (chapters N11 and N12).

Section N9.1: Weight and Projectile Motion

As noted in unit C, we define an object's weight to be the force that gravity exerts on that object. Weight is not the same as mass, but at a given position in a gravitational field, an object's weight is proportional to its mass:

$$\vec{F}_g = m\vec{g} \tag{N9.1}$$

where the constant of proportionality $\vec{g}$ at that location is called the **gravitational field vector** at that location: $\vec{g}$ expresses the strength and direction of the gravitational field at that point. At points near the earth's surface $|\vec{g}| = 9.8$ m/s^2 = 22 (mi/h)/s. An object is **freely falling** if its weight is the *only* force acting on it. Newton's second law implies that a freely falling object's acceleration is $\vec{a} = \vec{g}$: for this reason, $\vec{g}$ is often called the "acceleration of gravity."

We call an object a **projectile** and its motion **projectile motion** if:

1. The object remains close enough to the earth's surface so that $|\vec{g}| \approx$ constant.
2. The object's trajectory is short enough that the direction of $\vec{g}$ is also constant.

Note that $|\vec{g}|$ varies by less than $\pm 1\%$ within 30 km of the earth's surface, and the direction of $\vec{g}$ varies by less than $1°$ during a 100-km trajectory.

If the drag forces acting on an object are negligible compared to its weight, we say that its motion is **simple projectile motion**.

Section N9.2: Simple Projectile Motion

In the case of simple projectile motion, integrating both sides of $d\vec{v}/dt \equiv \vec{a}(t) = \vec{g}$ and $d\vec{r}/dt \equiv \vec{v}(t)$ with respect to time yields

The basic equations describing simple projectile motion

$$\begin{bmatrix} v_x(t) \\ v_y(t) \\ v_z(t) \end{bmatrix} = \begin{bmatrix} v_{0x} \\ v_{0y} \\ -|\vec{g}|t + v_{0z} \end{bmatrix} \tag{N9.6}$$

$$\begin{bmatrix} x(t) \\ y(t) \\ z(t) \end{bmatrix} = \begin{bmatrix} v_{0x}t + x_0 \\ v_{0y}t + y_0 \\ -\frac{1}{2}|\vec{g}|t^2 + v_{0z}t + z_0 \end{bmatrix} \tag{N9.7}$$

- **Purpose:** These equations express how the components $[v_x, v_y, v_z]$ and $[x, y, z]$ of a simple projectile's velocity $\vec{v}$ and position $\vec{r}$ vary with time t, where $|\vec{g}|$

is the gravitational field strength, and $[v_{0x}, v_{0y}, v_{0z}]$ and $[x_0, y_0, z_0]$ are the components of the object's initial velocity and position at time $t = 0$, respectively.
- **Limitations:** Gravity must be the only significant force acting on the object, and $\vec{g}$ must be an essentially constant vector during the object's trajectory.

Section N9.3: Some Basic Implications

Here are some of the important implications of equations N9.6 and N9.7:

1. The horizontal components of motion are unaffected by gravity.
2. An object's vertical motion is independent of its horizontal motion.
3. An object's trajectory is confined to a plane (which we usually choose to be the xz plane).

One can find the time t when a simple projectile reaches the peak of its trajectory by setting its vertical velocity component $v_z(t)$ to zero and solving for t. Similarly, one can determine when a simple projectile returns to the ground by solving $z(t) = 0$ for t. Since the latter is a quadratic equation, you will get two solutions for t, but one is usually unphysical (it yields a value for t that is before or after the time interval when the object is a simple projectile).

Section N9.4: A Projectile Motion Checklist

A checklist for simple projectile motion problems looks like this:

Simple Projectile Motion Checklist
Your main drawing should include:
- ☐ The object in its initial position.
- ☐ A labeled arrow showing its initial velocity.
- ☐ Coordinate axes with a clear origin.
- ☐ Labels defining symbols for other quantities.
Also, you should: ☐ Define when $t = 0$.
- ☐ Define symbols for other critical times.
- ☐ Use the master equations N9.6 and/or N9.7.

General Checklist
- ☐ Define symbols.
- ☐ Draw a picture.
- ☐ Describe your model and assumptions.
- ☐ List knowns and unknowns.
- ☐ Do algebra symbolically.
- ☐ Track units.
- ☐ Check your result.

Section N9.5: Drag and Terminal Speed

The drag force on a reasonably large object moving through air at a reasonably large speed $|\vec{v}|$ is usually accurately modeled by $F_D = \frac{1}{2}C\rho A|\vec{v}|^2$ (equation N9.15) where C is the drag coefficient, ρ is the density of air, and A is the object's cross-sectional area. An object falling vertically from rest will stop accelerating when it reaches a speed where the drag is equal to the object's weight: this **terminal speed** is

$$v_T = \sqrt{\frac{2m|\vec{g}|}{C\rho a}} \qquad (N9.16)$$

- **Purpose:** This equation describes the maximum speed v_T that an object dropped from rest will reach when falling through air, where A is the object's cross-sectional area, m is its mass, C is the object's drag coefficient, ρ is the density of air, and $|\vec{g}|$ is the gravitational field strength.
- **Limitations:** This equation does not apply to tiny, slowly moving objects.

An object falling from rest at time $t = 0$ will behave as a simple projectile for times $t \ll t_T \equiv v_T/|\vec{g}|$, but will essentially reach its terminal speed after a time $t \approx 2t_T$.

N9.1   Weight and Projectile Motion

One of the fundamental forces of nature that we experience daily is the *force of gravity*, the force that incessantly and insistently tugs us toward the center of the earth. Since this force is such a common part of our experience, it is worth studying in some detail.

As we saw in unit C, in physics we define an object's *weight* $\vec{F}_g$ to be *the force that gravity exerts on that object*. Note that this force expresses the interaction between the object and whatever nearby body (usually the earth) is massive enough to create an appreciable gravitational field. We also saw that near the earth's surface, the weight force on an object is given by

Defining the gravitational field vector $\vec{g}$

$$\vec{F}_g = m\vec{g} \qquad (N9.1)$$

where $\vec{g}$ is a *vector* that points toward the center of the earth and whose magnitude is $|\vec{g}| = 9.8$ N/kg $= 9.8$ m/s^2 (since 1 N $= 1$ kg·m/s^2). We can take this equation as the *definition* of $\vec{g}$. Since $\vec{g}$ expresses the effect of the gravitational interaction on an object of given mass at a certain point in space, and since $\vec{g}$ has the same value for all objects at that point, it expresses something basic about the character of the gravitational field that the earth creates at that point. We therefore call $\vec{g}$ the earth's **gravitational field vector** at that point. (We can define $\vec{g}$ near other celestial objects in a similar way.)

If the *only* force on an object is its weight $\vec{F}_g$, we say it is **freely falling**. As we saw in chapter N3, Newton's second law tells us that since the object's weight is strictly proportional to its mass, its acceleration is

The acceleration of a freely falling object is $\vec{g}$

$$\vec{a} = \frac{\vec{F}_{net}}{m} = \frac{\vec{F}_g}{m} = \frac{m\vec{g}}{m} = \vec{g} \qquad (N9.2)$$

implying that *every* object at a given point in space (independent of its mass, composition, or other characteristics) will fall with the *same* acceleration equal to the value of $\vec{g}$ at that point. Since the value of $\vec{g}$ at a point characterizes the common acceleration of all falling objects at that point, people sometimes call it the **acceleration of gravity** (even though it obviously isn't *gravity* that is accelerating!). I will use the term *gravitational field vector* not only because it more accurately describes the physical meaning of $\vec{g}$, but also because it underlines its analogy with the *electric field vector* $\vec{E}$ that we will discuss in unit E.

The simple projectile motion model

As long as (1) an object remains "sufficiently close" to the surface of the earth and (2) its trajectory is "sufficiently short" that the curvature of the earth is not significant, the magnitude and direction of $\vec{g}$ (and thus of the object's weight $\vec{F}_g$) will be approximately constant. If in addition (3) the object is "not significantly affected" by other forces except possibly air drag, then we call it a **projectile** and its motion **projectile motion**. When (4) we also neglect drag, we say that the object's motion is **simple projectile motion**. Newton's second law implies that such an object will have constant acceleration $\vec{a} = \vec{g}$.

When is drag important?

What do these restrictions mean in practice? Let's look at each in turn, starting with the last. A *freely falling* object is strictly defined to be an object that is not influenced by anything *except* its own weight. In practice, it is hard to isolate falling objects from the effects of air friction (drag). In some cases, these effects can be substantial: it is clear, for example, that a feather does not fall in the same way that a steel ball does. The statement that an object is "freely falling" thus could only strictly apply to objects in a vacuum. However, the description is *approximately* true in any case in which the drag force is small compared to its weight.

Whether this approximation is good depends on the object's *mass*, *shape*, *size*, and *speed*. For example, we would expect two objects of the same size and shape and moving at the same speed to experience the same drag. However the same drag force will change a more massive object's velocity *less* in a given time interval (according to Newton's second law) and thus will affect its motion less. A steel ball will thus behave more like a freely falling object than a ping-pong ball of the same size. An object's *shape* is also important: for example, a sheet of paper behaves more like a freely falling object if it is compressed into a tight ball than if it is left flat, because more air interacts with the flat sheet than with the ball. Finally, the object's *speed* is important: higher speeds produce higher drag forces. So the approximation works best if the object is small, massive, and moving relatively slowly.

What does it mean for an object to be "sufficiently close" to the earth's surface? We have seen that the effect of the earth's gravity diminishes with distance from the earth's center. However, one must travel a significant distance vertically before variations in $\vec{g}$ become noticeable. Empirically, the variation in the value of g is smaller than $\pm 1\%$ within a range of ± 30 km relative to sea level. Even at altitudes of several hundred kilometers (roughly the altitude where the International Space Station flies), the value of $|\vec{g}|$ is more than 90% of its value at sea level. So depending on what we would consider a significant variation in the value of $|\vec{g}|$ to be, "sufficiently close" could range from within a few kilometers to within hundreds of kilometers of the earth's surface.

When is an object "sufficiently close" to the earth?

The other restriction is that the object's trajectory be "sufficiently short" compared to the earth's curvature. This is because $\vec{g}$ points toward the earth's *center*, so if the object moves far enough horizontally, the direction of $\vec{g}$ could change significantly (relative to the distant stars, say). The earth is pretty large, though! Since the earth's circumference is about 40,000 km and the change in the direction of $\vec{g}$ for an object going around the earth will be $360°$, we see that we have to travel more than 100 km for the direction of $\vec{g}$ to change by more than $1°$. So we can consider the direction of $\vec{g}$ to be constant as long as the object's trajectory is not longer than 100 km or so.

When is its trajectory "sufficiently short"?

N9.2   Simple Projectile Motion

So, when the drag on an object is negligible and $\vec{g}$ is relatively constant in magnitude and direction, we can model the object as a *simple projectile* and take its acceleration to be $\vec{a} = \vec{g} =$ constant. Since this *is* a constant, we can easily integrate this by using the methods of chapter N3 to find the falling object's velocity and position as a function of time. Integrating the acceleration to find the velocity, we find that

Integrating $\vec{a}(t)$ to find $\vec{v}(t)$ and $\vec{r}(t)$

$$\vec{v}(t) - \vec{v}(0) = \int_0^t \vec{a}(t)\,dt = \int_0^t \vec{g}\,dt = \vec{g}\int_0^t dt = \vec{g}(t-0) = \vec{g}t$$

$$\Rightarrow \quad \vec{v}(t) = \vec{g}t + \vec{v}(0) = \vec{g}t + \vec{v}_0 \quad \text{[where } \vec{v}_0 \equiv \vec{v}(0)] \qquad \text{(N9.3)}$$

Similarly, integrating the velocity to find the position yields

$$\vec{r}(t) - \vec{r}(0) = \int_0^t \vec{v}(t)\,dt = \int_0^t [\vec{g}t + \vec{v}_0]\,dt = \vec{g}\int_0^t t\,dt + \vec{v}_0\int_0^t dt$$

$$= \vec{g}(\tfrac{1}{2}t^2 - \tfrac{1}{2}0^2) + \vec{v}_0(t-0) = \tfrac{1}{2}\vec{g}t^2 + \vec{v}_0 t$$

$$\Rightarrow \quad \vec{r}(t) = \tfrac{1}{2}\vec{g}t^2 + \vec{v}_0 t + \vec{r}_0 \quad \text{[where } \vec{r}_0 \equiv \vec{r}(0)] \qquad \text{(N9.4)}$$

Note how I have treated both $\vec{g}$ and $\vec{v}_0$ as simple constants in the integration.

In practical component language, if we define the z axis to be vertically upward and the x and y axes to be horizontal, then $\vec{g} = [0, 0, -|\vec{g}|]$. (The z component of $\vec{g}$ is negative because $\vec{g}$ points vertically downward. Note that $|\vec{g}|$ is always *positive*, no matter how $\vec{g}$ points.) If we also define

$$\vec{v}(0) \equiv \begin{bmatrix} v_{0x} \\ v_{0y} \\ v_{0z} \end{bmatrix} \quad \text{and} \quad \vec{r}(0) \equiv \begin{bmatrix} x_0 \\ y_0 \\ z_0 \end{bmatrix} \tag{N9.5}$$

then we can write equations N9.3 and N9.4 in column-vector form as follows:

Practical equations for the motion of a simple projectile

$$\begin{bmatrix} v_x(t) \\ v_y(t) \\ v_z(t) \end{bmatrix} = \begin{bmatrix} 0 \cdot t \\ 0 \cdot t \\ -|\vec{g}|t \end{bmatrix} + \begin{bmatrix} v_{0x} \\ v_{0y} \\ v_{0z} \end{bmatrix} \Rightarrow \begin{bmatrix} v_x(t) \\ v_y(t) \\ v_z(t) \end{bmatrix} = \begin{bmatrix} v_{0x} \\ v_{0y} \\ -|\vec{g}|t + v_{0z} \end{bmatrix} \tag{N9.6}$$

$$\begin{bmatrix} x(t) \\ y(t) \\ z(t) \end{bmatrix} = \begin{bmatrix} \frac{1}{2} 0 \cdot t^2 \\ \frac{1}{2} 0 \cdot t^2 \\ -\frac{1}{2}|\vec{g}|t^2 \end{bmatrix} + \begin{bmatrix} v_{0x}t \\ v_{0y}t \\ v_{0z}t \end{bmatrix} + \begin{bmatrix} x_0 \\ y_0 \\ z_0 \end{bmatrix} \Rightarrow \begin{bmatrix} x(t) \\ y(t) \\ z(t) \end{bmatrix} = \begin{bmatrix} v_{0x}t + x_0 \\ v_{0y}t + y_0 \\ -\frac{1}{2}|\vec{g}|t^2 + v_{0z}t + z_0 \end{bmatrix} \tag{N9.7}$$

- **Purpose:** These equations express how the components $[v_x, v_y, v_z]$ and $[x, y, z]$ of a simple projectile's velocity $\vec{v}$ and its position $\vec{r}$ vary with time t, where $|\vec{g}|$ is the gravitational field strength, and $[v_{0x}, v_{0y}, v_{0z}]$ and $[x_0, y_0, z_0]$ are the components of the object's initial velocity and position at time $t = 0$, respectively.
- **Limitations:** Gravity must be the only significant force acting on the object, and $\vec{g}$ must be essentially constant in magnitude and direction during the object's trajectory.

N9.3   Some Basic Implications

Gravity affects only the vertical component of an object's motion

Equations N9.6 and N9.7 imply three basic but important things about simple projectile motion that are worth underlining. First, notice that *gravity affects only the z component of a simple projectile's motion*: as it falls, a simple projectile will maintain whatever original horizontal velocity it had. For example, a package dropped from a plane will continue to move horizontally at the same rate as the plane (see figure N9.1a).

The component equations are independent

Second, *the component* equations in either vector equation *are completely independent of each other*. In particular, note that the projectile's vertical motion is completely disconnected from its horizontal velocity. This means, for example, that an object dropped from rest will have the same vertical motion as an object with a large initial horizontal velocity, as shown in figure N9.1b.

We can always define the horizontal motion to be along the x axis

Third, since a projectile's horizontal motion is unaffected by gravity, the projection of a projectile's motion on the horizontal (xy) plane will always be a straight line. It is thus generally possible to orient our reference frame so that its x axis lies in the direction of motion [that is, so that $v_y(t) = v_{0y} = 0$]. If we do this, the projectile's motion lies entirely in the xz plane, which simplifies things somewhat. I will commonly do this in the examples that follow.

When does a (simple) projectile reach its peak?

Most actual projectiles are launched from near the ground and follow a trajectory that looks like an arc, going first upward and then curving back downward to the ground. How do we determine when the object has reached its maximum altitude and when it reaches the ground?

If a projectile's initial velocity is at all upward, then $v_{0z} > 0$. The bottom line of equation N9.6 tells us that subsequently $v_z(t) = -|\vec{g}|t + v_{0z}$, so the projectile's

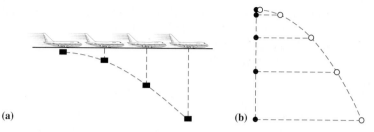

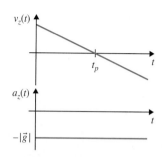

(a)

(b)

Figure N9.1

(a) A projectile's horizontal motion is unaffected by gravity: as an object falls, it will continue to move forward at whatever horizontal velocity it had originally. (b) A projectile's vertical motion is independent of its horizontal motion: objects fall vertically at the same rate no matter how they are moving horizontally.

Figure N9.2

Graphs of a projectile's z-velocity and z-acceleration near the peak of its trajectory.

z-velocity starts positive, but decreases linearly through zero and becomes negative, as shown in figure N9.2. I claim that the projectile is at its peak when its z-velocity passes through zero. Why? At an instant when $v_z(t)$ is positive, the projectile is still going upward, and so it will reach a still higher position later. At an instant where $v_z(t)$ is negative, it is moving downward and thus was at a higher position earlier. Only at the exact instant where $v_z(t) = 0$ is it neither climbing to nor coming down from a higher position: this must be the peak. So to find t_p ($\equiv$ time of the peak), we solve

$$0 = v_z(t_p) = -|\vec{g}|t_p + v_{0z} \quad \Rightarrow \quad t_p = \frac{v_{0z}}{|\vec{g}|} \tag{N9.8}$$

[Note, by the way, that even though the projectile's z-velocity is passing through zero at this point, its z-acceleration remains nonzero and equal to $-|\vec{g}|$ always. This can be seen clearly in figure N9.2: even as the value of $v_z(t)$ passes through zero, its slope is always constant and negative.]

If we have set up our reference frame so that the ground is at $z = 0$, then finding the time t_h ($\equiv$ time when the projectile hits the ground) is simply a matter of solving equation $z(t_h) = 0$ for t_h by using the quadratic equation:

When does a (simple) projectile hit the ground?

$$0 = z(t_h) = -\tfrac{1}{2}|\vec{g}|t_h^2 + v_{0z}t_h + z_0 \quad \Rightarrow \quad t_h = \frac{v_{0z} \pm \sqrt{v_{0z}^2 + 2|\vec{g}|z_0}}{|\vec{g}|} \tag{N9.9}$$

Exercise N9X.1

Check that the last result in equation N9.9 is correct.

Note that this equation yields *two* solutions. This is because if the projectile had been freely falling for *all time* and had sufficient energy to get up beyond $z = 0$ at all, then its parabolic trajectory would cross $z = 0$ *twice*, once going up and once going down. But since the projectile is *not* freely falling before it is launched, neither equation N9.8 nor equation N9.9 applies for times before the launch time, and so if either equation yields a solution before the launch time, the solution is not physically meaningful.

Exercise N9X.2

Are there values of v_{0z} and z_0 for which equation N9.9 yields *no* solution? If so, explain in physical terms *why* there is no solution.

Other implications

Other implications of equations N9.6 and N9.7 include the *parabolic* shape of a simple projectile's trajectory (see problem N9D.1) and the fact that the horizontal distance that a projectile launched from level ground will go is greatest if it is launched at a 45° angle (see problem N9D.2).

N9.4   A Projectile Motion Checklist

Projectile motion problem solutions usually have simpler structures than the constrained-motion problems we have seen in previous chapters. We already know the basic model we will use (the simple projectile model) and the master equations we will use (equations N9.6 and N9.7). Even so, one can do certain things to increase one's chances of solving the problem correctly.

The translation step

The drawing for a projectile motion solution should display

1. The initial position of the object of interest and its subsequent trajectory.
2. A labeled arrow indicating the object's initial velocity $\vec{v}_0$.
3. Coordinate axes with a clearly specified origin.
4. Labels defining symbols for other relevant quantities (particularly initial and final positions, important instants of time, and perhaps velocities).
5. A list of symbols with known values.

Establishing coordinate axes is very important for simple projectile motion problems because equations N9.6 and N9.7 are expressed in terms of components that have meaning only in the context of a well-defined reference frame with a clearly specified origin. Because time t is a very important variable in equations N9.6 and N9.7, defining symbols for important instants of time is also very important. As this is hard to do on a diagram, short prose descriptions are often a good way to define time symbols.

The conceptual model step

The point of the *modeling* part of a solution is to choose a model and assess its applicability and limitations. With simple projectile motion problems, the basic model is clear, so your main task is describing the approximations needed to make the situation fit the simple projectile model (and how good these approximations are) and the time interval during which the model applies. Establishing the latter helps us recognize whether a mathematically possible answer is in fact physically absurd. Again, a brief bit of prose is usually sufficient. The modeling part should end with the master equation(s) (equations N9.6 and/or N9.7).

So here is a checklist for projectile motion problems:

Simple Projectile Motion Checklist	**General Checklist**
Your main drawing should include:	☐ Define symbols.
☐ The object in its initial position.	☐ Draw a picture.
☐ A labeled arrow showing its initial velocity.	☐ Describe your model
☐ Coordinate axes with a clear origin.	and assumptions.
☐ Labels defining symbols for other quantities.	☐ List knowns and unknowns.
Also, you should: ☐ Define when $t = 0$.	☐ Do algebra symbolically.
☐ Define symbols for other critical times.	☐ Track units.
☐ Use the master equations N9.6 and/or N9.7.	☐ Check your result.

When you solve the problem, remember that each row in equations N9.6 and N9.7 is independent and must be satisfied simultaneously (as in a conservation of momentum problem). Often in projectile problems we will solve one of these independent equations for a time and then plug the result into a different equation to compute another quantity of interest.

Problem: Your friend, an unemployed actor, has a chance at a role in a Western movie if he can learn to slide a mug precisely down the length of a saloon bar. After practicing a lot, he thinks he can do it. But when asked to perform in front of the producer, he is a bit nervous, and the mug goes off the end of the bar (which is 1.4 m high) at a speed of 2.5 m/s. Does the mug compound your friend's problem by hitting the producer's foot, which is 2.0 m beyond the end of the bar?

Solution A sketch of the situation appears on the next page. Note that $t = 0$ is the instant that the mug leaves the edge of the bar; t_h is the instant when the mug hits the floor.

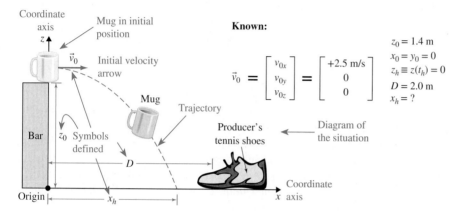

A mug is both small and relatively massive, so the simple projectile model should pretty accurately describe the situation for the short time the mug is in the air after it leaves the bar at $t = 0$ until it hits the floor (or the producer's foot) at $t = t_h$. I am also treating the mug as a point particle, essentially ignoring its size. If the mug's final position x_h turns out to be very close to D, I may need to do the calculation again, making some guesses as to the mug's actual size. I'm also assuming that the mug's initial velocity is exactly horizontal. Equation N9.7 tells us that

Discussion of model limitations

Definitions of $t = 0$ and other critical times

Other assumptions and approximations

$$\begin{bmatrix} x_h \\ 0 \\ 0 \end{bmatrix} = \begin{bmatrix} x(t_h) \\ y(t_h) \\ z(t_h) \end{bmatrix} = \begin{bmatrix} v_{0x}\,t_h + x_0 \\ v_{0y}\,t_h + y_0 \\ -\frac{1}{2}|\vec{g}|\,t_h^2 + v_{0z}\,t_h + z_0 \end{bmatrix} = \begin{bmatrix} |\vec{v}_0|\,t_h \\ 0 \\ -\frac{1}{2}|\vec{g}|\,t_h^2 + z_0 \end{bmatrix} \qquad (N9.10)$$

Master equation

The x and z components of this equation provide two useful equations in the two unknowns t_h and x_h. We know that $z(t_h) = 0$ (since the mug hits the floor at $t = t_h$ by definition), so the z component of equation N9.10 implies that

$$0 = z(t_h) = -\tfrac{1}{2}|\vec{g}|\,t_h^2 + z_0 \quad \Rightarrow \quad \tfrac{1}{2}|\vec{g}|\,t_h^2 = z_0 \quad \Rightarrow \quad t_h = \pm\sqrt{\frac{2z_0}{|\vec{g}|}} \qquad (N9.11a)$$

Algebra with symbols

Since the mug is not a projectile before $t = 0$, the negative solution is not relevant. Putting the positive result into the first line of equation N9.10 yields

$$x_h = |\vec{v}_0|\,t_h = |\vec{v}_0|\sqrt{\frac{2z_0}{|\vec{g}|}} = \left(2.5\ \frac{\text{m}}{\text{s}}\right)\sqrt{\frac{2(1.4\ \cancel{\text{m}})}{9.8\ \cancel{\text{m}}/\text{s}^{\cancel{2}}}} = 1.3\ \text{m} \qquad (N9.11b)$$

Calculations include and track units

This has the right units and the right sign (the mug should hit to the positive side of $x = 0$) and seems plausible. So the mug misses hitting the producer, whose shoes are $D = 2.0$ m from the end of the bar. (Let's just hope that the mug also does not splash the producer's shoes.)

Final check of plausibility

Example N9.2

Problem: During a soccer game, you kick the ball so that it leaves the ground at an angle of 33° traveling at a speed of 15 m/s. The ball reaches its maximum height just as it passes the goalie, who is standing well in front of the net. Is the ball high enough at this point to be out of the goalie's reach?

Definition of $t = 0$ and other critical times

Solution A drawing of the situation appears below. Let $t = 0$ be the instant the ball leaves the foot and t_p be the time when the ball reaches its peak. Note that the peak is defined to be where $v_z = 0$.

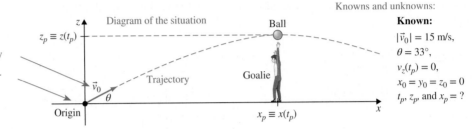

Knowns and unknowns:

Known:
$|\vec{v}_0| = 15$ m/s,
$\theta = 33°$,
$v_z(t_p) = 0$,
$x_0 = y_0 = z_0 = 0$
$t_p, z_p,$ and $x_p = ?$

Initial velocity arrow

Definition of coordinate axes and origin

Discussion of model limitations

Definitions of $t = 0$ and other critical times

Other assumptions and approximations

To use the simple projectile model, we have to assume that air resistance is negligible. (This may not be a very good assumption for this rapidly moving, fairly lightweight ball). If the model is good at all, it applies after the ball leaves the ground at $t = 0$ until it hits the ground or is caught: assume this occurs after $t = t_p$. I also am going to assume that the goalie can catch the ball if z_p is less than about 3.0 m (which is ≈ 9.5 ft): this is about how high I think I could reach (with a bit of a jump). Equations N9.6 and N9.7 at $t = t_p$ read

Master equations

$$\begin{bmatrix} v_x(t_p) \\ v_y(t_p) \\ v_z(t_p) \end{bmatrix} = \begin{bmatrix} |\vec{v}_0|\cos\theta \\ 0 \\ -|\vec{g}|t_p + |\vec{v}_0|\sin\theta \end{bmatrix} \qquad (N9.12a)$$

$$\begin{bmatrix} x(t_p) \\ y(t_p) \\ z(t_p) \end{bmatrix} = \begin{bmatrix} (|\vec{v}_0|\cos\theta)\,t_p + x_0 \\ 0 \\ -\frac{1}{2}|\vec{g}|t_p^2 + (|\vec{v}_0|\sin\theta)\,t_p + z_0 \end{bmatrix} \qquad (N9.12b)$$

Counting unknowns and equations

The x and z components of these equations provide four useful equations in our four unknowns t_p, z_p, $v_x(t_p)$, and x_p, so we can solve.

We know that $v_z(t_p)$ is zero, so we can solve the z component of equation N9.12a for t_p and then substitute it into the z component of equation N9.12b to determine the value of $z_p \equiv z(t_p)$ that we are particularly interested in. Doing this yields

Algebra with symbols

$$0 = v_z(t_p) = -|\vec{g}|t_p + |\vec{v}_0|\sin\theta \quad \Rightarrow \quad t_p = \frac{|\vec{v}_0|\sin\theta}{|\vec{g}|} \qquad (N9.13)$$

$$\Rightarrow z_p = z(t_p) = -\frac{1}{2}|\vec{g}|t_p^2 + (|\vec{v}_0|\sin\theta)t_p = -\frac{1}{2}|\vec{g}|\left(\frac{|\vec{v}_0|\sin\theta}{|\vec{g}|}\right)^2 + \frac{(|\vec{v}_0|\sin\theta)^2}{|\vec{g}|}$$

$$= \left(-\frac{1}{2}+1\right)\frac{|\vec{v}_0|^2\sin^2\theta}{|\vec{g}|} = \frac{|\vec{v}_0|^2\sin^2\theta}{2|\vec{g}|}$$

Calculations include and track units

$$= \frac{(15\text{m/s})^2\sin^2(33°)}{2(9.8\text{ m/s}^2)} = 3.4 \text{ m} \qquad (N9.14)$$

Final check of plausibility

This result has the right units, the right sign (the peak is above the ground!), and a reasonable magnitude. The kick is almost certainly out of the goalie's reach, though, so it's a gooaal!

N9.5 Drag and Terminal Speed

Up to now, we have been considering *simple* projectile motion, in which the only significant force acting on an object is its weight. Realistically, though, projectiles moving through air experience some drag. How would including drag affect our predictions about a projectile's motion?

In section N5.3, we saw that the drag force on a large object with cross-sectional area A moving at a reasonably large speed $|\vec{v}|$ through air has a magnitude of

$$|\vec{F}_D| = \tfrac{1}{2}C\rho A|\vec{v}|^2 \tag{N9.15}$$

where ρ is the density of air (which is about 1.2 kg/m³ at room temperature and standard pressure) and C is a unitless constant (called the *drag coefficient*) that depends on an object's shape.

First consider the simplest possible situation: an object falling vertically from rest. We can pretty easily determine *qualitatively* what will happen. At first (because $|\vec{v}|$ is very small) the magnitude of the drag force is negligible, and the object falls essentially freely downward (see figure N9.3a). But as its downward speed increases, the drag force also increases, decreasing the net force on the object and thus its downward acceleration (figure N9.3b). The downward speed of the object will continue to increase, however, until the drag force becomes essentially equal to the object's weight (figure N9.3c). As the object approaches the speed where this happens, the net force on it approaches zero. When the net force is essentially zero, the object no longer accelerates, but continues to fall with constant downward velocity. We call the magnitude of this final constant downward velocity the object's **terminal speed** v_T. You can easily show that its value must be

$$v_T = \sqrt{\frac{2m|\vec{g}|}{C\rho A}} \tag{N9.16}$$

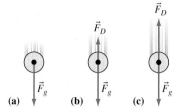

Figure N9.3
(a) Just after an object begins to fall, the magnitude of the drag force on it is very small. (b) As it begins to pick up downward speed, the drag force grows in magnitude. (c) When the magnitude of the drag force equals the magnitude of the object's weight, the object does not continue to accelerate, but instead maintains a *constant* downward velocity.

The terminal speed of an object falling with drag

Exercise N9X.3

Verify equation N9.16. (*Hint:* When is $|\vec{F}_D| = |\vec{F}_g|$?)

More mathematically, if we define the z axis to be positive upward, then the z component of Newton's second law implies that

$$ma_z = F_{net,z} = -m|\vec{g}| + \frac{1}{2}C\rho A v_z^2 = -m|\vec{g}| + m|\vec{g}|\left(\frac{C\rho A}{2m|\vec{g}|}\right)(v_z^2) = -m|\vec{g}|\left(1 - \frac{v_z^2}{v_T^2}\right)$$

$$\Rightarrow \quad a_z = -|\vec{g}|\left(1 - \frac{v_z^2}{v_T^2}\right) \tag{N9.17}$$

The acceleration of an object falling with drag

(Note that the drag force term is positive because it acts *upward* on an object falling downward.) You can see directly from this equation that the vertical acceleration becomes smaller and smaller until it becomes essentially zero as $|v_z|$ approaches the terminal speed v_T.

The next step would be to integrate both sides of this equation with respect to time to find v_z as a function of t. But here we run into a roadblock. To find v_z as a function of t, we have to integrate the right side of equation N9.17, and to do that, we already have to *know* how v_z depends on t!

This illustrates a typical problem that arises when we attempt to use forces to determine motion. *In principle*, once we know the forces acting on an object, we can use Newton's second law to find its acceleration $\vec{a}(t)$ and then integrate $\vec{a}(t)$ with respect to time to find the object's velocity $\vec{v}(t)$ and position $\vec{r}(t)$

Why we sometimes can't integrate Newton's second law

as functions of time. *In practice*, however, the expression for an object's acceleration almost always refers to the object's unknown velocity and/or position, making a straightforward integration of the acceleration impossible. For realistic problems, either we have to resort to some kind of trick to perform the integration, or we have to solve the problem numerically (either constructing a trajectory diagram by hand or using Newton or a similar computer application to do the job). Problem N9D.4 discusses the trick needed to solve this particular problem mathematically.

A useful guess about the time required to reach the terminal speed

But we can in fact make some useful estimates without much more mathematics. One very basic question we might like to answer is this: about how long will it take a falling object to reach its terminal velocity? If there is no drag force acting on the object, the z component of equation N9.6 implies that an object falling from rest satisfies $v_z = -|\vec{g}|t$, so the time it would take a drag-*free* object to reach speed v_T would be

$$t_T = \frac{v_T}{|\vec{g}|} \tag{N9.18}$$

Now, drag will cause the real object to fall more slowly than a free object, so we would *not* expect it to actually reach the terminal speed by this time. However, the time required is also not likely to be many times this value: surely after, say, 10 times the time required for a drag-free object to reach v_T, the real object must finally be falling at pretty close to this speed as well.

Checking the guess against the exact solution

The exact mathematical solution to this problem is shown in figure N9.4. One can see that by time t_T, the object has only reached a speed of about $\frac{3}{4}v_T$, but the object essentially reaches its terminal speed after a time $t \approx 2t_T$. For very early times ($t < 0.2t_T$ or so), the object's downward speed increases at essentially the free-fall rate of $v_z = -|\vec{g}|t$, so this provides a good estimate of the time period during which the simple projectile model might apply.

A human example

As a quick example, a human being's terminal speed $v_T \approx 60$ m/s. About how long would a skydiver have to fall to reach that terminal speed? According to the argument above, this will occur at approximately a time

$$t \approx 2t_T = \frac{2v_T}{|\vec{g}|} = \frac{120 \text{ m/s}}{9.8 \text{ m/s}^2} = 12 \text{ s} \tag{N9.19}$$

If the object is *not* falling from rest, the problem becomes even *more* difficult. In such cases, using the Newton application (at sixideas.pomona.edu/resources.html), which can handle *any* situation, becomes a virtual necessity. The important issue to understand here is that even in very straightforward circumstances, it may not be easy to solve Newton's second law for the trajectory of an object (a problem we will struggle with for the rest of the unit).

How long will it take this person to reach terminal speed? (Credit: germanskydiver/123RF)

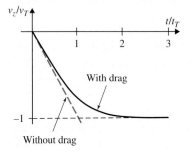

Figure N9.4

A graph of the mathematical solution for the z-velocity of an object falling from rest with drag (this solution is discussed in problem N9D.4). The straight line shows that at very early times ($t \ll t_T$) the object's downward speed increases at approximately the free-fall rate. However, the object essentially reaches its terminal speed by $t \approx 2t_T$, where $t_T = v_T/|\vec{g}|$.

TWO-MINUTE PROBLEMS

N9T.1   You are driving 5 ft or so behind a pickup truck (don't actually try such a stupid tailgating stunt, of course!). A jostled crate tips off the back of the truck with only a very small backward velocity. The crate will not hit your car until after it hits the road, regardless of your speed. True (T) or false (F)? (Ignore air resistance.)

N9T.2   A person standing in the cabin of a jet plane drops a coin. This coin hits the floor of the cabin at a point directly below where it was dropped (as seen in the cabin) no matter how fast the plane is moving. T or F?

N9T.3   A tennis ball is dropped from rest at the exact same instant and height that a bullet is fired horizontally. Which hits the ground first (ignoring air resistance)?
A. The bullet hits first.
B. The ball hits first.
C. Both hit at the same time.

N9T.4   As a projectile moves along its parabolic trajectory, which of the following remain constant (ignoring air resistance, and defining the z axis as pointing upward)?
A. Its speed
B. Its velocity
C. Its x-velocity and y-velocity
D. Its z-velocity
E. Its acceleration
F. Its x-velocity, y-velocity, and acceleration
T. Some other combination of the given quantities

N9T.5   Suppose we throw a baseball with an initial speed of 12 m/s in a direction 60° upward from the horizontal. What is the baseball's speed at the peak of its trajectory? (*Hint:* You do not need to do a *lot* of calculation here.)
A. 12 m/s
B. 10.4 m/s
C. 6 m/s
D. 3 m/s
E. 0 m/s
F. Other (specify)

N9T.6   Suppose you serve a tennis ball with an initial speed of 10 m/s in a direction 10° below the horizontal. What is its speed at the peak of its trajectory?
A. 10 m/s
B. 9.8 m/s
C. 1.7 m/s
D. 0 m/s
E. There is no "peak" to this tennis ball's trajectory.
F. Other (specify)

N9T.7   An orbiting satellite experiences no forces other than the force of gravity. It therefore exhibits simple projectile motion. T or F?

N9T.8   Imagine you throw a tennis ball vertically into the air. At the exact top of its trajectory it is at rest. What is the magnitude of its acceleration at this point?
A. 9.8 m/s^2
B. -9.8 m/s^2
C. $0 < |\vec{a}| < 9.8 \text{ m/s}^2$
D. 0
E. Other (specify)

N9T.9   Two balls have the same size and surface texture, but one is twice as heavy as the other. How many times larger is the terminal speed of the more massive ball falling through air than that of the lighter ball?
A. The balls fall with the *same* speed in air.
B. The massive ball's terminal speed is $[2]^{1/2}$ times larger than the other's.
C. The massive ball's terminal speed is 2 times larger than the other's.
D. The massive ball's terminal speed is 4 times larger than the other's.
E. The massive ball's terminal speed is some other multiple of the other's (specify).

N9T.10   Two balls have the same weight and surface texture, but one has twice the diameter of the other. How many times larger is the terminal speed of the smaller ball falling through air than that of the bigger ball?
A. The balls fall with the *same* speed in air.
B. The smaller ball's terminal speed is $[2]^{1/2}$ times larger than the other's.
C. The smaller ball's terminal speed is 2 times larger than the other's.
D. The smaller ball's terminal speed is 4 times larger than the other's.
E. The smaller ball's terminal speed is some other multiple of the other's (specify).

N9T.11   Suppose a baseball travels 425 ft (130 m) from where the batter hits it to where it lands in the stands. The ball's velocity makes an angle of 40° with respect to the horizontal as the ball leaves the bat. Suppose we first calculate the ball's initial speed ignoring drag and find it to be $|\vec{v}_0|$. Including drag might plausibly require a higher initial speed for the ball to travel the same horizontal distance. On the other hand, by slowing down the ball's vertical motion, including drag will increase the ball's flight time, allowing the ball the same distance with a slower horizontal velocity. So if we include drag, for the ball to fly the same horizontal distance will we need
A. An initial speed greater than $|\vec{v}_0|$.
B. An initial speed less than $|\vec{v}_0|$.
C. An initial speed equal to $|\vec{v}_0|$.
D. More information to answer for sure.
(I suggest checking your reasoning with the Newton app.)

HOMEWORK PROBLEMS

Basic Skills

N9B.1   A ballistic missile traveling 3200 km between its launch point and final destination freely falls during much of its trajectory, most of which is above the atmosphere. Is this missile a simple projectile or not? Explain.

N9B.2   A package dropped from an airplane flying at an altitude of 1.0 km will take how long to reach the ground? (Ignore air resistance.)

N9B.3   How long will it take a stone thrown with a vertical velocity of 22 m/s to reach the peak of its trajectory?

N9B.4   If a fireworks rocket has an initial upward speed of 58 m/s when launched, for how long will it coast upward before reaching its peak? (Ignore air resistance and assume that the rocket engine burns out very shortly after launch.)

N9B.5   If an object is launched from the ground with an initial z-velocity of 25 m/s, how much time will pass before it returns to the ground? (Ignore air resistance.)

N9B.6   If an object is launched from a point 10 m above the ground with an initial z-velocity of 25 m/s, how much time will pass before it returns to the ground? (Ignore air resistance.)

N9B.7   Estimate the terminal speed for a ping-pong ball whose diameter is ≈1.5 in. and whose mass is ≈2.5 g. (For a sphere, C is roughly 0.5.)

N9B.8   A person's terminal speed in air is typically about 60 m/s. If so, what is the value for CA for a falling person? (Assume that $m ≈ 60$ kg.)

Modeling

N9M.1   Estimate the speed at which the drag on a 150-g steel ball becomes equal to about 1% of its weight. (The density of steel is about 7900 kg/m^3, and C for a sphere is about equal to 0.5.)

N9M.2   **(a)** Estimate the drag coefficient for a falling spread-eagled person in air, using data given in section N9.5.
(b) Roughly how far has a person fallen by the time the person reaches terminal speed, within ±10% or so? (*Hint:* Look at figure N9.4. Remember that an integral is equal to the area under the curve of the integrand.)

N9M.3   Suppose your Frisbee is lodged in a tree. The frisbee is 12 m above your head.
(a) How fast would you have to throw a baseball to dislodge the Frisbee?

(b) Check your work using a calculation based on conservation of energy.
(c) Which method is easier to use in this kind of problem?

N9M.4   A police officer is chasing a burglar across a rooftop. Both are running at a speed of 6.0 m/s. Before the burglar approaches the edge of the roof, the burglar needs to make a decision about whether to jump the gap to the next building, whose roof is 6.2 m away but 3.5 m lower. Will the burglar be able to land on the next building on his or her feet (assuming that the burglar's initial velocity is 6.0 m/s horizontally when the jump begins)?

N9M.5   You are standing on a cliff 32 m tall overlooking a flat beach. The edge of the water is 25 m from the base of the cliff. How fast would you have to throw a stone horizontally to reach the water?

N9M.6   In serving a tennis ball, the server hits the ball horizontally from a height of about 2.2 m. The ball has to travel over the net (0.9 m high) a distance of 12 m away. What is the minimum initial speed that the tennis ball can have if it is to make it over the net?

N9M.7   You are the pilot of a Coast Guard rescue plane. Your job is to drop a package of emergency supplies to a person floating in the ocean. If your plane is flying directly toward the person at an elevation of 520 m at a speed of 85 m/s (about 190 mi/h), about how far ahead of the person should you release the package if it is to land near the person? Assume the package's motion is not significantly affected by air friction (probably not a very good assumption here).

N9M.8   A batter hits a fly ball with an initial velocity of 37 m/s at an angle of 32° from the horizontal. How much time does the outfielder have to get to the appropriate position to catch the ball?

N9M.9   During volcanic eruptions, chunks of solid rock can be blasted out of a volcano: these projectiles are called *volcanic blocks*. Imagine that during an eruption of Mount St. Helens (a volcano in Washington state) a block lands near an observing station that is located 2.5 km east of and 900 m below the summit. Assume that air drag is negligible and that the block was ejected from the summit at a 35° angle from the horizontal. Find
(a) its initial speed and
(b) its time of flight.
(c) Imagine that this particular block is a cube of rock about 1 m on a side: such a block would have a mass of about 3000 kg. Is drag really negligible for this block at the launch speed you calculated for part (a)?
(d) Use the Newton application to find (by trial and error) the block's *actual* initial velocity and time of flight. (*Hint:* Calculate b in the expression $|\vec{F}_D|/m = b|\vec{v}|^2$.)

N9M.10 Consider dropping a baseball from rest at an altitude of 300 m. Assume that the baseball has a drag coefficient of 0.3, a mass of 0.145 kg, and a radius of 3.7 cm (roughly the real values for a baseball).

(a) What is the terminal speed v_T of this baseball?

(b) Calculate the characteristic time $t_T = v_T/|\vec{g}|$.

(c) Set up the Newton application to model this situation. Let's agree that the ball has essentially reached its terminal velocity when its acceleration falls below $|\vec{g}|/20$. About how long must the baseball fall to reach terminal velocity according to this criterion? Express your answer as a multiple of the characteristic time t_T. (*Hints:* Equation N9.17 might be helpful as you set up the program. A time step of 0.1 s works pretty well.)

N9M.11 (Baseball physics)[†]

(a) Use the Newton app to calculate the trajectory of a batted baseball that starts at $y \approx 1$ m (where y is the vertical axis) and has an initial velocity of 50 m/s in a direction that is 37° above the horizontal (note that $\cos 37° = \frac{4}{5}$, $\sin 37° = \frac{3}{5}$). Use a time step of 0.2 s. Assume zero drag. Look at a close-up view of the construction to verify that the (blue) acceleration arrows are indeed vertical. (1) How far does the baseball travel horizontally? (2) What is the baseball's height above the ground at the peak of its trajectory? (3) How long is it in flight? Print out a copy of the trajectory, and record your answers on the printout.

(b) Now add drag, using equation N9.15 to model the drag. Note that the drag always acts opposite to the object's velocity. Note also that the Newton app requires you to enter terms in the expression for the ball's *acceleration*, so you need to solve Newton's second law for acceleration to determine the values of these terms. The radius of a real baseball is 3.7 cm, its mass is 0.145 kg, and its drag coefficient is about 0.3. Use the same initial conditions as in part (a). Look at a close-up view and verify that the acceleration arrows are no longer vertical, but tip backward, as might be expected when there is a backward force on the ball. (1) How far does the baseball travel now? (2) What is its peak height? (3) How long does it take to reach the ground? (4) Is drag an important effect here? Write your responses to these questions on a printout of the trajectory, and show how you calculated the coefficient for the drag acceleration term that you typed into the app.

(c) The baseball could also be spinning. It turns out that a baseball spinning in the correct way (so that the bottom part of the ball moves in the same direction as the ball's center of mass) experiences a lift force in addition to the drag force. This lift force has a magnitude of $|\vec{F}_L| = kf|\vec{v}|^2$, where f is the frequency (in rotations per second) of the ball's rotation, $|\vec{v}|$ is the speed of the ball's center of mass, and k is a constant whose experimental value is roughly 1.1×10^{-5} N·s³/(m² rev). The direction of the force is perpendicular to the velocity on the upward side (this is exactly vertically upward *only* if the ball's velocity is exactly horizontal, so *don't* set the direction of the acceleration term contributed by this force to be in the *y*-direction). Add this effect into the simulation, assuming the baseball spins at 20 turns per second. (1) How far does the ball travel now? (2) What is its peak height? (3) How long does it take to reach the ground? (4) Could such a spin turn a long fly ball into a home run, do you think? Record your answers on a printout of the trajectory, and show how you calculated the acceleration coefficient for the spin lift term that you typed into the app.

(d) Finally, check out what happens if you vary the size of the time step (this is something one should always do with computer models). (1) What happens if you choose Δt equal to 0.1 s? How about 0.05 s? Does the distance the ball travels, its peak height, or the time of flight change much? (2) What if you choose something larger, such as 0.5 s or 1 s? (3) What do you think is the largest time step you could choose to get answers that are within 2% or so of the true result? (Don't do new printouts here. Just record your answers on one of the previous printouts.)

N9M.12 A soccer ball has a mass of about 0.45 kg and a radius of about 10 cm.

(a) Do some runs, using the Newton application to determine the optimal angle (±1°) for the goalie to kick the ball to achieve the maximum range in still air with a launch speed of 30 m/s. Use a time step of 0.05 s, and do *not* ignore drag. The optimal angle when drag is not significant is 45° (see problem N9D.2), so start with that angle.

(b) Explain physically why the optimal angle should be *less* than 45° when drag is significant.

(c) On the other hand, does choosing the optimal angle (as opposed to 45°) in this case give you much improvement in the range? Submit printouts justifying your conclusions.

Derivations

N9D.1 The formula for a parabola in the zx plane is

$$z(x) = Ax^2 + Bx + C \qquad (N9.20)$$

where A, B, and C are constants. Prove that the trajectory of a simple projectile has this form.

[†]Thanks to Phil Krasicky of Cornell University for developing this application of the Newton program. See A. F. Rex, "The Effect of Spin on the Flight of Batted Baseballs," *American Journal of Physics*, **53:** 1073–1075 (1985) for greater discussion of the general issue.

N9D.2   A projectile's range R is the horizontal distance that it travels between its launch and the time it hits the ground.

(a) Show that if the projectile is launched from the ground, with velocity $\vec{v}_0$ at an angle of θ with respect to the horizontal, then its range will be

$$R = \frac{2|\vec{v}_0|^2 \sin\theta \cos\theta}{|\vec{g}|} \qquad (N9.21)$$

(b) Prove that a projectile launched from the ground with a given initial speed $|\vec{v}_0|$ will travel the farthest if it is launched at an angle of $45°$ with respect to the horizontal. (*Hint:* Consider the trigonometric identity that says that $\sin 2\theta = 2 \sin\theta \cos\theta$.)

N9D.3   Prove that the ranges of two projectiles launched from the ground with the same initial speed $|\vec{v}_0|$ at an angle of $45° \pm \phi$ are the same. [*Hints:* See problem N9D.2. Note also that $\sin(\pi/2 + \phi) = \cos\phi$.]

N9D.4   (Requires some sophisticated calculus.) One can integrate equation N9.17 as follows.

(a) Using the fact that $a_z = dv_z/dt$, show that you can rearrange equation N9.17 to read

$$\frac{dv_z}{1 - v_z^2/v_T^2} = |\vec{g}|\,dt \qquad (N9.22)$$

(b) Take the integral of both sides of this equation, evaluating the integral on the left from $v_z(0)$ to $v_z(t)$ and the integral on the right from 0 to t. (Feel free to take advantage of the labors of some brave and selfless mathematician by looking up the integral of the left side in a table of integrals or use WolframAlpha to give it to you. You should find that the result involves an inverse hyperbolic tangent of v_x/v_T.)

(c) Rearrange the result to show that

$$\frac{v_z(t)}{v_T} = \tanh\!\left(\frac{-|\vec{g}|t}{v_T}\right) \qquad (N9.23)$$

This is the function graphed in figure N9.4.

(d) Integrate equation N9.23 to find an expression for the z-position of the object as a function of time. Again, feel free to look up the integral you need in a table of integrals. (You should get something involving the natural logarithm of a hyperbolic cosine.)

Rich-Context

N9R.1   Abel Knaebble, professional stunt man and amateur physicist, wants to jump the Grand Canyon in his new, super-streamlined motorcycle (which is completely immune to the effects of air friction). He simply plans to ride his motorcycle off one rim, so his initial velocity will be entirely horizontal. Assume that the width of the canyon is 8.0 km at the proposed jump site, that the mass of his fully loaded cycle (including Abel himself) is 380 kg, and that the rim of the canyon from which he will jump is 320 m higher than the far side.

(a) How fast will he need to take off to successfully make it to the other side?

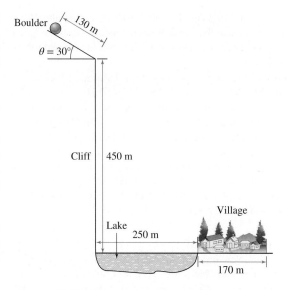

Figure N9.5
A village at risk? (See problem N9R.2.)

(b) Discuss some of the physical impracticalities of this proposed stunt. (The more the better, but discuss at least two of the impracticalities you consider to be the most important.)

N9R.2   An essentially spherical boulder the size of several houses sits precariously on a slope above a village, as shown in figure N9.5. If this boulder ever gets loose, will it hit the village? (*Hint:* First estimate the speed with which the boulder will roll off the cliff. Show that the total center-of-mass kinetic plus rotational energy of a rolling spherical object is $\frac{7}{10}m|\vec{v}|^2$: see chapter C11.)

N9R.3   You are the physics consultant working on an action movie set where an SUV must start from rest, accelerate along a pier, and jump over a stretch of water to land in a departing ferry (the director assures you it will all be very exciting). The drawing below shows the situation.

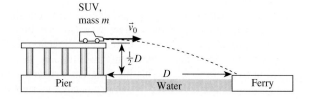

The SUV has four-wheel drive (see problem N5D.1), and the coefficients of friction between the tires and the pier's surface are $\mu_s = 3/4$ and $\mu_k = 1/2$. Calculate the length L (as a fraction or multiple of D) of the run along the pier that you will need for the SUV to make it to the ferry. (Ignore air drag.)

ANSWERS TO EXERCISES

N9X.1   The quadratic formula implies that

If $ax^2 + bx + c = 0$ then $x = \dfrac{-b \pm \sqrt{b^2 - 4ac}}{2a}$ (N9.24)

In the case at hand, $a = -|\vec{g}|/2$, $b = v_{0z}$, and $c = z_0$. Substituting these into the above yields equation N9.9.

N9X.2   If z_0 is negative and $2|\vec{g}||z_0| > v_{0z}^2$, then the quantity inside the square root is negative and the equation has no solution. Having z_0 being negative means that the projectile begins below $z = 0$. Conservation of energy means that the projectile in this case must have an initial kinetic energy

of $\frac{1}{2} m v_{0z}^2 > m|\vec{g}||z_0| \Rightarrow v_{0z}^2 = 2|\vec{g}||z_0|$ just to get up to $z = 0$. Therefore, if $v_{0z}^2 < 2|\vec{g}||z_0|$, the projectile *never* makes it up to $z = 0$, and thus there *should* be no real solutions to the equation.

N9X.3   The object reaches its terminal velocity when the drag acting on it is equal in magnitude to its weight, so

$$mg = \tfrac{1}{2} C\rho A v_T^2 \quad \Rightarrow \quad v_T^2 = \dfrac{2m|\vec{g}|}{C\rho A} \qquad \text{(N9.25)}$$

Taking the square root of this yields equation N9.16.

Oscillatory Motion

N10

CORE

Chapter Overview

Introduction

This is the second in the series of chapters exploring how we can use Newton's second law and knowledge of the forces acting on an object to predict its motion. In this chapter, we explore the important case of *oscillatory motion*, where an object moves in one dimension in response to a force that seeks to push it back toward an equilibrium point.

Section N10.1: Mass on a Spring

Consider an object with mass m connected to a fixed point with an ideal spring, and imagine that the mass moves in one dimension along the x axis on a level, frictionless surface. If $x = 0$ is the object's position when the spring is relaxed, then we can show from the spring potential energy formula that

$$F_{Sp,x} = -k_s x \qquad \text{(N10.2)}$$

- **Purpose:** This expression describes the x-force $F_{Sp,x}$ that an ideal spring exerts on an object whose x-position is x, where k_s is the spring's **spring constant**.
- **Limitations:** The spring must be ideal, the object must be in equilibrium at position $x = 0$, and the object must be moving in one dimension.
- **Note:** This equation is called **Hooke's law**.

Section N10.2: Solving the Equation of Motion

We call any object moving in one dimension whose *net x*-force has the form of equation N10.2 a **simple harmonic oscillator (SHO)**. Newton's second law then implies that the object's position obeys the **(simple) harmonic oscillator equation**:

$$\frac{d^2 x}{dt^2} = -\omega^2 x \qquad \text{(N10.4)}$$

- **Purpose:** This equation describes the consequences of Newton's second law for an object of mass m experiencing a *net* force given by Hooke's law, where x is the mass's x-position relative to its equilibrium position, t is time, and ω is a constant $= (k_s/m)^{1/2}$, where k_s is the constant appearing in Hooke's law.
- **Limitations:** It assumes that the object moves much slower than light in *one dimension* and that the net force on the object is given by Hooke's law.

One can show by substitution that the following function satisfies the simple **differential equation** given by equation N10.4:

$$x(t) = A\cos(\omega t + \theta) \qquad \text{(N10.6)}$$

- **Purpose:** This equation describes the most general solution to the harmonic oscillator equation, where $x(t)$ is the oscillating object's x-position at time t, A is a constant with units of distance called the oscillation's **amplitude,** θ is a constant with units of angle called the oscillation's **initial phase,** and ω is a constant with units of angle/time called the oscillation's **phase rate** or **angular frequency.**
- **Limitations:** This equation assumes that the harmonic oscillator equation adequately describes the motion of the object.
- **Notes:** Again, $\omega \equiv (k_s/m)^{1/2}$, where m is the object's mass and k_s is the spring constant appearing in Hooke's law. The values of A and θ are determined by the object's position and velocity at time $t = 0$.

The oscillation's **period** (the time required to complete 1 cycle) is $T = 2\pi/\omega$, and the oscillation's **frequency** (the number of oscillations per unit time) is $f = \omega/2\pi$. The SI unit for the latter quantity is the **hertz,** abbreviated as Hz, where $1\ \text{Hz} \equiv 1$ cycle/s.

Section N10.3: The Oscillator as a Model

Any oscillating object behaves as a simple harmonic oscillator if:

1. It has an **equilibrium position** at which the net force on the object is zero.
2. It experiences a **restoring force** if it is displaced from equilibrium.
3. The magnitude of the restoring force is directly proportional to the displacement.

For sufficiently small displacements, almost any restoring force satisfies these conditions, so almost any oscillating system will behave as an SHO in that limit.

Section N10.4: A Mass Hanging from a Spring

Even though the net force on an object hanging from a spring includes gravity, if we orient our x axis vertically and shift the origin so that $x = 0$ is where the net force (including gravity) is zero, then the force law for displacements from this new origin becomes $F_{\text{net},x} = -k_s x$. Therefore, such an object is a simple harmonic oscillator.

Section N10.5: An Analogy to Circular Motion

Whether our oscillating object moves vertically or horizontally, its x-position is the same as the x coordinate of an object fixed to the edge of a steadily rotating circular disk whose radius R is equal to the oscillation amplitude A. We conventionally think of the disk as rotating counterclockwise, and we take counterclockwise angles to be positive. An oscillation's initial phase θ then corresponds to the angle of the disk at $t = 0$ relative to its orientation where x is maximum.

Mathematically, one can solve for A and θ from initial conditions by noting that $v_x(t) = dx/dt = -A\omega\sin(\omega t + \theta)$, setting $x(0)$ and $v_x(0)$ equal to the initial conditions, and solving these two equations for A and θ. The disk analogy often helps one to keep signs straight.

Section N10.6: The Simple Pendulum

A **simple pendulum** consists of an object (the pendulum **bob**) swinging from an inextensible massless string of length L. Newton's second law implies that the angle ϕ that the pendulum makes with the vertical obeys the differential equation that in the limit of small angles reduces to $d^2\phi/dt^2 = -(|\vec{g}|/L)\phi$. This is the harmonic oscillator equation with ϕ replacing x and $|\vec{g}|/L$ replacing ω^2. Therefore, the solution to this equation is $\phi(t) = A\cos(\omega t + \theta)$ with $\omega = (|\vec{g}|/L)^{1/2}$, and the pendulum's period is

$$T = \frac{2\pi}{\omega} = 2\pi\sqrt{\frac{L}{|\vec{g}|}} \qquad \text{for small angles} \qquad \text{(N10.29)}$$

N10.1   Mass on a Spring

Conventional definition of
coordinate axes

Imagine an ideal, massless spring with one end connected to a fixed point (such as a wall) and the other connected to a movable object with mass m, as shown in figure N10.1a. Assume the object is free to slide in one dimension on a horizontal frictionless surface. In this case, the object's weight will be exactly canceled by the normal force due to its interaction with the surface, and the net force acting on the object will be the force supplied by the spring.

We conventionally define the x axis to coincide with the line along which the object moves. Let r be the object's separation from the fixed point at a given time, and let r_0 be the same when the spring is relaxed. If we define our reference frame origin so that $x = r - r_0$, then $x = 0$ corresponds to the object's position when the spring is relaxed, and $|x|$ expresses the distance that the spring is either stretched or compressed.

According to equation N2.4c, the force exerted by an ideal spring is

$$\vec{F}_{Sp} = -k_s(r - r_0)\hat{r} \tag{N10.1}$$

where k_s is the **spring constant** that characterizes the spring's stiffness and $\hat{r}$ specifies the direction in which r increases. In our coordinate system, $x = r - r_0$, and $\hat{r}$ is the $+x$ direction, so the x component of the force exerted on the object by the spring is given by the simple linear formula

The force law for a spring (Hooke's law)

$$F_{Sp,x} = -k_s x \tag{N10.2}$$

- **Purpose:** This equation describes the x-force $F_{Sp,x}$ that an ideal spring with **spring constant** k_s exerts on an object whose x-position is x.
- **Limitations:** The spring must be ideal, and the object must move in only one dimension and be in equilibrium at position $x = 0$.
- **Note:** We call this equation **Hooke's law**.

(Robert Hooke first stated this law in 1678.) Note that the force acts in the negative x direction when x is positive and in the positive x direction when x is negative: in both cases, this tends to push the object back toward $x = 0$. This is completely consistent with what we know qualitatively about the behavior of springs.

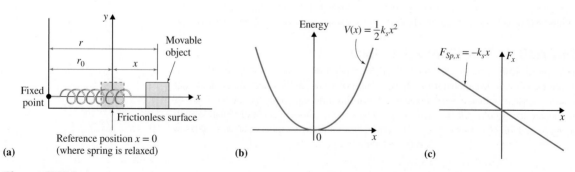

Figure N10.1
(a) The archetype of the simple harmonic oscillator: a movable object allowed to move in one dimension that is connected to a fixed point by an ideal, massless spring whose relaxed length is r_0. (b) A graph of the system's potential energy as a function of x. (c) A graph of the x-force on the movable object as a function of x.

Exercise N10X.1

The units of the spring constant k_s were given in chapter C9 as joules per meter squared (J/m^2). These units were appropriate when we were using k_s to calculate potential energies. What would be the appropriate units for k_s in the context of equation N10.1? Show that your units are equivalent to J/m^2.

N10.2   Solving the Equation of Motion

We call any object moving in one dimension whose net x-force has the form of equation N10.2 a **simple harmonic oscillator (SHO)**. In *any* such situation, Newton's second law reads

$$F_{Sp,x} = ma_x \quad \Rightarrow \quad a_x = \frac{F_{Sp,x}}{m} = \frac{-k_s x}{m} = -\omega^2 x \qquad \text{(N10.3)}$$

where $\omega^2 \equiv k_s/m$. (Why I have defined the *square* of ω to equal k_s/m will become clear shortly.) Since $a_x = dv_x/dt = d^2x/dt^2$, equation N10.3 becomes

$$\frac{d^2x}{dt^2} = -\omega^2 x \qquad \text{(N10.4)}$$

- **Purpose:** This equation describes the consequences of Newton's second law for an object of mass m experiencing a *net* force given by Hooke's law, where x is the mass's x-position relative to its equilibrium position, t is time, ω is a constant $= \sqrt{k_s/m}$, and k_s is the spring constant appearing in Hooke's law.
- **Limitations:** This equation assumes that the object moves much slower than light in one dimension and that the net force on the object is given by Hooke's law.

The harmonic oscillator equation

This important equation is called the **(simple) harmonic oscillator equation** or **SHO equation**.

How can we determine the object's motion in this case? When we studied projectile motion in chapter N9, we simply integrated both sides of Newton's second law twice to find the projectile's position as a function of time. We cannot do the same thing in this case, because in order to integrate both sides of equation N10.4 with respect to time, we have to know how x depends on time. Unfortunately, this is what we are trying to *find*. As I mentioned in section N9.5 on drag, this commonly occurs when we are trying to determine an object's motions from forces, and we ultimately have to find some trick to get around this problem.

The harmonic oscillator equation is an example of a simple **differential equation**. A differential equation sets a function, $x(t)$ in this case, in an equation having terms that also involve derivatives of that function. Differential equations cannot generally be solved in any straightforward manner. Often, the only method is to guess what the solution is, plug the guess into the differential equation, and check to see whether the guessed solution satisfies the equation. If the guess works, you have solved the equation; if not, you try another guess. (A course in differential equations basically makes you a more intelligent guesser!)

So the trick we will use here is to guess a possible solution and see if it works. What intelligent guess can we make in this case? We know that an object on the end of a spring will oscillate, so we expect $x(t)$ to be some function that repeats in time, something like $\sin(bt)$ or $\cos(bt)$, where b is some constant. The differential equation $d^2x/dt^2 = -\omega^2 x$ tells us that the second time derivative of our function for $x(t)$ should be equal to a negative constant times that same function. But both $\sin(bt)$ and $\cos(bt)$ have that characteristic as well! Note that (according to appendix NA on derivatives):

$$\frac{d}{dt}\sin(bt) = b\cos(bt) \qquad \frac{d^2}{dt^2}\sin(bt) = \frac{d}{dt}b\cos(bt) = -b^2\sin(bt) \qquad \text{(N10.5}a\text{)}$$

$$\frac{d}{dt}\cos(bt) = -b\sin(bt) \qquad \frac{d^2}{dt^2}\cos(bt) = -\frac{d}{dt}b\sin(bt) = -b^2\cos(bt) \qquad \text{(N10.5}b\text{)}$$

So if we set $x(t)$ equal to either one of these functions, we could satisfy the harmonic oscillator equation $d^2x/dt^2 = -\omega^2 x$ as long as we identify the constant b as being equal to $\omega = \sqrt{k_s/m}$.

Now, the simple equation $x(t) = \sin(\omega t)$ can't be right, as $x(t)$ has units of meters while the sine function always produces a unitless number by definition (it is the ratio of two sides of a right triangle). So a viable solution must be something like $x(t) = A\sin(\omega t)$, where A is a constant with units of meters.

However, as we have already pointed out, the function $x(t) = A\cos(\omega t)$ will *also* be a solution to this equation. The most general solution to the harmonic oscillator equation is in fact

The most general solution to
the simple harmonic oscillator
equation

$$x(t) = A\cos(\omega t + \theta) \qquad \text{(N10.6)}$$

- **Purpose:** This equation describes the most general solution to the harmonic oscillator equation, where $x(t)$ is the oscillating object's x-position at time t; A is a constant with units of distance called the oscillation's **amplitude**; θ is a constant with units of angle called the oscillation **initial phase**; and ω is a constant with units of angle/time called the oscillation's **phase rate** or **angular frequency**.
- **Limitations:** This equation assumes that the harmonic oscillator equation adequately describes the motion of the object.
- **Notes:** Again, $\omega = \sqrt{k_s/m}$, where m is the object's mass and k_s is the spring constant appearing in Hooke's law. The values of A and θ are determined by the object's position and velocity at time $t = 0$.

The object's motion is thus described by a cosine wave that cycles back and forth between the limits $x = \pm A$ as shown in figure N10.2. Note that the *total* distance that the oscillating object travels from one extreme to the other is $2A$.

As shown in figure N10.2, the constant θ specifies how far along a given cycle the oscillation is at time $t = 0$ (measured from the first peak to the left of the origin). For example, if $\theta = 0$, the object is at $+A$ at $t = 0$; if $\theta = \pi/2$, then the object is at $x = 0$ at $t = 0$; and so on. Different values of θ essentially shift the oscillation back and forth along the ωt axis of figure N10.2.

Note that if $\theta = \pi/2$, then $\cos(\omega t + \theta)$ is equivalent to $-\sin(\omega t)$ (if you shift the peak to the left of the origin back a full quarter-cycle from the origin, you can see that it is like an upside-down sine function). Similarly, if $\theta = \pi$, $\cos(\omega t + \theta) = -\cos\theta$; and if $\theta = 3\pi/2$, then $\cos(\omega t + \theta) = +\sin\theta$. The point is that equation N10.6 embraces *both* the sine and cosine solutions to the

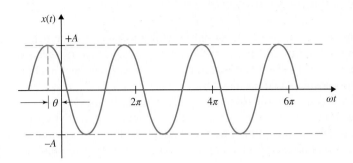

Figure N10.2
The solution to the harmonic oscillator equation is $x(t) = A\cos(\omega t + \theta)$. This function oscillates between the limits $\pm A$ and goes through one complete oscillation every time that ωt increases by 2π. The initial phase θ specifies how far the function is along its cycle at time $t = 0$ (a positive value of θ thus corresponds to shifting the wave crest left from zero by the magnitude of θ). The value of ω is determined by the values of k_s and m, while the values of A and θ are determined by initial conditions.

harmonic oscillator equation and everything in between. Since we can also always change the sign of the function by choosing the right initial phase θ, we conventionally choose θ so that the amplitude A is positive.

The constant ω specifies the rate at which the quantity $\omega t + \theta$ (which is sometimes called the oscillation's *phase*) increases with time. Since the object goes through one complete cycle every time that $\omega t + \theta$ increases by 2π, the larger ω is, the more cycles the object will complete in a given time.

An oscillation's **period** T is defined to be the time it takes the object to go through one complete oscillation of the cosine function. This will happen when the value of ωt (and thus $\omega t + \theta$) increases by 2π rad, that is,

The period of an oscillation

$$\omega(t + T) = \omega t + 2\pi \qquad \text{(N10.7)}$$

Subtracting ωt from both sides, we see that

$$\omega T = 2\pi \quad \Rightarrow \quad T = \frac{2\pi}{\omega} = 2\pi\sqrt{\frac{m}{k_s}} \qquad \text{(N10.8)}$$

The **frequency** f of an oscillation is defined to be the number of *cycles* completed per unit time. Since exactly 1 cycle is completed in time T by definition, the frequency f is given by

The frequency of an oscillation

$$f = \frac{1 \text{ cycle}}{T} = \frac{\omega}{2\pi}\text{cycle} = \frac{\text{cycle}}{2\pi}\sqrt{\frac{k_s}{m}} \qquad \text{(N10.9)}$$

[The Greek letter ν (nu) is also commonly used for frequency, but since this letter is easy to confuse with the letter v used for velocity, I'll avoid it here.] The standard unit for frequency f in the SI system is the **hertz** (abbreviation: Hz), where 1 Hz $\equiv$ 1 cycle/s, whereas the standard SI units for phase rate ω are simply s^{-1} (which we can also think of as being radians per second).

Note that an oscillation's frequency f and/or period T is determined by the spring constant k_s and the object's mass m ($\omega = \sqrt{k_s/m}$), but the oscillation's amplitude A and initial phase θ are determined by the oscillator's initial state at time $t = 0$ (as we will shortly see).

Exercise N10X.2

Show by direct substitution that $x(t) = A\cos(\omega t + \theta)$ does indeed solve the harmonic oscillator equation (equation N10.4).

N10.3   The Oscillator as a Model

The simple harmonic oscillator model has many applications

The simple harmonic oscillator model is not only useful for describing the behavior of objects connected to springs, but also for an astounding range of physical systems (from ocean waves to electrical oscillations to atomic vibrations), making it one of the most useful models in all physics.

We have already seen part of the reason why this model is so useful in section C9.5. There, I argued that whenever the graph of an interaction's potential energy function $V(r)$ has a valley, we can approximate the valley's bottom by the parabolic harmonic oscillator potential energy function. In this section, I want to present a different way of saying the same thing.

Characteristics of an oscillating system

The most important characteristics of *any* oscillating object are as follows: (1) It has a position or configuration (called its **equilibrium position**) where the force on it is zero; (2) if it is displaced from that position, it experiences a force (called a **restoring force**) that pushes it *back toward* the equilibrium position; and (3) this force (at least for small displacements) grows in magnitude as the object's displacement increases. Any object satisfying these criteria will oscillate about its equilibrium position if it is displaced and then released.

Clearly, an object attached to a spring has these characteristics. The special characteristic of a simple harmonic oscillator is that the magnitude of the restoring force is strictly proportional to the distance that the object is displaced. Many oscillating systems do *not* share this characteristic. For example, the force on an atom in a solid is a complicated function of its position as a result of the complicated electrostatic interactions between the atom and its neighbors.

For small displacements, $F_x = -k_s x$ in many cases

Even so, calculus tells us that near $x = 0$, we can approximate the curve of almost any physically reasonable (that is, differentiable) function $F_x(x)$ that is positive for $x < 0$ and negative for $x > 0$ by a straight line $F_x = -k_s x$, where k_s is the value of $-dF_x/dx$ evaluated at $x = 0$. This is a good approximation when x is sufficiently small in magnitude, as shown in figure N10.3.

Thus, an object responding to almost *any* restoring force function $F_x(x)$ will find that for small displacements $F_x \approx -k_s x$. In this *small oscillation limit*, therefore, almost any oscillating object behaves as a simple harmonic oscillator. This is why the harmonic oscillator model is so useful and important.

N10.4   A Mass Hanging from a Spring

An easy way to construct a practical simple harmonic oscillator is to simply hang an object vertically at the end of a spring. This is much easier than constructing the system illustrated in figure N10.1 (frictionless surfaces are hard to come by). But is a hanging system really the same as that in figure N10.1?

Figure N10.3

Almost *any* function $F_x(x)$ that is negative for $x > 0$ and positive for $x < 0$ can be approximated for small |x| by a straight line $-k_s x$ for an appropriately chosen value of k_s.

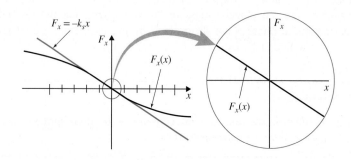

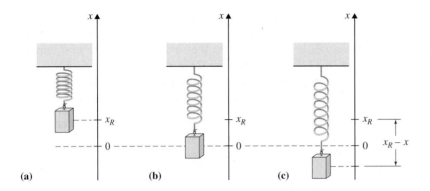

Figure N10.4
A system in which an object hangs from a spring. (a) The object located at the position x_R where the spring is relaxed. (b) The position of the object when the spring tension exactly balances the object's weight. This position is defined to be $x = 0$. (c) Here the spring is stretched farther. The distance by which the spring is stretched when the object is at x is $x_R - x$.

The complication here is that the net force on the object is a sum of the vertical component of the spring force and the force of gravity. If we set up our coordinate system so the x axis is along the direction of oscillation (vertical in this case) and set $x = 0$ to be the object's position when the spring is relaxed, then the net x-force on the object is

The net force in this case includes gravity

$$F_{net,x} = -k_s x - m|\vec{g}| \qquad (N10.10)$$

This is *not* the same as the harmonic oscillator force law (equation N10.2).

But it turns out that this force law really does work out to be the simple harmonic oscillator force law if we choose our coordinate system cleverly. Let's choose a coordinate system whose x axis is vertical (and let's choose the $+x$ direction to be upward), but let's define $x = 0$ not to be the position of the hanging object when the spring is relaxed, but rather to be its position at *equilibrium*, where the spring tension force exactly balances the object's weight. (The spring will be somewhat stretched at this position!) Let x_R be the position of the object when the spring is relaxed: the distance s by which the spring is stretched when the object is at any given position x is then

We can get rid of this term by redefining the origin

$$s = x_R - x \quad \text{(assuming that } x < x_R) \qquad (N10.11)$$

(see figure N10.4). Note that since the $+x$ direction is upward, as x becomes more positive, the object is moving *up*, and the distance that the spring is stretched will *decrease*: this is why x is *subtracted* in equation N10.11.

Since the spring exerts an upward force proportional in magnitude to s, the net force acting on the object at position x in this reference frame is

$$F_{net,x} = +k_s s - m|\vec{g}| = +k_s(x_R - x) - m|\vec{g}| = k_s x_R - k_s x - m|\vec{g}| \qquad (N10.12)$$

Note that $k_s s$ is *positive* here because the force is upward (in the $+x$ direction) when the spring is stretched by s. (As a check, note that this formula implies that as the object gets lower, x decreases and the upward force exerted by the spring increases, as we would expect.) At $x = 0$, the net force on the movable object is supposed to be zero, so equation N10.12 tells us that at $x = 0$,

$$0 = F_{net,x} = k_s(x_R - 0) - m|\vec{g}| \quad \Rightarrow \quad k_s x_R = m|\vec{g}| \qquad (N10.13)$$

If we substitute this into equation N10.12, the two constant terms cancel, so

$$F_{net,x} = -k_s x \qquad (N10.14)$$

the right force law for a simple harmonic oscillator. Choosing the right origin thus enables us to cancel the gravitational force with a constant term in the spring tension force making the resulting net force directly proportional to x.

So a hanging mass behaves as a harmonic oscillator

The point of all this is that an object hanging from a spring, even though it is not physically the same thing as a mass oscillating horizontally back and forth on a frictionless surface, will *behave* in exactly the same manner. The gravitational force here simply has the effect of displacing the equilibrium point ($x = 0$ in the reference frame we've been using) downward from the position where the spring is relaxed. The hanging oscillator even oscillates at the same frequency it would if it were horizontal!

Example N10.1

Problem: You see a couple of neighborhood kids bouncing on the hood of your car. When one jumps on the hood, you see the car's front end oscillate with a period of about 1.5 s. After you yell at the kids, you naturally get to wondering about the spring constant k_s of the car's suspension. Estimate k_s from what you saw.

Solution A typical car's weight $\approx$ 3000 lb, corresponding to a mass of roughly 1500 kg. Let's say that the front suspension effectively suspends about one-half of this mass, or about 800 kg when the roughly 40-kg mass of the child is included. According to equation N10.8, the period of oscillation is $T = 2\pi\sqrt{m/k_s}$, so

$$k_s = \frac{4\pi^2(800 \text{ kg})}{(1.5 \text{ s})^2}\left(\frac{1 \text{ N}}{1 \text{ kg·m/s}^2}\right) = 14{,}000 \text{ } \frac{\text{N}}{\text{m}} \tag{N10.15}$$

Here is a way to check this estimate. If you were to sit on the hood (assume your weight is about 700 N $\approx$ 155 lb), the hood should sink until the compressed springs exert an upward force equal to your weight, that is, by about $|\Delta x| = \Delta F/k_s = (700 \text{ N})/(14{,}000 \text{ N/m}) \approx 0.05$ m = 5 cm. This seems entirely credible.

N10.5   An Analogy to Circular Motion

An oscillating object's x component behaves like that of an object on a rotating disk

According to the discussion in section N10.4, equation N10.6 should accurately describe the oscillatory motion of an object hanging from a spring. Note that the motion described by this equation is mathematically equivalent to the x coordinate of an object attached to a rotating disk (see figure N10.5). This provides a very useful way to visualize the meaning of the oscillation's amplitude A and initial phase θ: the amplitude corresponds to the radius of the circular motion, and θ to the disk's angle at time $t = 0$ relative to its position when the object's x-position is maximum. If we imagine the disk to be rotating counterclockwise, a negative phase angle θ means the disk is oriented clockwise of this maximal position at $t = 0$, while a positive phase angle means it is oriented counterclockwise of this position.

Finding A and θ from initial conditions with the help of this analogy

So, for example, suppose we know that an object hanging from a spring is moving upward through the equilibrium point $x = 0$ at time $t = 0$. If we look at figure N10.5, we can see that this will only happen if the corresponding object on the disk is 90° clockwise of its maximum upward position, because only then will that object be moving *upward* through $x = 0$. Therefore, the initial phase θ for the oscillation must be 90° clockwise or $-\pi/2$. Similarly, we can immediately see that

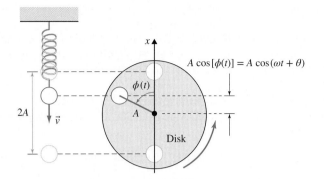

Figure N10.5
The vertical motion of a hanging oscillator is mathematically identical to the projection on the x axis of the motion of an object in uniform circular motion with angular speed ω. Such a diagram helps us visualize the meaning of the initial phase θ.

if the object is passing through its lowest point at $t = 0$, the initial phase must be $\pm 180°$ or $\theta = \pm\pi$ (either sign is acceptable in this case).

Examples N10.2 and N10.3 illustrate how one can calculate both the oscillation's amplitude A and its initial phase θ from initial conditions. Even when the initial phase is *not* a multiple of $\pi/2$ (see example N10.3), the rotating disk analogy in figure N10.5 can help us determine the sign of the initial phase and check its value.

Problem: An object of mass 0.52 kg hangs from a spring with spring constant $k_s = 130$ N/m. The object is measured to pass through its equilibrium point with a speed of 1.0 m/s. Define the time when it passes the equilibrium point going downward to be $t = 0$. **(a)** What is initial phase θ? **(b)** What is the amplitude of its oscillation?

Example N10.2

Solution **(a)** If we study figure N10.5, we can see that the object will pass downward through the origin only when the corresponding object on the disk is 90° counterclockwise of its maximum upward position. Since this is the situation at time $t = 0$, we must have $\theta = +\pi/2$.

(b) We can calculate the object's velocity at any time t by calculating the time derivative of equation N10.6:

$$v_x(t) = \frac{dx}{dt} = \frac{d}{dt} A \cos(\omega t + \theta) = -A\omega \sin(\omega t + \theta) \qquad (N10.16)$$

At time $t = 0$, this becomes

$$v_x(0) = -A\omega \sin\theta \quad \Rightarrow \quad A = \frac{-v_x(0)}{\omega \sin\theta} = \frac{-v_x(0)}{\sin\theta} \sqrt{\frac{m}{k_s}} \qquad (N10.17)$$

where in the last step, I used $\omega = \sqrt{k_s/m}$. Since we know that $\theta = \pi/2$ from part (a), and all other quantities are given, we can calculate A:

$$A = \frac{-(-1.0 \text{ m/s})}{\sin\frac{1}{2}\pi} \sqrt{\frac{0.52 \text{ kg}}{130 \text{ N/m}} \left(\frac{1 \text{ N}}{1 \text{ kg·m/s}^2}\right)} = 0.062 \text{ m} \qquad (N10.18)$$

This is positive (as it should be) and has the right units.

Example N10.3

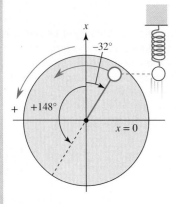

Figure N10.6

The rotating disk analogy shows that an object that at $t = 0$ is in the $+x$ direction with respect to the origin and is moving in the $+x$ direction must have a phase angle θ of $-32°$, not $+148°$.

Problem: Suppose an object hanging from a spring oscillates with a period of $T = 2.0$ s. At $t = 0$, the object is 5 cm above the equilibrium point and is moving upward at 10 cm/s. What are A and θ for this oscillation?

Solution The equation $x(t) = A\cos(\omega t + \theta)$ gives us the object's position as a function of time, while equation N10.16 gives us its velocity. At time $t = 0$, these equations reduce to

$$x(0) = A\cos\theta \quad \text{and} \quad v_x(0) = -A\omega\sin\theta \quad \text{(N10.19)}$$

We are given T, so equation N10.8 allows us to calculate ω: $\omega = 2\pi/T$. Since we are also given $x(0)$ and $v_x(0)$, we have two equations in our two unknowns A and θ, so we have enough information to solve the problem. We can eliminate the unknown A by dividing the second of equations N10.19 by the first:

$$-\frac{v_x(0)}{\omega x(0)} = \frac{\cancel{A}\sin\theta}{\cancel{A}\cos\theta} = \tan\theta \quad \Rightarrow \quad \theta = \tan^{-1}\left[\frac{-v_x(0)}{\omega x(0)}\right] \quad \text{(N10.20)}$$

Substituting in the numbers, we get

$$\tan\theta = \frac{-v_x(0)}{\omega x(0)} = \frac{T}{2\pi}\left[\frac{-v_x(0)}{x(0)}\right] = \frac{-(2.0 \cancel{s})(10 \cancel{cm/s})}{2\pi(5 \cancel{cm})} = -\frac{2}{\pi} \quad \text{(N10.21)}$$

So $\theta = \tan^{-1}(-2/\pi) = -32°$ or $+148°$ [because $\tan(\theta + \pi) = \tan\theta$, either one of these results would satisfy the equation]. However, we can see from the rotating disk analogy in figure N10.6 that if the object is above the equilibrium point and moving upward at $t = 0$, only the first solution makes any sense. We can substitute this value for θ back into either of equations N10.19 to solve for A:

$$A = \frac{x(0)}{\cos\theta} = \frac{5.0 \text{ cm}}{\cos(-32°)} = +5.9 \text{ cm} \quad \text{(N10.22)}$$

The amplitude is positive, as it should be, has the right units, and is plausibly comparable to the value of $x(0)$.

Note that equation N10.20, in combination with figure N10.5, provides a *general* way to compute θ from initial conditions when $x(0) \neq 0$.

Exercise N10X.3

A 1.0-kg object hangs from a spring whose spring constant is 100 N/m. You take the mass, pull it down 10 cm, and then give it an initial downward speed of 0.50 m/s. What are the values of A and θ here?

N10.6   The Simple Pendulum

A *simple pendulum* is an example of a system that has nothing to do with a mass connected to a spring, and yet (for small oscillations) it obeys the same mathematics as a simple harmonic oscillator does. A **simple pendulum** consists of an object (called a **bob**) that swings back and forth at the end of a massless and inextensible string of length L tied to a fixed point (as shown in figure N10.7).

The bob is confined to a circular path by the string, so this is an example of *nonuniform circular motion*. According to the results of section N7.3, the bob's acceleration at any instant in this situation is

$$\vec{a}(t) = \frac{d|\vec{v}|}{dt}\hat{v} - \frac{|\vec{v}|^2}{L}\hat{r} \tag{N10.23}$$

where $|\vec{v}|$ is the bob's speed, $\hat{v}$ is a unit vector in the direction of its velocity, and $\hat{r}$ points directly away from the center of the object's circular motion.

Since the bob's velocity changes direction at the extreme points, it is awkward to use $\hat{v}$ as a unit vector in this case. Instead, let's define the unit vector $\hat{\phi}$ as the direction in which the bob is moving when it moves counterclockwise (the direction in which ϕ increases). Note that $\hat{\phi} = \pm\hat{v}$, where the plus sign applies when the bob moves counterclockwise and the minus sign when it moves clockwise. Now, if the bob's angle changes by $d\phi$ during a short interval dt, the bob travels an arc length $L|d\phi|$ during dt, so its speed is $|\vec{v}| = L|d\phi/dt|$, implying that

$$\frac{d|\vec{v}|}{dt} = L\frac{d}{dt}\left|\frac{d\phi}{dt}\right| = +L\frac{d}{dt}\left(\pm\frac{d\phi}{dt}\right) = \pm L\frac{d^2\phi}{dt^2} \tag{N10.24}$$

where the plus sign applies if $d\phi/dt$ is positive, which is again when the bob is moving counterclockwise. Combining this with $\hat{\phi} = \pm\hat{v}$, we see that in *both* cases, equation N10.23 becomes

$$\vec{a}(t) = L\frac{d^2\phi}{dt^2}\hat{\phi} - \frac{|\vec{v}|^2}{L}\hat{r} \tag{N10.25}$$

Now, Newton's second law tells us that $\vec{a} = \vec{F}_{net}/m$, where $\vec{F}_{net}$ is the net force on the bob and m is the bob's mass. The bob is acted on by two forces, a gravitational force due to the earth and the tension force exerted by the string. The latter always acts in the $-\hat{r}$ direction, so only the gravitational force can possibly have a component in the $\hat{\phi}$ direction. When the string makes an angle ϕ with the vertical, the component of $\vec{F}_g$ in that direction is $F_{g,\phi} = -m|\vec{g}|\sin\phi$, so

$$L\frac{d^2\phi}{dt^2} = \frac{F_{g,\phi}}{m} = -|\vec{g}|\sin\phi \quad \Rightarrow \quad \frac{d^2\phi}{dt^2} = -\frac{|\vec{g}|}{L}\sin\phi \tag{N10.26}$$

The pendulum equation

Now, for small ϕ in radians (less than a few tenths of a radian), $\sin\phi \approx \phi$. (Try this on a calculator and see for yourself!) Using this, we get

$$\frac{d^2\phi}{dt^2} = -\frac{|\vec{g}|}{L}\phi \qquad \text{for small angles} \tag{N10.27}$$

Now, compare this equation to the SHO equation $d^2x/dt^2 = -\omega^2 x$ (equation N10.4). We see that equation N10.27 is mathematically identical to the SHO equation, with ϕ here playing the role of x in the SHO equation and $\sqrt{|\vec{g}|/L}$ here playing the role of ω in the SHO equation. We don't need to solve this differential equation yet again: all we need to do is use the answers we found before, substituting ϕ for x and $\sqrt{|\vec{g}|/L}$ for ω. So the angle of the pendulum bob depends on time as follows (by analogy to equation N10.6):

The low-angle limit is equal to the oscillator equation

$$\phi(t) \approx A\cos(\omega t + \theta) \qquad \text{for small angles} \tag{N10.28}$$

where $\omega = \sqrt{|\vec{g}|/L}$ here, A expresses the absolute value of the angle at the extreme points, and θ is the initial phase. The oscillation period is

So the solution must be analogous as well

$$T = \frac{2\pi}{\omega} = 2\pi\sqrt{\frac{L}{|\vec{g}|}} \tag{N10.29}$$

independent of the (small) swing angle. This is an interesting and almost counterintuitive result!

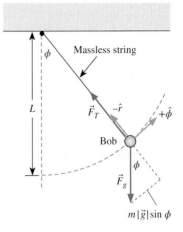

Figure N10.7
The simple pendulum. Note that ϕ is conventionally taken to increase in the counterclockwise direction.

TWO-MINUTE PROBLEMS

N10T.1   If you double the amplitude of a harmonic oscillator, the oscillator's period
A. Decreases by a factor of 2.
B. Decreases by a factor of $\sqrt{2}$.
C. Does not change.
D. Increases by a factor of $\sqrt{2}$.
E. Increases by a factor of 2.
F. Changes by some other factor (specify).

N10T.2   If you double the amplitude of a harmonic oscillator, the object's maximum speed does what? (Use the answers for problem N10T.1)

N10T.3   If you double the spring constant of a harmonic oscillator, the oscillation frequency does what? (Use the answers for problem N10T.1.)

N10T.4   A glider on an air track is connected by a spring to the end of the air track. If it takes 0.30 s for the glider to travel the distance of 12 cm from one turning point to the other, its amplitude is
A. 12 cm
B. 6 cm
C. 24 cm
D. 36 cm
E. 3.6 cm
F. We are not given enough information to answer.

N10T.5   Consider the glider described in problem N10T.4. Its phase rate is
A. $0.30 \ \text{s}^{-1}$
B. $0.15 \ \text{s}^{-1}$
C. $0.60 \ \text{s}^{-1}$
D. $3.77 \ \text{s}^{-1}$
E. $0.096 \ \text{s}^{-1}$
F. Some other result (specify)

N10T.6   Suppose the x-position of an oscillating object is given by $x(t) = A \cos(\omega t + \theta)$. What is the object's *speed* at time $t = 0$?
A. A
B. Zero
C. $|A \cos \theta|$
D. $|A \sin \theta|$
E. $|A \omega \sin \theta|$
F. Something else (specify)
T. We are not given enough information to answer.

N10T.7   A glider on an air track is connected by a spring to the end of the air track. If it is pulled 3.5 cm in the $+x$ direction away from its equilibrium point and then released from rest at $t = 0$, what is the initial phase θ?
A. 0
B. $\pi/4$
C. $\pi/2$
D. π (more choices $\longrightarrow$)

E. $3\pi/2$
F. Some other result (specify)

N10T.8   A glider on an air track is connected by a spring to the end of the air track. If it is pulled 3.5 cm in the $-x$ direction away from its equilibrium point and then released from rest at $t = 0$, what is the initial phase θ?
A. 0
B. $\pi/4$
C. $\pi/2$
D. π
E. $3\pi/2$
F. Some other result (specify)

N10T.9   A mass hanging from the end of a spring has a phase rate of $\omega = 6.3 \ \text{s}^{-1} \approx 1$ cycle/s. Let's define $t = 0$ to be when the mass passes $x = 0$ going up. If its speed as it passes is 1.0 m/s, what is its amplitude A?
A. 0
B. 0.16 m
C. 1.0 m
D. 6.3 m
E. We are not given enough information to answer.
F. Some other result (specify)

N10T.10   Suppose we know the value of ω for a simple harmonic oscillator. What further information do we need to know to fix the amplitude A and the initial phase θ in the solution to the simple harmonic oscillator equation? (If more than one item is needed, feel free to indicate more than one answer.)
A. The oscillating object's initial position
B. The oscillating object's initial velocity
C. The object's mass
D. The oscillation's period
E. The spring's spring constant
F. We do not need additional information.

N10T.11   To double the period of a pendulum, you need to multiply its length by a factor of:
A. $\frac{1}{2}$
B. 2
C. $\sqrt{\frac{1}{2}}$
D. $\sqrt{2}$
E. 4
F. Some other result (specify)

N10T.12   For a swinging pendulum, the *maximum* angle that the pendulum's string makes with the vertical corresponds to what in the pendulum solution $\phi = A \cos(\omega t + \theta)$?
A. A
B. ϕ
C. θ
D. Something else (specify)

HOMEWORK PROBLEMS

Basic Skills

N10B.1 An oscillating object repeats its motion every 3.3 s.
(a) What is the period of this oscillation?
(b) What is its frequency?
(c) What is its phase rate?

N10B.2 An object of mass 0.30 kg hanging from a spring is observed to have an oscillation frequency of 2.2 Hz. What is the spring's spring constant k_s?

N10B.3 An object of mass 0.36 kg hanging at the end of a spring oscillates with an amplitude of 4.8 cm and a frequency of 1.2 Hz. What is the spring's spring constant k_s?

N10B.4 A magnesium atom (mass $\approx$ 24 proton masses) in a crystal is measured to oscillate with a frequency of roughly 10^{13} Hz. What is the effective spring constant of the forces holding that atom in the crystal?

N10B.5 An object of mass 0.30 kg hanging from a spring is pulled 2.5 cm below its equilibrium position and then is released from rest. What is θ for this oscillation in equation N10.6? Explain your reasoning.

N10B.6 A pendulum is observed to swing with a period of 2.0 s. How long is it?

N10B.7 What will be the natural oscillation period of a 30-kg child on a swing whose seat is 3.2 m below the bar where the chains from the seat are attached?

N10B.8 Suppose a pendulum hangs vertically at rest. We pull the bob to the right so that the string makes an angle of 15° with respect to the vertical and then release the bob from rest. It swings with a period of exactly 1.0 s. What are the values of A, θ, and L in this case?

Modeling

N10M.1 An object of mass 0.25 kg extends a spring by a distance of 5.0 cm when it hangs from the spring at rest. If it is then set in vertical motion, what will be its period of oscillation?

N10M.2 An object of mass 0.30 kg hanging from a spring is lifted 2.5 cm above its equilibrium position and is dropped from rest. What is the amplitude of the subsequent oscillation? Explain your reasoning.

N10M.3 Any real spring has mass. Do you think that this mass would make the actual period of a real harmonic oscillator longer or shorter than the period predicted by equation N10.8? Explain your reasoning.

N10M.4 An object of mass 1.0 kg hanging from the end of a spring (whose spring constant is 120 N/m) is observed to pass downward through position $x = 2.5$ cm traveling at a speed of 0.60 m/s.
(a) What is the amplitude of oscillation?
(b) What is the initial phase θ for this oscillation? Please explain your reasoning.

N10M.5 An object of mass 0.60 kg hangs from the end of a spring. Imagine that the object is lifted upward and held at rest at the position where the spring is not stretched. The object is then released. It is observed that the lowest point in the object's subsequent oscillation is 12 cm below the point where it was released.
(a) What is the amplitude of the oscillation?
(b) What is the spring constant of the spring?
(c) What is the object's maximum speed?
Explain your reasoning and show your work for each step.

N10M.6 When a 55-kg friend of yours sits on a trampoline, your friend sinks about 45 cm below the trampoline's normal level surface. If your friend were to bounce gently on the trampoline (never leaving its surface), what would be your friend's period of oscillation? (*Hint:* Model the trampoline as if it were a harmonic oscillator. Do you think this will be a good model?)

N10M.7 If you stand on a pogo stick, you note that its spring-loaded foot is pushed in about 18 cm. If you bounce *gently* up and down on it (so that the foot never leaves the ground), what is your approximate oscillation frequency?

N10M.8 A certain bungee cord has a relaxed length of 20 m. When Kelly hangs from the cord, it becomes 22.45 m long. What would Kelly's oscillation period be after being set into a vertical oscillation of amplitude 1.0 m while hanging from this cord?

N10M.9 Do you think that a pendulum swinging through a large angle will have a longer or shorter period than the period predicted by equation N10.29? Explain your reasoning carefully. (Possibly helpful hints: consider extreme cases, or compare $\sin\phi$ to ϕ.)

N10M.10 The net x-force on a 1.0-kg oscillating object moving along the x axis is $F_x = -b(x - a^2x^3)$, where $b = 120$ and $a = 5$ in appropriate SI units.
(a) What are the units of a and b?
(b) Show that for small oscillations, this force law reduces to the harmonic oscillator law $F_x \approx -k_s x$ where k_s depends on the value of a and/or b.
(c) What is the frequency of such small oscillations?
(d) How small is "small"? For what range of x will $F_x = -k_s x$ to within 1%?

N10M.11   Suppose the x-force on a 1.0-kg oscillating object in a certain situation is given by $F_x = -a \sin(bx)$, where $a = 100$ and $b = 10$ in appropriate SI units.

(a) What are the SI units of a and b?

(b) Show that for small x, this force law reduces to $F_x \approx -k_s x$, where k_s can be calculated from a and b.

(c) Find the frequency of small oscillations.

N10M.12   Suppose we have a block of mass m connected to an ideal spring (with an unknown spring constant), which is connected to an ideal string whose other end is connected to a fixed point on the ceiling. Let L be the object's distance below the ceiling when the spring is relaxed, and let ΔL be the additional amount that the block stretches the string when the block hangs at rest. Define $z = 0$ to be the block's vertical position when it hangs at rest. Now imagine that we drop the block from rest at the ceiling. Our goal is to find the block's lowest position. The parts will guide you through the process. Define $t = 0$ to be the instant that the string begins to stretch.

(a) Draw diagrams of the situation, one showing the object hanging at rest $z = 0$ and how that position is related to L and ΔL, and one showing the situation at $t = 0$.

(b) Use conservation of energy to determine the block's z-velocity $v_{0z}(0)$ at time $t = 0$.

(c) Describe the time interval during which the simple harmonic oscillator (SHO) model describes the block's motion. When does it start? When does it stop?

(d) During the time that the SHO model applies, the block's vertical position is $z(t) = A \cos(\omega t + \theta)$. Explain how the block's lowest position is related to A.

(e) Determine $\omega^2 = k_s/m$ from the distance that the block stretches the string when the block is at rest.

(f) We know that at time $t = 0$, $z(0) = A \cos\theta$ and $v_{0z} = -A\omega \sin\theta$. Explain why.

(g) Since we know $z(0)$, v_{0z}, and ω, we can solve these two equations for the two unknowns A and θ. Solve for A. [*Hint:* Start by noting that $A^2 = (A \cos\theta)^2 + (A \sin\theta)^2$.]

(h) So what is the block's lowest position?

Derivation

N10D.1   Consider an object moving along the x axis under the influence of a force whose x component is $F_x = a[(b - x)^2 - b^2]$, where a and b are positive constants. Note that this force component is negative when $b > x > 0$ and positive when $x < 0$, so it qualifies as a restoring force. Argue that for small $|x|$, this formula becomes $F_x = -k_s x$, and find k_s in terms of a and/or b. (*Hint:* Multiply out the square and drop a term that becomes small when $x \ll b$.)

N10D.2   Yet another way to find the amplitude from initial conditions is to use conservation of energy.

(a) Consider an oscillating object of mass m sliding on a level, frictionless surface. The object is connected to a spring whose other end is connected to a fixed point. Explain why we can treat the spring-object system as an isolated system. (*Hint:* What work do external forces on the system do?)

(b) Argue that the velocity of an oscillating object at either of its extreme positions must be zero.

(c) Argue that the system's conserved total energy E is given by $E = \frac{1}{2} k_s A^2$, where k_s is the spring's spring constant and A is the oscillation amplitude, as long as we assume that $x = 0$ is the equilibrium point.

(d) Argue, therefore, that at $t = 0$, conservation of energy and $k_s/m \equiv \omega^2$ imply that

$$A^2 = \left[\frac{v_x(0)}{\omega}\right]^2 + [x(0)]^2 \qquad (N10.30)$$

If we know $v_x(0)$ and $x(0)$, we can find A this way. We can then use $x(0) = A \cos\theta$ to determine θ.

N10D.3   We can actually determine the motion of a harmonic oscillator from *conservation of energy* instead of solving the harmonic oscillator equation. Here's how.

(a) Show that a bit of manipulation of the conservation of energy formula for the harmonic oscillator yields

$$A^2 = x^2(t) + \left[\frac{v_x(t)}{\omega}\right]^2 \qquad (N10.31)$$

where $A = \sqrt{2E/k_s}$, E is the system's total energy, and $\omega = \sqrt{k_s/m}$ as usual. (*Hint:* See problem N10D.2.)

(b) Argue that this means that at any instant of time, we can find an angle ψ such that

$$x(t) = A \cos\psi \quad \text{and} \quad v_x(t) = -\omega A \sin\psi \qquad (N10.32)$$

Thus, these expressions give x and v_x for *all* times, with ψ depending in some unknown way on time. [*Hint:* $|x|$ will always be less than or equal to A, so we can *define* $x(t) = A \cos\psi$. Then, solve equation N10.31 for the other term.]

(c) Use the definition $v_x = dx/dt$ and the chain rule to show that the most general possible expression for ψ is

$$\psi = \omega t + \theta \qquad (N10.33)$$

where θ is some constant. Therefore

$$x(t) = A \cos\psi = A \cos(\omega t + \theta) \quad !! \qquad (N10.34)$$

Rich-Context

N10R.1   How do you measure the mass of an astronaut in orbit? (You can't just use a scale!) For the 1973–1979 Skylab program (Skylab was a space station that pre-dated the current International Space Station), NASA engineers designed a Body Mass Measuring Device (BMMD): see figure N10.8. This is essentially a chair of mass m mounted on a spring with a carefully measured spring constant $k_s = 605.6$ N/m. (The other end of the spring is connected to the Skylab itself, which has a mass much larger than the astronaut or the chair, and so remains essentially fixed as the astronaut oscillates.) The period of oscillation of the empty chair is measured to be 0.90149 s. When an astronaut is sitting in the chair, the period is 2.12151 s. What is the astronaut's mass? Please describe your reasoning! (Adapted from Halliday and Resnick, *Fundamentals of Physics,* 3/e, New York: Wiley, 1988, p. 324.)

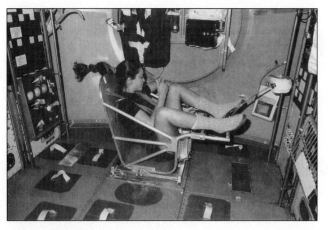

Figure N10.8
Skylab astronaut Tamara Jernigan uses the Body Mass Measuring Device (BMMD). (Credit: NASA)

N10R.2 Known mobster and gambler Larry the Loser is found dead, hanging at about the 7-story level from a bungee cord tied to the top of the 15-story (45-m) Prudential Building in downtown Chicago. Larry has a mass of 90 kg, and when the bungee cord is cut down, it is found to have a relaxed length of 5 stories. Detective Lestrade from the Chicago Police Department thinks that Larry must have been bungee-jumping from the top of the building and just unluckily hit the ground, but *you* know that

Larry was *murdered*. What is your evidence? (*Notes:* A bungee cord will not exert any force on the jumper until the jumper has fallen a distance equal to the bungee cord's relaxed length. Also, the force a bungee cord exerts actually exceeds $k_s|x|$ in magnitude when it is stretched a lot.)

Advanced

N10A.1 Imagine that the force law expressing the interaction between a hydrogen atom and a chlorine atom in an HCl molecule is approximately

$$F_x = -a\left[\left(\frac{b}{r}\right)^2 - \left(\frac{c}{r}\right)^3\right] \qquad (N10.35)$$

where F_x is the x-force on the hydrogen atom, a is a constant having units of force and b and c are constants having units of distance, and r is the distance measured from the chlorine atom (which is so massive compared to the hydrogen atom that we will consider it fixed).
 (a) What is the equilibrium position r_0 for the hydrogen atom? (Express your answer in terms of b and c.)
 (b) Define $x = r - r_0$ and show that for small x this force law becomes approximately $F_x = -k_s x$. (One approach is to find the first few terms of a Taylor series expansion of the exact force law. Alternatively, one can use the approximation $(1 + q)^n \approx 1 + nq$ if $q \ll 1$.)
 (c) What is the frequency of small oscillations of the hydrogen atom in terms of its mass m, and a, b, and c?

ANSWERS TO EXERCISES

N10X.1 The appropriate units for k_s in equation N10.3 are N/m. Since 1 J = 1 N·m, this is the same as J/m^2.

N10X.2 If we define $u = \omega t + \theta$, then $x(t) = A \cos u$, and

$$\frac{d}{dt}A\cos u = -A\sin u\frac{du}{dt} = -\omega A\sin u \qquad (N10.36)$$

$$\text{since} \quad \frac{du}{dt} = \frac{d}{dt}(\omega t + \theta) = \omega + 0 = \omega \qquad (N10.37)$$

Taking the time derivative again, we get

$$\frac{d^2x}{dt^2} = -\omega A\frac{d}{dt}\sin u = -\omega^2 A\cos u = -\omega^2 x \qquad (N10.38)$$

So this equation does indeed solve the SHO equation.

N10X.3 Using equation N10.20, we get

$$\theta = \tan^{-1}\left[\frac{-v_x(0)}{\omega x(0)}\right] = \tan^{-1}\left[\frac{-(-0.5 \text{ m/s})}{(10/\text{s})(-0.1 \text{ m})}\right]$$

$$= \tan^{-1}(-0.5) = -26.6° = -0.46 \text{ radians} \qquad (N10.39)$$

As a fraction of an oscillation, this result corresponds to shifting the cosine wave by $26.6°/360° = 0.074 = 7.4\%$ of a complete oscillation. But the result for the phase shift θ could *also* be $180° - 26.6° = 153.4°$ (43% of an oscillation), since $\tan\theta = \tan(\theta + \pi)$. It is in fact the latter value that gives the correct positive value for the amplitude:

$$A = \frac{x(0)}{\cos\theta} = \frac{-0.1 \text{ m}}{\cos(153.4°)} = +0.11 \text{ m} \qquad (N10.40)$$

Kepler's Laws

Chapter Overview

Introduction

In this chapter and the next, we will explore how to use the gravitational force law discussed in chapter N2 to predict the motions of the planets and other celestial objects. We will see that Newton's laws work just as well for celestial physics as terrestrial physics, thus providing a *universal* model for (macroscopic) mechanics.

Section N11.1: Kepler's Laws

Johannes Kepler used Tycho Brahe's collection of extremely accurate naked-eye observations of the planets to develop three empirical laws of planetary motion, which we call **Kepler's laws of planetary motion**:

1. All planets move in ellipses, with the sun at one focus.
2. A line from the sun to the planet sweeps out equal areas in equal times.
3. If a is the planet's **semimajor axis** (one-half the ellipse's greatest width), then $T^2 \propto a^3$ where T is the planet's **period**.

Newton's triumph was to offer an *explanation* of these empirical laws, using a theory that also applied to terrestrial physics.

Section N11.2: Orbits Around a Massive Primary

Consider an isolated (or freely falling) system of two interacting objects with masses M and m. A reference frame connected to the system's center of mass (CM) will be inertial. If in addition we have $M \gg m$, we call the massive object the **primary** and the other its **satellite**, and the following simplifications apply in the CM frame:

1. The primary is essentially at rest at the origin.
2. The primary's kinetic energy is negligible.
3. The objects' angular momenta are parallel and are each conserved.
4. The primary's angular momentum is negligible.
5. The objects' separation is essentially equal to the satellite's distance from the origin.

$M \gg m$ is a good approximation for almost all pairs of objects in the solar system.

Section N11.3: Kepler's Second Law

Conservation of angular momentum implies (1) that the orbit of either object around the system's center of mass must lie in a *plane* perpendicular to the fixed angular momentum vector and (2) Kepler's second law. This section provides a detailed argument supporting both assertions.

Section N11.4: Circular Orbits and Kepler's Third Law

From now on, we will assume that $M \gg m$ unless otherwise specified. In this approximation, the primary basically provides a fixed gravitational field to which the satellite responds. **Newton's law of universal gravitation** states that the magnitude of the gravitational force that the primary exerts on the satellite is

$$|\vec{F}_g| = \frac{GMm}{r^2} \qquad \text{(N11.9)}$$

- **Purpose:** This equation specifies the magnitude of the gravitational force $\vec{F}_g$ that an object of mass M exerts on an object of mass m (or vice versa) when their centers of mass are separated by a distance r. $G = 6.67 \times 10^{-11}$ N·m²/kg² is the **universal gravitational constant**.
- **Limitations:** This equation applies to point masses or spheres, but not to irregularly shaped objects. Also, the equation does not apply to *extremely* strong gravitational fields (*much* stronger than any fields in our solar system) or to objects moving at close to the speed of light.

Newton's second law, the law of universal gravitation, and the expression for an object's acceleration in circular motion imply that for a satellite in a circular orbit

$$|\vec{v}| = \sqrt{\frac{GM}{R}} \qquad \text{and} \qquad T^2 = \frac{4\pi^2}{GM}R^3 \qquad \text{(N11.11, N11.13)}$$

- **Purpose:** These equations specify the orbital speed $|\vec{v}|$ and **period** T of a satellite in a circular orbit of radius R around a primary with mass M where G is the universal gravitational constant.
- **Limitations:** The orbit must be circular, and the primary must be much more massive than the satellite. The limitations on equation N11.9 also apply here.
- **Note:** The second equation is Kepler's third law for circular orbits.

Section N11.5: Circular Orbit Problems

Circular orbit problems are very much like the constrained-motion problems in chapter N7, except that we use equation N11.11 or N11.13 or the *magnitude* of Newton's second law as the master equation. In the conceptual model section, you really only need to (1) describe the interacting objects, (2) check that one is much more massive than the other, and (3) check that the satellite's orbit is essentially circular.

Section N11.6: Black Holes and Dark Matter

This section discusses how astrophysicists have used Kepler's third law to show that black holes exist and to discover the existence of dark matter.

Section N11.7: Kepler's First Law and Conic Sections

In this section, we use the Newton application to show that Newton's second law and the law of universal gravitation predict that a satellite's orbit is a **conic section** (a circle, ellipse, parabola, or hyperbola) described by

$$r(\theta) = \frac{b}{1 + \varepsilon\cos\theta}, \quad \text{with} \quad a \equiv \frac{b}{|1 - \varepsilon^2|} \qquad \text{(N11.23)}$$

- **Purpose:** This equation describes a *conic section* by specifying the distance r the curve is from the origin (the **focus**) at a given angle θ measured from the line that connects the origin to the curve's nearest point. The constant ε is the curve's **eccentricity**, b (= r at $\theta = 90°$) is a constant, and a is the semimajor axis.
- **Limitations:** None: this is a definition.
- **Notes:** The point of the orbit closest to the origin is $r_c = b/(1 + \varepsilon) = a|1 - \varepsilon|$, and for an ellipse, the farthest point is $r_f = b/(1 - \varepsilon) = a(1 + \varepsilon)$. For a parabola and hyperbola, r goes to infinity at angle $\theta_\infty = \cos^{-1}(-1/\varepsilon)$.

N11.1   Kepler's Laws

In the year 1600, Johannes Kepler came to Prague to join the research staff at an observatory operated by Tycho Brahe, an astronomer who was both the official "imperial mathematician" of the Holy Roman Empire and a friend of Galileo. The following year, Brahe died, and Kepler succeeded him as imperial mathematician and as director of the observatory. His new position gave him complete access to Tycho Brahe's extraordinary collection of careful astronomical observations of the planets, the result of a lifetime of work by perhaps the greatest naked-eye astronomer who ever lived. (The telescope was not invented until about 1608.)

In 1609, Kepler published his *Astronomia Nova* ("New Astronomy") in which he stated two empirical laws that he induced from Brahe's planetary observations. In modern language, these laws state that

1. The orbits of the planets are *ellipses*, with the sun at one focus.
2. The line from the sun to a planet sweeps out *equal areas in equal times*.

(I'll define the *focus* of an ellipse later.) Kepler offered no theoretical explanation for these laws, he simply presented them as being *descriptive* of planetary orbits (according to Brahe's observational data). These laws represented a rather radical departure from the accepted wisdom of the time: up to then, most astronomers assumed that the motions of the planets could be described in terms of combinations of uniform circular motions (although it was becoming clear that such schemes had to be extraordinarily complex to fit the best observational data available).

Ten years later, in his book *Harmonice Mundi* ("Harmonics of the World"), Kepler stated a third empirical law:

3. The square of a planet's **period** T (the time that it takes to complete one orbit) is proportional to the cube of the orbit's **semimajor axis** a.

(An ellipse's *semimajor axis* is defined to be one-half its widest diameter.) The three numbered laws above are known as **Kepler's three laws of planetary motion**. Figure N11.1 illustrates these laws.

Again let me emphasize that these laws are entirely *empirical*: they do not so much explain as *describe* the motion of the planets. However, because Kepler supported them so carefully with observational data of extraordinary quality, these laws became widely accepted in spite of their radical character.

For more than six decades after Kepler's third law was published, the scientific community could not say *why* these laws were true. Isaac Newton's amazing triumph was to show that each of these laws follows directly from his second law of motion and the assumption that the force of gravity between two objects depends on the inverse square of their separation. In other words, Newton offered an *explanation* of Kepler's laws in terms of physical principles that applied equally well to terrestrial motion: one simply had to accept that the planets were endlessly falling around the sun.

Let me emphasize how radical *this* suggestion was at the time! Before Newton, scholars had believed that the laws of physics pertaining to the motion of heavenly bodies (where unceasing motion in approximate circles seemed to be the rule) were completely distinct from the laws pertaining to terrestrial motion (where objects generally come to rest rather quickly). It took Newton's genius not only to see that this very credible division between celestial and terrestrial physics was in fact not necessary at all (which, granted, was *beginning* to occur to others as well), but also to provide a complete theoretical perspective that unified terrestrial and celestial physics in a manner that demonstrably

Kepler's laws

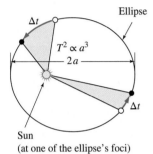

Figure N11.1
Kepler's laws illustrated. The colored regions show the area swept out by a line connecting the planet and the sun during equal time intervals Δt during different parts of the orbit: their areas are equal by Kepler's second law.

Kepler's laws are empirical

Newton offered an explanation of these laws

worked. The simplicity, beauty, and extraordinary predictive power of Newton's ideas were so compelling that it brought the physics community to its first real consensus on a grand theoretical structure for physics. This first consensus (as discussed in chapter C1) in some sense marks the birth of physics as a scientific discipline.

Our goal in this chapter is to show that Kepler's laws are a consequence of the laws of mechanics that we have been studying. We will not quite follow the same path Newton did in proving this, since computers and the law of conservation of angular momentum give us more powerful tools than Newton had at his disposal. (Using these tools will enable us to do in a few pages what it took Newton scores of pages to show in the *Principia*.)

Our goal is to prove that Kepler's laws follow from principles we have studied

N11.2 Orbits Around a Massive Primary

The first step toward understanding what Newton's laws say about planetary motion is to take advantage of the simplifications that result when (1) our system of interest consists of a very massive object interacting with a much lighter object and (2) we choose the origin of our reference frame to be the system's center of mass.

Consider an isolated system consisting of an object of mass M interacting with a smaller object of mass m (see figure N11.2). If the system is *really* isolated, then its center of mass will move at a constant velocity and thus can be used as the origin of an inertial reference frame. In practical situations, it actually is more likely that the system is *freely falling* in some external gravitational field (for example, the earth and moon falling around the sun), but we can *still* treat the system's center of mass as the origin of an inertial reference frame if we ignore the external gravitational field (as we saw in chapter N8).

General situation: an isolated pair of interacting objects

If we define the origin to be the system's center of mass, then the definition of the center of mass means that

Results in a frame based on the system's CM

$$0 = \frac{m\vec{r} + M\vec{R}}{M + m} \quad \Rightarrow \quad M\vec{R} = -m\vec{r} \quad \Rightarrow \quad \vec{R} = -\frac{m}{M}\vec{r} \qquad \text{(N11.1)}$$

This implies that in *all* circumstances (no matter how the two objects move and/or interact with each other) the positions $\vec{R}$ and $\vec{r}$ of the objects relative to the center of mass are opposite and have magnitudes proportional to each other: if r gets bigger or smaller, then so does R (proportionally).

Taking the time derivative of both sides of equation N11.1 yields

$$\vec{V} = -\frac{m}{M}\vec{v} \quad \Rightarrow \quad |\vec{V}| = \frac{m}{M}|\vec{v}| \qquad \text{(N11.2)}$$

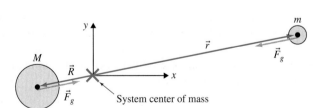

Figure N11.2

An isolated system of two objects interacting gravitationally. Note that if we define the origin of our reference frame to be attached to the system's center of mass, the positions of the objects are opposite. The gravitational force exerted on each object by the interaction points directly toward the other.

The fact that $|\vec{V}| = (m/M)|\vec{v}|$ means that

$$\frac{1}{2}M|\vec{V}|^2 = \frac{1}{2}M\left(\frac{m}{M}|\vec{v}|\right)^2 = \left(\frac{m}{M}\right)\left(\frac{1}{2}m|\vec{v}|^2\right) \quad \Rightarrow \quad K_M = \frac{m}{M}K_m \qquad (\text{N11.3})$$

where K_M and K_m are the kinetic energies of the massive and light objects, respectively. Thus (as we've seen before), the more massive object's kinetic energy in this frame is smaller than that of the lighter object by the factor m/M.

Chapter C7 argued that a particle's angular momentum $\vec{L}$ about a given origin is given by the "cross product" $\vec{L} = \vec{r} \times m\vec{v}$, where $\vec{r}$ is the particle's position relative to the origin, m is its mass, and $\vec{v}$ is its velocity. If you did not read this chapter, all you need to know for our present purposes is that this cross product produces a *vector* whose direction is perpendicular to both $\vec{r}$ and $m\vec{v}$ and whose magnitude is $|\vec{L}| = |\vec{r}||m\vec{v}||\sin\theta|$ (where θ is the angle between $\vec{r}$ and $\vec{v}$).

Given this, we can express the larger object's angular momentum $\vec{L}_M$ around the system's center of mass in terms of $\vec{L}_m$ as follows:

$$\vec{L}_M \equiv \vec{R} \times M\vec{V} = \left(-\frac{m}{M}\right)\vec{r} \times M\left(-\frac{m}{M}\right)\vec{v} = +\frac{m}{M}(\vec{r} \times m\vec{v}) = \frac{m}{M}\vec{L}_m \qquad (\text{N11.4})$$

The system's total angular momentum $\vec{L} = \vec{L}_m + \vec{L}_M = (1 + m/M)\vec{L}_m$ around its center of mass will be conserved *if and only if* the light object's angular momentum is conserved around that center of mass.

Implications when $M \gg m$

Equations N11.1 through N11.4 apply to *any* isolated (or freely falling) system of two interacting objects described in a reference frame whose origin is the system's center of mass. If, in addition, we have $M \gg m$, then

1. The massive object's position is $\vec{R} \approx 0$ (by equation N11.1).
2. This object is essentially at rest at the origin: $\vec{V} \approx 0$ (by N11.2).
3. Its kinetic energy is negligible: $K_M \approx 0$ (by N11.3).
4. Its angular momentum is negligible: $\vec{L}_M \approx 0$ (by N11.4).
5. The objects' separation $\approx$ distance of the lighter object from the origin.

In such a case, the massive object (which we call the system's **primary** under these circumstances) is essentially at rest at the origin, and the lighter object (which we call a **satellite**) orbits it. The primary then provides an essentially fixed origin for the gravitational force exerted on the satellite.

This approximation is very useful in the solar system

This approximation holds very well in the solar system. Even Jupiter's mass is more than 1000 times smaller than the sun's mass, and the earth's mass is more like 330,000 times smaller. Similarly, the moons that orbit the major planets typically have masses much smaller than their primary: even our own moon (the largest major-planet moon in the solar system compared to its primary) has 81 times less mass than the earth.

The fact that planetary masses are so small compared to the sun has another important implication. When computing the orbit of one planet, we can ignore the gravitational effects of the others (to an excellent degree of approximation). This is because the sun is so much more massive than anything else that the gravitational force it exerts is by far the greatest influence on each planet's motion. Therefore, as an excellent approximation, we can pretend that each planet orbits the sun as if it were alone.

N11.3   Kepler's Second Law

Equation N11.4 and $\vec{L} = \vec{L}_M + \vec{L}_m$ imply that the angular momentum of *either* object in an isolated interacting pair is conserved along with that of the whole. This has two important consequences: (1) the object's orbit lies in a plane, and (2) its position vector sweeps out equal areas in equal times.

The first of these consequences follows from the definition of the angular momentum, which says that $\vec{L}$ for *either* object is defined to be $\vec{L} \equiv \vec{r} \times m\vec{v}$. By definition of the cross product, this means that $\vec{L}$ is always perpendicular to $\vec{r}$. But if $\vec{L}$ has a fixed orientation in space (because it is conserved), then the object's position vector $\vec{r}$ must always lie in the fixed plane perpendicular to $\vec{L}$. Therefore, the object's orbit lies in a certain fixed plane.

The orbit of either object must be in a fixed plane in space

Conservation of angular momentum also implies Kepler's second law as follows. Imagine that in an infinitesimal time dt, the object moves a certain infinitesimal angle $d\phi$ (shown greatly exaggerated in figure N11.3) as it moves in its orbit around the system's center of mass. The area swept out by the line between the object and the system's center of mass is the colored pie slice in figure N11.3. If $d\phi$ is very small, the shape of the slice is very nearly triangular, so its area is $dA \approx \frac{1}{2}(\text{base})(\text{height})$. Now, as shown in the drawing, the height h of the triangle is very nearly equal to the arc length $r\,d\phi$, and this approximation gets better and better as $d\phi \to 0$. So if $d\phi$ is small,

The position vector of either object sweeps out equal areas in equal times

$$dA \approx \tfrac{1}{2}r(r\,d\phi) = \tfrac{1}{2}r^2\,d\phi \qquad (\text{N11.5})$$

Dividing both sides by dt and taking the limit as dt and $d\phi$ go to zero yields

$$\frac{dA}{dt} = \frac{1}{2}r^2\frac{d\phi}{dt} \qquad (\text{N11.6})$$

Now note from the figure that $(\vec{v}\,dt)\sin\theta = h \approx r\,d\phi$. The magnitude of the object's angular momentum is thus

$$|\vec{L}| = |\vec{r}\,||m\vec{v}\,||\sin\theta| = rm\frac{r\,d\phi}{dt} = mr^2\frac{d\phi}{dt} \qquad (\text{N11.7})$$

This means that

$$\frac{dA}{dt} = \frac{1}{2}r^2\frac{d\phi}{dt} = \frac{1}{2m}\left(mr^2\frac{d\phi}{dt}\right) = \frac{|\vec{L}|}{2m} \qquad (\text{N11.8})$$

Since $|\vec{L}|$ is conserved, dA/dt is constant, meaning that *the object's radius vector sweeps out equal areas in equal times*. Note that this applies to *either* object, independent of the objects' relative masses or the nature of their interaction.

Now, Kepler's second law actually says that the line between the planet and the *sun* sweeps out equal areas in equal times, not the line between the planet and the system's *center of mass*. But if the sun is much more massive than the planet, it essentially *is* located at the system's center of mass, and Kepler's second law is essentially correct.

Application to Kepler's second law

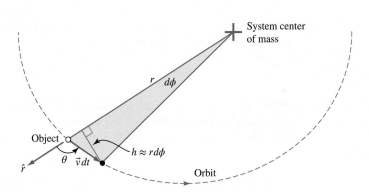

Figure N11.3

The colored region shows the area swept out by the line between the object and the system's center of mass as the object moves a tiny angle $d\phi$ around the center of mass (the object's initial position is the white dot and its final position is the black dot). This area is almost equal to $\frac{1}{2}rh$, and h in turn is approximately equal to the arclength $r\,d\phi$.

N11.4   Circular Orbits and Kepler's Third Law

We will assume $M \gg m$ from now to the end of the unit

From here through the next chapter, we will (unless otherwise specified) assume the primary-satellite approximation, in which the mass of the primary is much greater than the mass of the satellite. To simplify notation, when we refer to the satellite's kinetic energy and/or angular momentum from now on, we will drop the subscripts on K_m and $\vec{L}_m$ and simply use K and $\vec{L}$.

Circular orbits at constant speed are possible

If the primary is much more massive than its satellite, then it is possible for the satellite to follow an essentially circular orbit at constant speed *around the primary*. Let us see whether such an orbit is consistent with what we know about uniform circular motion and the gravitational interaction.

We know that an object moving in a circular trajectory at a constant speed is accelerating toward the center of its circular path, and that the magnitude of its acceleration is $|\vec{a}| = |\vec{v}|^2/R = $ a *constant,* where R is the radius of the object's orbit and $|\vec{v}|$ is its constant orbital speed. Newton's second law tells us that this acceleration must be caused by some force that is directed toward the center of the satellite orbit and which has a *constant* magnitude.

The gravitational force $\vec{F}_g$ exerted on the satellite by its gravitational interaction with its primary satisfies these criteria. It is directed toward the center of the satellite's trajectory (to the extent that we can consider the massive object to be at rest). As we saw in chapter N2, its magnitude is given by Newton's **law of universal gravitation**:

$$|\vec{F}_g| = \frac{GMm}{r^2} \qquad \text{(N11.9)}$$

- **Purpose:** This equation specifies the magnitude of the gravitational force $\vec{F}_g$ that an object of mass M exerts on an object of mass m (or vice versa) when their centers of mass are separated by a distance r. $G = 6.67 \times 10^{-11}$ N·m^2/kg^2 is the **universal gravitational constant**.
- **Limitations:** This equation applies to point masses or spheres, but not to irregularly shaped objects. Also it does not apply to *extremely* strong gravitational fields (*much* stronger than any fields in our solar system) or to objects moving at close to the speed of light.

(We derived this in chapter N2 from the gravitational potential energy formula.) Therefore, in the case of a truly circular orbit (where $r = R = $ constant) the magnitude of the force will be GMm/R^2, which is a constant, as needed.

This means that it is at least *plausible* that the gravitational force exerted on the satellite due to its interaction with the primary can hold the satellite in a circular orbit. This does not mean that orbits *always* are circular: indeed, orbits generally are *not.* But it does mean that a circular orbit is a possibility. In fact, the orbital radii of most major objects in the solar system are constant to within a few percent. Most artificial satellites orbit the earth with approximately circular orbits as well. Therefore, out of all the kinds of possible orbits, this "special case" is a very good approximate model for a broad variety of realistic situations.

Implications of Newton's second law for circular orbits

Let's see what Newton's second law can tell us quantitatively about such orbits. If we assume that the gravitational force is the *only* force acting on the satellite, then $|\vec{F}_{net}| = |\vec{F}_g| = GMm/R^2$. Therefore, the magnitude of Newton's second law tells us that

$$m|\vec{a}| = |\vec{F}_{net}| = \frac{GMm}{R^2} \qquad \text{(N11.10)}$$

Dividing both sides by m and substituting $|\vec{v}^2|/R$ for $|\vec{a}|$ yields

$$\frac{|\vec{v}|^2}{R} = \frac{GM}{R^2} \quad \Rightarrow \quad |\vec{v}| = \sqrt{\frac{GM}{R}} \qquad \text{(N11.11)}$$

- **Purpose:** This equation specifies the orbital speed $|\vec{v}|$ of a satellite in a circular orbit of radius R around a primary with mass M, where G is the universal gravitational constant.
- **Limitations:** The orbit must be circular, and the primary must be much more massive than the satellite. The limitations on equation N11.9 also apply here.

(Note that $|\vec{v}|$ is appropriately constant.)

We can use this information to determine how long it will take the satellite to go once around its circular orbit. The orbit's period T in this case is the time it takes the satellite to travel a distance $2\pi R$ at a constant speed $|\vec{v}|$:

$$T = \frac{2\pi R}{|\vec{v}|} \qquad \text{(N11.12)}$$

If we square equation N11.12 and use equation N11.11, we can show that

$$T^2 = \frac{4\pi^2}{GM}R^3 \qquad \text{(N11.13)}$$

- **Purpose:** This equation specifies the orbital period T of a satellite in a circular orbit of radius R around a primary with mass M, where G is the universal gravitational constant.
- **Limitations:** The orbit must be circular, and the primary must be much more massive than the satellite. The limitations on equation N11.9 also apply here.

Exercise N11X.1

Verify equation N11.13.

Kepler's third law states that *the square of a planet's period is proportional to the cube of its semimajor axis*. When an orbit is circular, its semimajor axis (one-half of the distance measured across the widest part of the orbit) *is* its radius R. So we see here that in a few lines we have *derived* Kepler's third law (for the special case of circular orbits anyway). Moreover, this derivation even gives us the constant of proportionality between T^2 and R^3!

Kepler's third law for circular orbits

Exercise N11X.2

The earth orbits the sun in an approximately circular orbit once each year. If it were 4 times as far from the sun, how long would it take to orbit once?

N11.5   Circular Orbit Problems

In this section, we'll explore some examples of how equations N11.11 and N11.13 can answer many questions about orbiting objects and their primaries. Since these examples involve circularly "constrained" motion (although only because most orbits *happen* to be roughly circular, not because they *must* be), we can adapt the checklist we used for constrained-motion problems. But we need some modifications. For example, we don't need to write Newton's laws in component form (equation N11.11 or N11.13 will probably be more helpful), so we don't need to define coordinate axes. A free-body diagram of an orbiting object will generally show only one force vector, so it is pretty pointless. The only important issues to be addressed in the modeling part are the validity of the circular orbit approximation and the assumption that the primary is indeed much more massive than the satellite.

A checklist for solving circular orbit problems

So an adapted problem-solving checklist for orbital motion problems might look as follows:

Circular Orbit Problem Checklist
- ☐ Draw the two interacting objects.
- ☐ Draw a labeled arrow for the orbiting object's velocity vector.
- ☐ Check that one object is very massive.
- ☐ Check that the satellite's orbit is circular (or state this as an assumption).
- ☐ Apply equation N11.11 and/or N11.13 as your master equation(s).

General Checklist
- ☐ Define symbols.
- ☐ Draw a picture.
- ☐ Describe your model and assumptions.
- ☐ List knowns and unknowns.
- ☐ Use a master equation.
- ☐ Do algebra symbolically.
- ☐ Track units.
- ☐ Check your result.

The examples that follow illustrate solutions following this outline. You should check for yourself that all of the checklist elements are present in these example solutions.

Example N11.1

Problem: Consider a spaceship in a circular orbit at an altitude of 250 km above the earth's surface. What is its orbital speed? Its orbital period?

Solution

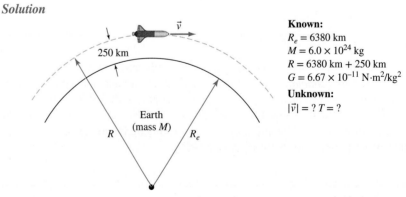

Known:
$R_e = 6380$ km
$M = 6.0 \times 10^{24}$ kg
$R = 6380$ km $+ 250$ km
$G = 6.67 \times 10^{-11}$ N·m²/kg²

Unknown:
$|\vec{v}| = ?\ T = ?$

According to the problem statement, the ship is following a circular orbit around the earth. The earth is much more massive than a spaceship, so equations N11.11 and N11.13 should apply.

Equation N11.11 tells us that a circularly orbiting object's speed is

$$|\vec{v}| = \sqrt{\frac{GM}{R}} = \sqrt{\frac{(6.67 \times 10^{-11} \,\cancel{N} \cdot m^2/\cancel{kg^2})(6.0 \times 10^{24} \,\cancel{kg})}{6{,}380{,}000 \,\cancel{m} + 250{,}000 \,\cancel{m}}\left(\frac{1 \,\cancel{kg} \cdot m/s^2}{1 \,\cancel{N}}\right)}$$

$$= 7770 \text{ m/s} = 7.8 \text{ km/s} \qquad\qquad\qquad \text{(N11.14)}$$

If we know the speed, equation N11.12 more quickly yields the period than equation N11.13 does: the orbit's period is

$$T = \frac{2\pi R}{|\vec{v}|} = \frac{2\pi(6380 \,\cancel{km} + 250 \,\cancel{km})}{7.77 \,\cancel{km}/\cancel{s}}\left(\frac{1 \text{ min}}{60 \,\cancel{s}}\right) = 89.4 \text{ min} \qquad \text{(N11.15)}$$

This is a bit longer than the result we found in example N7.5. But this is plausible: since the ship orbits some distance above the earth's surface, not only is gravity a bit weaker than at the earth's surface but the distance around the orbit is a bit longer. Both effects will make an orbit a bit longer than one just barely above the earth's surface.

Problem: Ganymede, Jupiter's largest moon, has a nearly circular orbit whose radius is 1.07 Gm. The moon goes around Jupiter once every 7 d, 3 h, and 43 min. What is Jupiter's mass?

Solution Here is a drawing of the situation:

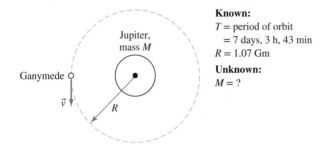

Known:
T = period of orbit
  = 7 days, 3 h, 43 min
R = 1.07 Gm

Unknown:
$M = ?$

We are told that Ganymede's orbit is circular, and we can safely assume that Ganymede is much less massive than Jupiter. Since we are given R and T, we can solve equation N11.13 for the mass M of Jupiter:

$$T^2 = \frac{4\pi^2}{GM}R^3 \quad \Rightarrow \quad M = \frac{4\pi^2 R^3}{GT^2} \qquad\qquad \text{(N11.16)}$$

The period T, expressed in seconds, is

$$T = 7 \,\cancel{d}\left(\frac{24 \,\cancel{h}}{1 \,\cancel{d}}\right)\left(\frac{3600 \text{ s}}{1 \,\cancel{h}}\right) + 3 \,\cancel{h}\left(\frac{3600 s}{1 \,\cancel{h}}\right) + 43 \,\cancel{min}\left(\frac{60 \text{ s}}{1 \,\cancel{min}}\right) = 6.18 \times 10^5 \text{ s} \qquad \text{(N11.17)}$$

Substituting this into equation N11.16 yields

$$M = \frac{4\pi^2(1.07 \times 10^9 \,\cancel{m})^3}{(6.67 \times 10^{-11} \,\cancel{N} \cdot \cancel{m^2}/kg^2)(6.18 \times 10^5 \,\cancel{s})^2}\left(\frac{1 \,\cancel{N}}{1 \text{ kg} \cdot \cancel{m}/\cancel{s^2}}\right) = 1.9 \times 10^{27} \text{ kg} \qquad \text{(N11.18)}$$

This seems plausible. Note that we can determine an object's mass by observing the motion of a smaller object orbiting it (this is indeed the standard method of determining the masses of astronomical objects).

Example N11.3

Problem: Astronomical measurements show that Jupiter orbits the sun in a nearly circular orbit once every 11.86 years. How does the radius of Jupiter's orbit compare with that of the earth?

Solution

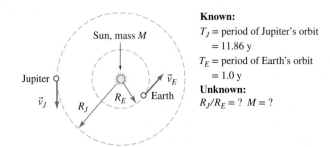

Known:
T_J = period of Jupiter's orbit
     = 11.86 y
T_E = period of Earth's orbit
     = 1.0 y
Unknown:
R_J/R_E = ? M = ?

Jupiter and earth both orbit the sun, which is much more massive than either planet. We are ignoring the comparatively tiny interaction between earth and Jupiter. To do this problem, we have to *assume* that both the earth's orbit and Jupiter's orbit are circular (which is pretty closely true). If this is so, then equation N11.13 applies to both orbits, so

$$T_J^2 = \frac{4\pi^2}{GM}R_J^3 \quad \text{and} \quad T_E^2 = \frac{4\pi^2}{GM}R_E^3 \tag{N11.19}$$

Since we are really looking for the ratio R_J/R_E, the fastest way to find that ratio is to divide the first expression by the second:

$$\frac{T_J^2}{T_E^2} = \frac{(4\pi^2/GM)R_J^3}{(4\pi^2/GM)R_E^3} \quad \Rightarrow \quad \frac{R_J}{R_E} = \left(\frac{T_J}{T_E}\right)^{2/3} = \left(\frac{11.86\,\cancel{y}}{1.0\,\cancel{y}}\right)^{2/3} = 5.20 \tag{N11.20}$$

This is consistent with the result stated in the inside front cover.

N11.6   Black Holes and Dark Matter

In 1994, H. C. Ford and R. J. Harms announced that they had discovered a giant black hole at the center of the galaxy known as M87. This was the first time anyone had presented an apparently iron-clad argument for the existence of a giant black hole at the center of any galaxy.

A black hole is an object so dense that light cannot escape it. So how can we hope to "discover" an object that doesn't emit any light? If you can find something orbiting the black hole and you can measure the orbit's period and radius, Kepler's third law allows you to compute the black hole's mass!

Using Kepler's third law to find a black hole

Ford and Harms used the Hubble Telescope to take spectra from very small regions of a gas cloud near the center of M87. By measuring the Doppler shifts of the spectra obtained, they were able to determine that gas that was a radius R = 60 light-years (where 1 ly = 0.946×10^{16} m) from the center of the cloud was orbiting the center with a speed of 450 km/s. If we assume that the orbit is roughly circular, $T = 2\pi R/|\vec{v}|$. Plugging the latter equation into Newton's version of Kepler's third law and solving for M, we get

$$\frac{4\pi R^3}{GM} = T^2 = \left(\frac{2\pi R}{|\vec{v}|}\right)^2 \quad \Rightarrow \quad M = \frac{4\pi R^3|\vec{v}|^2}{G(2\pi R)^2} = \frac{R|\vec{v}|^2}{G} \tag{N11.21}$$

Substituting in $R = 60(0.946 \times 10^{16}$ m) and $|\vec{v}| = 450$ km/s yields 1.7×10^{39} kg, which is about 10^9 solar masses (1 solar mass $= 2.0 \times 10^{30}$ kg). This is just one of many bits of evidence that Ford and Harms cite.

How do we know that this is a black hole? No other explanation works! A cluster containing a billion stars within a radius of 60 ly would emit lots of light that is not seen. No plausible model *other* than the black hole model explains how a billion solar masses could fit within a sphere 60 ly in radius and yet be consistent with all the available data collected by Ford and Harms.

For several decades, astronomers also have been collecting evidence that more than 80% of the mass associated with matter in the universe is *dark matter*, that is, matter that is *not* in stars that emit light or in dust clouds that emit radio signals, reflect light, or obstruct light. Consistent evidence for this dark matter comes from a variety of studies of the orbital motion of stars within galaxies, satellite galaxies around large galaxies, and galaxies in galactic clusters. Many of these studies use Newton's version of Kepler's third law to determine the mass of the unseen matter by its gravitational effects.

Dark matter and Kepler's third law

In one study, D.N.C. Lin, B. F. Jones, and A. R. Klemona carefully measured the transverse movement of the *Large Magellanic Cloud* (LMC) relative to background objects. The LMC is a small galaxy that is a companion to and presumably orbits our own much larger galaxy. Lin and his collaborators measured the transverse velocity of the LMC to be about 200 km/s. Since the LMC is about 170,000 ly $(1.6 \times 10^{21}$ m) from our galaxy, if it were in a circular orbit, that would imply that our galaxy's mass is

$$M = \frac{R|\vec{v}|^2}{G} = \frac{(1.6 \times 10^{21} \text{ m})(200,000 \text{ m/s})^2}{6.67 \times 10^{-11} \text{ N·m}^2/\text{kg}^2} \left(\frac{1 \text{ N}}{1 \text{ kg·m/s}^2} \right) = 1.0 \times 10^{42} \text{ kg} \tag{N11.22}$$

which is about 500 billion solar masses (a more accurate calculation based on the LMC's real trajectory actually yields 600 billion solar masses). *Visible* matter in our galaxy amounts to about 100 billion solar masses.

What is this dark matter? No one knows! Recent experiments have strongly suggested it *cannot* be dwarf stars, large planets, or black holes; mounting evidence suggests it cannot be ordinary matter made of protons, neutrons, and electrons at all. The mystery is still unsolved, but part of the point here is that we would not even know this major fraction of the universe existed if it weren't for Newton's form of Kepler's third law!

N11.7 Kepler's First Law and Conic Sections

Kepler's first law states that planetary orbits are *ellipses*. An **ellipse** is one of a family of curves that mathematicians call **conic sections** (the curves that result from slicing a cone through various angles); the family also includes circles, **parabolas**, and **hyperbolas**. The general formula for a conic section is

Conic sections

$$r = \frac{b}{1 + \varepsilon \cos\theta} \tag{N11.23}$$

- **Purpose:** This equation describes a *conic section* by specifying the distance r the curve is from the origin (the **focus**) at a given angle θ measured from the line that connects the origin to the curve's nearest point. The constant ε is the curve's **eccentricity** and b ($= r$ at $90°$) is a constant.
- **Limitations:** None: this is a definition.

Figure N11.4

The features of an ellipse. We measure the value of r for any angle θ from the origin, which we call a *focus* of the ellipse. The constant b is the value of r at $\theta = \pi/2$, $r_c = r(0)$ is the distance from the focus to the closest point on the ellipse, and $r_f = r(\pi)$ is the distance to the farthest point. The *major axis* is the ellipse's greatest width: half of that width is a, the semimajor axis. (The ellipse technically has *two* foci located symmetrically along the ellipse's major axis.)

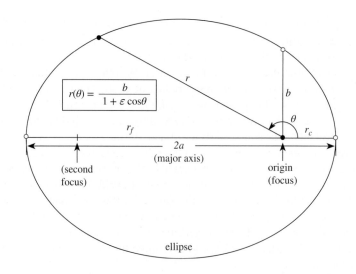

$$r(\theta) = \frac{b}{1 + \varepsilon \cos\theta}$$

2a
(major axis)

(second focus)

origin (focus)

ellipse

The eccentricity ε specifies the curve's *shape*. If $\varepsilon = 0$, then the expression reduces to $r = b$, and the curve is a circle of radius $R = b$. If ε is nonzero but less than 1, then the curve is an ellipse, whose closest point to the origin (focus) is $r_c = r(0) = b/(1 + \varepsilon)$ and whose farthest point from the origin is $r_f = r(\pi) = b/(1 - \varepsilon)$: see figure N11.4. If $\varepsilon = 1$, then the farthest point is at infinity, and the curve becomes a parabola. If $\varepsilon > 1$, then the curve is a hyperbola. Only ellipses have a finite farthest point, but *all* conic sections have a closest point. So we can calculate a given conic section's eccentricity ε as follows:

$$r_c = \frac{b}{1 + \varepsilon} \quad \Rightarrow \quad \varepsilon = \frac{b}{r_c} - 1 \tag{N11.24}$$

An ellipse's semimajor axis a

I'll talk more about hyperbolas shortly, but let's focus for the moment on ellipses. We define an ellipse's semimajor axis a to be half of its widest diameter. As figure N11.4 illustrates, the widest diameter is $r_c + r_f$, so

$$a = \frac{1}{2}(r_f + r_c) = \frac{1}{2}\left(\frac{b}{1 + \varepsilon} + \frac{b}{1 - \varepsilon}\right) = \frac{b}{2}\left(\frac{1 - \varepsilon + 1 + \varepsilon}{1 - \varepsilon^2}\right) = \frac{b}{1 - \varepsilon^2} \tag{N11.25}$$

This means that $b = a(1 - \varepsilon^2)$ and therefore that

$$r_c = \frac{b}{1 + \varepsilon} = \frac{a(1 - \varepsilon^2)}{1 + \varepsilon} = a\frac{(1 + \varepsilon)(1 - \varepsilon)}{1 + \varepsilon} = a(1 - \varepsilon) \tag{N11.26}$$

and similarly $r_f = a(1 + \varepsilon)$. These results will be useful in what follows.

Using the Newton app to "prove" Kepler's first law

Newton's second law and the law of universal gravitation together imply that if a satellite's orbital velocity at a given distance r from its primary is in the range $\sqrt{GM/r} < |\vec{v}| < \sqrt{2GM/r}$ (where M is the primary's mass), then its orbit will be an *ellipse* (at the lower limit, the orbit is circular and at the upper limit, it is parabolic). Proving this *mathematically* is not easy, but we can use the Newton app to show this is true.

The gravitational force acting on the satellite is directed toward the primary and has a magnitude equal to $|\vec{F}_g| = GMm/r^2$, where m is the satellite's mass. Therefore, the satellite's acceleration is directed toward the center with a magnitude $|\vec{a}| = |\vec{F}_g|/m = GM/r^2$. We can set up the Newton application to model such an acceleration by typing the numerical value of $-GM$ followed by "/r^2" in the setup edit field for an acceleration in the "radially outward direction" (we need the value of $-GM$ as opposed to GM because the acceleration is actually radially *inward*, not outward). We also need to specify the satellite's initial position and velocity.

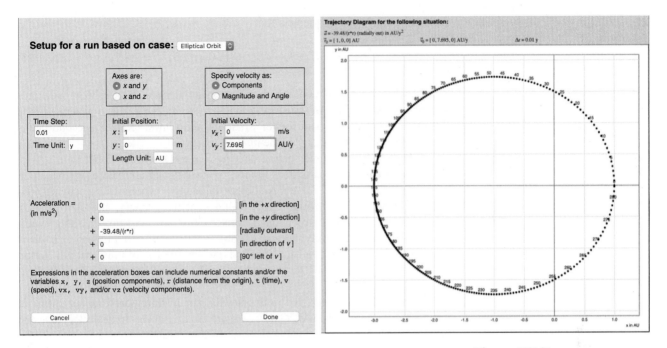

Figure N11.5
The Newton setup and resulting
trajectory for a hypothetical asteroid in our solar system. (Credit:
Courtesy of Thomas Moore)

One can use SI units for these quantities, but for objects in the solar system, it is much more convenient to measure distances in **astronomical units** (where 1 AU is the mean distance between the earth and the sun), and times in years. Kepler's third law allows us to determine the value of GM in these units: since the earth's orbit is nearly circular

$$T^2 = \frac{4\pi^2}{GM}R^3 \quad \Rightarrow \quad GM = 4\pi^2 \frac{R^3}{T^2} = 4\pi^2 \frac{(1 \text{ AU})^3}{(1 \text{ y})^2} = 39.48 \frac{\text{AU}^3}{\text{y}^2} \qquad \text{(N11.27)}$$

This value of GM applies to *any* object that orbits the sun. We can also conveniently describe the object's initial position in AU and its speed in AU/y (since the orbiting earth covers a distance 2π AU during the year, its orbital speed is 6.28 AU/y). Choosing the time step to be smaller than about 1/50 of the object's expected orbital period generally yields good accuracy.

Figure N11.5*a* shows the Newton setup for an asteroid whose initial position is $r_0 = r_c = 1$ AU from the sun but whose perpendicular velocity is 7.695 AU/y (too fast to be in a circular orbit at that radius). Note that choosing the initial velocity to be perpendicular to the object's initial position vector ensures that the initial position is either the *closest* point in the trajectory or the *farthest*, since only at these extreme points (where r is minimum or maximum) is $dr/dt = 0$, meaning there is no component of the velocity in the radial direction. If we also choose the speed to be *larger* than the circular orbit speed at this point, r_0 will be the minimum value of r, that is, r_c.

The resulting orbit (figure N11.5*b*) has $r_c = 1.0$ AU and $b = 1.5$ AU (the value of r when the orbit crosses the vertical axis). According to equation N11.24, $\varepsilon = 1.5 - 1 = 0.5$. If the trajectory really is an ellipse, then according to equation N11.25, we should have $a = R/(1 - 0.5^2) = (1.5 \text{ AU})/(0.75) = 2.0$ AU, and $r_f = a(1 + \varepsilon) = (2.0 \text{ AU})(1.5) = 3.0$ AU. These results agree with the plotted trajectory, where you can see that $r_f = 3.0$ AU and the orbit is 4.0 AU across at the widest point. Indeed, if you calculate r at various arbitrary values of θ for the given values of b and ε, you will find that the asteroid's plotted trajectory matches what the formula $r = b/(1 + \varepsilon \cos\theta)$ predicts at *all* angles.

Moreover, the plot shows that the orbit takes 283 steps of 0.01 y to complete, which corresponds to a period $T = 2.83$ y. Kepler's third law says that $T^2 \propto a^3$ for

"Proving" Kepler's third law for elliptical orbits

a given primary, and so predicts that

$$\frac{T_a^2}{T_e^2} = \frac{a_a^3}{a_e^3} \quad \Rightarrow \quad T_a = T_e\left(\frac{a_a}{a_e}\right)^{3/2} = (1.0 \text{ y})(2.0)^{3/2} = 2.83 \text{ y} \qquad \text{(N11.28)}$$

exactly what we see. Therefore, the Newton application tells us that Newton's second law and the law of universal gravitation do indeed seem to generate orbits that are consistent with both Kepler's first and third laws! (Kepler's second law is also obeyed because angular momentum is conserved.)

Hyperbolic orbits

A *hyperbola* is a conic section obeying $r = b/(1 + \varepsilon\cos\theta)$ but with $\varepsilon > 1$. The value of r in such a case goes to infinity when $\varepsilon\cos\theta$ approaches -1, meaning that θ approaches $\pm\cos^{-1}(-1/\varepsilon)$. An object following a hyperbolic trajectory would therefore come in from infinity, pass within r_c of the primary, and then go back out to infinity, never to be seen again. Kepler's first law does not state that hyperbolic orbits are a possibility, because an object following such a trajectory is obviously not a "planet." But Newton's second law and the law of universal gravitation predict that objects moving fast enough will follow hyperbolic instead of elliptical trajectories.

To see that this is true, set up the Newton application as before, but put in an initial velocity larger than $\sqrt{2GM/r_0}$. Figure N11.6 shows the trajectory that the Newton application generates for an object whose initial position is 1.0 AU in the $+x$ direction from the origin and a perpendicular initial velocity of 9.93 AU/y. I chose this speed because you can see that it makes $b = 2.5$ AU almost exactly. Equation N11.24 still applies, so we see that $\varepsilon = (b/r_c) - 1 = 2.5 - 1 = 1.5$. Although a hyperbola obviously has no finite r_f and therefore no finite "widest diameter," we can *define* a hyperbola's "semimajor axis" a so that

How we define *a* for a hyperbolic orbit

$$a \equiv \frac{b}{|1 - \varepsilon^2|} = \frac{b}{\varepsilon^2 - 1} \quad \Rightarrow \quad r_c = \frac{b}{1 + \varepsilon} = a\frac{\varepsilon^2 - 1}{\varepsilon + 1} = a(\varepsilon - 1) \qquad \text{(N11.29)}$$

in analogy with the elliptical case, where equations N11.25 and N11.26 tell us that $a = b/(1 - \varepsilon^2)$ and $r_c = a(1 - \varepsilon)$. In the hyperbolic orbit shown in figure N11.6, we have $a = (2.5 \text{ AU})/(1.5^2 - 1) = 2.0$ AU. You can also easily check that the trajectory shown in figure N11.6 obeys the conic-section formula $r = b/(1 + \varepsilon\cos\theta)$ at all angles.

In 2017, astronomers discovered an object, later named 1I/'Oumuamua, in a clearly hyperbolic orbit around the sun ($\varepsilon = 1.2$). This was the first object known to approach the sun from outside the solar system. In 2019, astronomers observed a second such object, named 2I/Borisov, following an even more extreme hyperbolic orbit ($\varepsilon = 3.36$). Both these interstellar visitors showed that Newton's predictions continue to have fresh applications even within the solar system.

Exercise N11X.3

Note that the trajectory approaches a straight line as r becomes large. Calculate the angle that this line makes with the horizontal (perhaps by calculating its slope near the upper left edge), and compare with $\theta_\infty = \cos^{-1}(-1/\varepsilon)$ (the angle at which r goes to infinity).

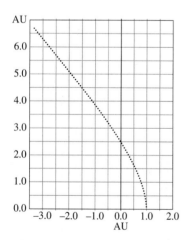

AU

6.0
5.0
4.0
3.0
2.0
1.0
0.0
    −3.0 −2.0 −1.0   0.0   1.0   2.0
                      AU

Figure N11.6

The trajectory generated by the Newton application for a rapidly moving object in our solar system.

In summary, we see that Newton's second law and the law of universal gravitation imply quite generally that a satellite's orbit around a primary will be a *conic section* (that is, either a circle, an ellipse, a parabola, or a hyperbola), consistent with (but also generalizing) Kepler's first law.

TWO-MINUTE PROBLEMS

N11T.1 Kepler's second law implies that as a planet's distance from the sun increases in an elliptical orbit, its orbital speed
A. Increases.
B. Decreases.
C. Remains the same.
D. Changes in a way we cannot determine.

N11T.2 The sun's mass is about 1000 times that of Jupiter, and the radius of Jupiter's orbit is about 1100 times the sun's radius. The center of mass of the sun/Jupiter system is inside the sun. True (T) or false (F)?

N11T.3 Two stars, one with radius r and the other with radius $3r$, orbit each other so that their centers of mass are $25r$ apart. Assume that the stars have the same uniform density. This system's center of mass is inside the larger star. T or F?

N11T.4 The speed of a satellite in a circular orbit of radius R around the earth is 3.0 km/s. The speed of another satellite in a different circular orbit around the earth is one-half this value. What is the radius of that satellite's orbit?
A. $4R$
B. $2R$
C. $\sqrt{2}\,R$
D. $R/\sqrt{2}$
E. $R/2$
F. $R/4$
T. Some other multiple of R (specify)

N11T.5 A satellite orbits the earth once every 2.0 h. What is the orbital period of another satellite whose orbital radius is 4.0 times larger?
A. 4.0 h
B. 8.0 h
C. 16 h
D. 64 h
E. Some other period (specify)

N11T.6 The radius of the earth's (almost circular) orbit around the sun is 150,000,000 km, and it takes 1 y for the earth to go around the sun. Imagine that a certain satellite goes in an almost circular orbit of radius 15,000 km around the earth (this radius is 10,000 times smaller than the earth's orbital radius around the sun). What is the period of this orbit?
A. 10^{-6} y
B. 10^{-4} y
C. 10^{-3} y
D. 10^{6} y
E. These periods are not related in any simple way.

N11T.7 Imagine that we launch a spaceship from the earth so its speed is 6 AU/y *relative to the earth* in the direction that the earth orbits the sun. This spaceship's subsequent orbit around the sun is:
A. Circular
B. Elliptical
C. Parabolic
D. Hyperbolic

N11T.8 Consider an orbit where $r_c = 1$ AU and $b = 2$ AU. What kind of orbit is this?
A. Circular
B. Elliptical
C. Parabolic
D. Hyperbolic

N11T.9 Consider an asteroid in an elliptical orbit whose closest point to the sun is $r_c = 1$ AU and farthest point is 7 AU from the sun. Its orbital period is
A. 1 yr
B. $(3.5)^{2/3}$ y = 2.31 y
C. 4 yr
D. $8^{3/2}$ y = 22.6 y
E. Some other period (specify)

N11T.10 Kepler's 2nd law applies to hyperbolic orbits. (T or F?) Kepler's 3rd law applies to hyperbolic orbits. (T or F?)

HOMEWORK PROBLEMS

Basic Skills

N11B.1 A satellite in a circular orbit around the earth has a speed of 3.00 km/s. What is the radius of this orbit?

N11B.2 The circular orbit of a satellite going around the earth has a radius of 10,000 km. What is the satellite's orbital speed?

N11B.3 *Geostationary* satellites are placed in a circular orbit around the earth at such a distance that their orbital period is exactly equal to 24 h (this means that the satellite seems to hover over a certain point on the earth's surface). What is the radius of such a geocentric orbit?

N11B.4 What is the orbital speed of the moon? The radius of the moon's orbit is roughly 384 Mm.

N11B.5 What is the earth's speed as it orbits the sun?

N11B.6 Triton is the largest moon of Neptune (it is roughly the same size as the earth's moon). It orbits Neptune once every 5.877 days at a distance of roughly 354 Mm from Neptune's center. What is the mass of Neptune?

N11B.7 The radius of Neptune's nearly circular orbit is about 30 times larger than that of earth's orbit. What is the period of Neptune's orbit?

N11B.8 The radius of Venus's nearly circular orbit is 0.723 times that of earth's orbit. What is the period of Venus's orbit?

N11B.9 A neutron star is an astrophysical object having a mass of roughly 2.8×10^{30} kg (about 1.4 times the mass of the sun) but a radius of only about 12 km. If you were in a circular orbit of radius 320 km (about 200 mi), how long would it take you to go once around the star?

N11B.10 An asteroid orbiting the sun has a semimajor axis of 3.0 AU. What is the period of its orbit in years?

N11B.11 A space probe is put in an orbit around the sun that is 0.5 AU from the sun at the closest and 2.0 AU at the farthest. What is its period in years?

N11B.12 Satellite *A* travels around the earth in a circular orbit of radius *R*. Another satellite orbits in an elliptical orbit that is *R* from the earth at its closest point and 3*R* from the earth at its farthest point. How does the period of satellite *B* compare to that of satellite *A*? Show your work.

Modeling

N11M.1 Is the center of mass of the earth/moon system inside the earth?

The earth–moon system viewed from space (see problem N11M.1). (Credit: NASA)

N11M.2 Imagine that the sun, earth, and Jupiter are aligned so that all three are in a line. What is the magnitude of the gravitational force exerted by Jupiter on the earth compared to that exerted by the sun on the earth?

N11M.3 What is the magnitude of the gravitational force exerted on the earth by the moon compared to that exerted on the earth by the sun?

N11M.4 During a certain 5-day time period, the line connecting a comet with the sun changes angle by about 3.2°. Assume that the comet's distance from the sun is roughly 130 million km during this time. During a 5-day time period some time later, the angle of this line changes by 0.80°. How far is the comet from the sun now? Explain.

N11M.5 Consider a light object and a massive object connected by a spring. Both objects are floating in deep space.
  (a) Will the path of the light object around the massive object lie in a plane?
  (b) Will it obey Kepler's second law?
      Explain your reasoning in both cases.

N11M.6 Not all orbit problems are astrophysical! In 1911, Ernest Rutherford proposed that all atoms consist of a tiny nucleus surrounded by comparatively lightweight electrons in circular orbits. He assumed that each electron was held in its orbit by the force of electrostatic attraction between the negatively charged electron (whose charge is $-e$ where $e = 1.6 \times 10^{-19}$ C) and the positively charged nucleus. We now know that this model is simplistic, but it was an important first step in understanding atomic structure.

The potential energy formula that describes the electrostatic interaction between two particles with charges q_1 and q_2 is $V_e = kq_1q_2/r$, where k is the Coulomb constant = 8.99×10^9 J·m/C^2. Since this has the same mathematical form as the formula for gravitational potential energy $V_g = -GMm/r$ but with kq_1q_2 replacing $-GMm$, the electrostatic force law must be $|\vec{F}_e| = k|q_1q_2|/r^2$ in analogy to the gravitational force law $|\vec{F}_g| = GMm/r^2$.

So let's consider hydrogen, where the nucleus consists of a single proton of charge $+e$ (but still has a mass almost 2000 times larger than the electron). Assume that the electron's orbit is circular with a radius equal to the hydrogen atom's measured radius of 5.4×10^{-11} m. Show that the electron's orbital speed is about 1/137 times the speed of light. [This is indeed the measured value of the electron's average speed, so Rutherford's model is good enough to get at least this right. It also illustrates the surprising range of Newton's physics, from the celestial to the microscopic!]

N11M.7 In Newton's time, people knew that the moon was about 60 earth radii from the earth.
  (a) According to the law of universal gravitation, about how many times smaller is the earth's gravitational field strength at the radius of the moon's orbit than it is on the earth's surface?
  (b) Use this information and the earth's radius (rather than its mass) to estimate the period of the moon's orbit in days.
  (c) Compare with the moon's observed orbital period.
(This was one of the ways Newton argued for his hypothesis of an inverse-square law for gravitation.)

N11M.8 In a certain binary star system, a small red star with a mass $m \approx 0.22$ solar masses orbits a bright white-hot star with a mass $M \approx 4.2$ solar masses. (These masses are estimated from the stars' color and luminosity.) The red star is observed to eclipse the other every 482 days.
  (a) What is the approximate distance D between the centers of these stars if we use the $M \gg m$ approximation? What other assumptions (if any) do you have to make to solve the problem?
  (b) Problem N11A.1 discusses a version of Kepler's third law that applies even when M and m are comparable. Use this result to calculate the distance between the stars' centers and compare it to the result of part (a).

Derivations

N11D.1 Suppose two objects with masses M and m are connected by a spring with zero relaxed length, so that the attractive force that each exerts on the other is $F = k_s r$. Assume that $M \gg m$, and that the satellite orbits in a circular orbit of radius R around its primary. Find an expression (analogous to Kepler's third law) that gives the period of the orbit T as a function of the orbital radius R, the spring constant k_s, and whatever else you need.

N11D.2 Imagine that in a different universe the magnitude of the force of gravity exerted by one object on another were given by

$$|\vec{F}_g| = \frac{BMm}{r^3} \qquad \text{(N11.30)}$$

where r is the separation between the two objects and B is some constant. What would Kepler's third law for circular orbits be in this universe?

N11D.3 The force law for the magnitude of the force between two quarks separated by a distance r turns out to be very roughly $|\vec{F}| = b$, where b is a constant. If we pretend that Newton's laws apply to quarks, and imagine that one quark is much lighter than the other and is in a circular orbit around it, what would be the equation for the lighter quark's period T as a function of its orbital radius R?

Rich-Context

N11R.1 Consider a spherical asteroid made mostly of iron (whose density is 7.9 g/cm^3) whose radius is 2.2 km. Could you run fast enough to put yourself in orbit around this asteroid?

N11R.2 Astrophysicists believe that a collision between galaxies might lead to the resulting combined galaxy having a *pair* of supermassive black holes at its center. Interactions between the black holes and the stars and gas in the galactic center will quickly circularize the black holes' orbits around their common center of mass.

(a) Suppose that the two black holes have *equal* masses M and that their orbits are circular and have a radius R_0 around the system's center of mass. Show that the speed of each black hole in its orbit is given by

$$|\vec{v}| = \frac{1}{2}\sqrt{\frac{GM}{R_0}} \qquad \text{(N11.31)}$$

(*Hint:* Equation N11.11 does not apply in this case because the black holes have comparable masses.)

(b) Interactions with the surrounding galactic matter will cause the distance between the black holes to shrink very gradually with time. Imagine that over a period of 60 My (60 million years) the orbital radius shrinks by one-half to $R = \frac{1}{2}R_0$. Compute the change in the black-hole binary system's total energy (kinetic + potential) during this time (express your answer in terms of G, M, and R_0).

(c) Has the binary lost energy to or gained energy from its surroundings? Does this make intuitive sense?

(d) If $M = 2.0 \times 10^{36}$ kg and the initial separation of the black holes is $R_0 \approx 10$ ly $\approx 10^{17}$ m, roughly what is the average rate at which energy is transferred, in watts? (For comparison, the rate at which the sun radiates energy is about 3.9×10^{26} W.)

N11R.3 Suppose you wanted to put a satellite in such a circular orbit that it appeared on the western horizon every Monday morning at 6 a.m. and was never at any other time on the western horizon. What radius should the orbit have? (Don't forget to account for the earth's rotation!)

Advanced

N11A.1 Consider an isolated two-object system interacting gravitationally such that M is not necessarily much greater than m. Assume that the smaller object's orbit around the system's center of mass is circular and has radius r. Argue that both objects have the same orbital period T around the system's center of mass and that

$$T^2 = \frac{4\pi^2 D^3}{G(M + m)} \qquad \text{(N11.32)}$$

where D is the separation of the two objects.

ANSWERS TO EXERCISES

N11X.1 Substituting equation N11.11 into equation N11.12 yields

$$T = \frac{2\pi R}{\sqrt{GM/R}} = 2\pi R\sqrt{\frac{R}{GM}} = 2\pi\sqrt{\frac{R^3}{GM}} \qquad \text{(N11.33)}$$

Squaring both sides of this gives us equation N11.13.

N11X.2 According to equation N11.13, $T \propto R^{3/2}$. If we increase R by a factor of 4, then $R^{3/2}$ increases by a factor of $(\sqrt{4})^3 = 2^3 = 8$. So the orbit would last 8 y.

N11X.3 The formula predicts that the angle where r goes to infinity is $\theta = \cos^{-1}(-1/\varepsilon) = \cos^{-1}(-1/1.5) = 132°$. To check this, note that as the trajectory's y value increases from 5.5 AU to 6.5 AU, its x value decreases from -2.3 AU to about -3.2 AU. So the tangent of the angle of this line is opposite over adjacent $= 1/(-0.9) = -1.11$. For comparison, $\tan(132°) = -1.12$, so this is pretty close. [Note that if you calculate $\tan^{-1}(-1.11)$, most calculators will deliver $-48°$, but $132° = 180° - 48°$ is also a valid solution to $\tan^{-1}(-1.11)$.]

N12
OPTIONAL

Orbits and Conservation Laws

Chapter Overview

Introduction

In this final chapter of the unit, we will bring many of the tools we have developed in this unit (and in unit C) to bear on the problem that represents the crowning triumph of the Newtonian synthesis. In particular, we will see how we can use conservation of energy and angular momentum as tools to help us understand orbital behavior and determine an orbit's characteristics.

Section N12.1: Overview and Review

This section reviews the conic section results from chapter N11. We see that we generally only need a pair of numbers to completely describe an orbit.

Section N12.2: Conservation Laws and Elliptical Orbits

Both the energy E and the angular momentum magnitude $L \equiv |\vec{L}|$ of a primary/satellite system will be conserved throughout the orbit. This section discusses how one can use this fact to link the values of E and L for a given system to the values of a (or b) and ε that determine the size and shape of the orbit.

Section N12.3: Conservation Laws and Hyperbolic Orbits

Similar equations link the values of E and L for a given system to the values of a (or b) and ε for a hyperbolic orbit. One of the most important basic consequences of the equations is that

$$\text{If} \quad E < 0 \quad \text{then} \quad \varepsilon < 1 \text{ (the orbit is } elliptical)$$
$$\text{If} \quad E > 0 \quad \text{then} \quad \varepsilon > 1 \text{ (the orbit is } hyperbolic)$$

Section N12.4: Solving Orbit Problems

Table N12.1 summarizes the equations that link various features of an orbit with the energy and angular momentum of the system. Given the equations in table N12.1, one can solve an astonishing range of orbit problems given only two pieces of information about the orbit (such as the satellite's distance from the primary and its velocity at a given time, or the points of the orbit that are nearest or farthest from the primary). This section contains a variety of examples that illustrate approaches to solving such problems.

Table N12.1   Table of useful equations for solving orbit problems

Item	Equation(s)	Symbols
Formula for a conic section, Formula for a	$r = \dfrac{b}{1 + \varepsilon \cos\theta}$, $\qquad a = \dfrac{b}{\|1 - \varepsilon^2\|}$	r = distance from the primary b = scale constant ε = eccentricity θ = angle from the axis between the primary and the orbit's closest point
Definitions of E and L	$\dfrac{2E}{m} = \|\vec{v}\|^2 - \dfrac{2GM}{r}$ $\quad$ and $\quad$ $\dfrac{L}{m} = r\|\vec{v}\|\sin\phi$	E = total system energy L = system's angular momentum
Formulas for ε and b	$\varepsilon^2 = 1 + \dfrac{1}{(GM)^2}\left(\dfrac{L}{m}\right)^2\left(\dfrac{2E}{m}\right)$ $\quad$ and $\quad$ $b = \dfrac{1}{GM}\left(\dfrac{L}{m}\right)^2$	G = gravitational constant m = satellite's mass M = primary's mass $\|\vec{v}\|$ = satellite's speed at an instant

Item	Elliptical Case	Hyperbolic Case	
Connection between a and E	$\dfrac{2E}{m} = -\dfrac{GM}{a}$	$\dfrac{2E}{m} = +\dfrac{GM}{a}$	ϕ = angle between $\vec{r}$ and $\vec{v}$ a = semimajor axis (for ellipse, half the widest diameter)
Location of extremes	$r_c = a(1 - \varepsilon)$ $\quad a = \tfrac{1}{2}(r_f + r_c)$ $r_f = a(1 + \varepsilon)$ $\quad \varepsilon = \dfrac{r_f - r_c}{r_f + r_c}$	$r_c = a(\varepsilon - 1)$ $\theta_\infty = \cos^{-1}(-1/\varepsilon)$	r_c = distance from primary to the closest point on the orbit r_f = distance from primary to the farthest point on the orbit
Other useful relations	$T^2 = \dfrac{4\pi^2}{GM}a^3$ (Kepler's third law)	$\tan\theta_\infty = -\sqrt{\varepsilon^2 - 1}$ $= -\dfrac{1}{GM}\left(\dfrac{L}{m}\right)\sqrt{\dfrac{2E}{m}}$	T = orbital period θ_∞ = value that θ approaches as $r \to \infty$ in a hyperbola

N12.1   Overview and Review

We saw in chapter N11 that orbits around a massive primary are either elliptical or hyperbolic. Armed with this knowledge, we can use conservation of energy and angular momentum to calculate many things about an orbit without having to solve Newton's second law directly. In this chapter, we will discover just how useful these laws can be in helping us think about, quantify, and understand orbital motion.

Before we begin, let's briefly review what we learned about orbits in section N11.7. We saw that the orbit of an object orbiting a massive primary at the origin (the orbit's "focus") is described by the formula for a conic section

$$r = \frac{b}{1 + \varepsilon \cos \theta} \tag{N12.1}$$

where ε and b are two constants that completely specify the orbit's shape and size. The eccentricity ε describes the orbit's *shape* ($\varepsilon = 0$ means the orbit is circular, $0 < \varepsilon < 1$ that it is elliptical, $\varepsilon = 1$ that it is parabolic, and $\varepsilon > 1$ that it is hyperbolic). Within the elliptical range, the ellipse becomes more elongated as ε increases; within the hyperbolic range, the angle between the hyperbola's incoming and outgoing legs increases. The constant b specifies something about the orbit's *size* for a given shape.

In section N11.7, we also defined an orbit's semimajor axis a to be

$$a \equiv \frac{b}{|1 - \varepsilon^2|} \tag{N12.2}$$

For an ellipse, this quantity corresponds to half the ellipse's widest diameter (and reduces to the simple radius of the circular orbit when $\varepsilon = 0$). The semimajor axis is important because Kepler's third law links the orbit's period to this quantity. For a hyperbola, it is less clear what the value of a describes physically, but we will see that it is still useful. In both cases, a is linked to r_c, the distance from the origin to the orbit's closest point as follows:

$$r_c = a|1 - \varepsilon| \tag{N12.3a}$$

and to the distance r_f of an *elliptical* orbit's farthest point as follows:

$$r_f = a(1 + \varepsilon) \tag{N12.3b}$$

For elliptical orbits, we also have

$$a = \tfrac{1}{2}(r_f + r_c) \quad \text{and} \quad \varepsilon = \frac{\tfrac{1}{2}(r_f - r_c)}{a} = \frac{r_f - r_c}{r_f + r_c} \tag{N12.4}$$

so knowing the pair r_f and r_c is the same as knowing a and ε and vice versa.

In what follows, we will see that we can link these orbital shape-and-size constants b, ε, a, r_c, and r_f to the total energy E and angular momentum $\vec{L}$ of the orbiting object, and that this will give us powerful insight into the physical meaning and implications of all of these quantities.

N12.2   Conservation Laws and Elliptical Orbits

Now consider an object of mass m in an *elliptical* orbit around a massive primary with mass M. Note that the orbit's closest and farthest points, the orbiting object's velocity is entirely perpendicular to the radial direction (see figure N12.1), since at these extreme points, the object's velocity is switching from being outward to inward (or vice versa), so its radial component is passing through zero. Therefore, the magnitude

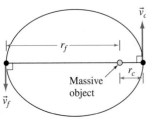

Figure N12.1

An orbiting object's velocity vectors are perpendicular to the radial direction at the orbit's extreme points.

of the orbiting object's angular momentum around the primary is simply $L \equiv |\vec{L}| = |\vec{r} \times m\vec{v}| = rm|\vec{v}| \sin 90° = rm|\vec{v}|$ at these points. Conservation of angular momentum and conservation of energy for these extreme points then imply that

$$r_c|\vec{v}_c| = \frac{L}{m} = r_f|\vec{v}_f| \qquad \text{(N12.5a)}$$

$$|v_c|^2 - \frac{2GM}{r_c} = \frac{2E}{m} = |\vec{v}_f|^2 - \frac{2GM}{r_f} \qquad \text{(N12.5b)}$$

Conservation of energy and angular momentum for an elliptical orbit

The ratios L/m and $2E/m$ will appear in many of the equations that follow. I recommend you think about them less as ratios and more as if they were single numbers whose symbols happen to be L/m and $2E/m$, respectively. I will place these ratios in parentheses in most of the equations that follow so you can more easily view each as a unit.

Multiplying both sides of the *left* equality in equation N12.5b by $r_c^2 = a^2(1-\varepsilon)^2 = a^2(1-2\varepsilon+\varepsilon^2)$ (see equation N12.3a) and substituting L/m for $r_c|\vec{v}_c|$ (equation N12.5a) yields

$$\left(\frac{L}{m}\right)^2 - 2GMa(1-\varepsilon) = \frac{2E}{m}a^2(1-2\varepsilon+\varepsilon^2) \qquad \text{(N12.6)}$$

Similarly, multiplying both sides of the right equality in equation N12.5b by $r_f^2 = a^2(1+\varepsilon)^2 = a^2(1-2\varepsilon+\varepsilon^2)$ and using $r_f|\vec{v}_f| = L/m$ (equation N12.5a) yields

$$\left(\frac{L}{m}\right)^2 - 2GMa(1+\varepsilon) = \frac{2E}{m}a^2(1+2\varepsilon+\varepsilon^2) \qquad \text{(N12.7)}$$

Subtracting equation N12.7 from equation N12.6 then yields

$$\left(\cancel{\frac{L}{m}}\right)^2 - 2GMa(1-\varepsilon) - \left(\cancel{\frac{L}{m}}\right)^2 + 2GMa(1+\varepsilon) = \frac{2E}{m}a^2(\cancel{1} - 2\varepsilon + \cancel{\varepsilon^2} - \cancel{1} - 2\varepsilon - \cancel{\varepsilon^2})$$

$$\Rightarrow \quad 4GMa\varepsilon = -\frac{2E}{m}a^2 4\varepsilon \quad \Rightarrow \quad E = -\frac{GMm}{2a} \qquad \text{(N12.8)}$$

The link between energy and the semimajor axis

This surprising result tells us that *the total conserved energy of an object in an elliptical orbit depends on the orbit's semimajor axis alone* (*not* on the object's angular momentum or the orbit's eccentricity), and that the energy gets larger (less negative) as the semimajor axis increases. This very powerful equation is useful in a number of contexts.

If we *add* equations N12.6 and N12.7, we get

$$2\left(\frac{L}{m}\right)^2 - 4GMa = \frac{2E}{m}a^2(2+2\varepsilon^2) \qquad \text{(N12.9)}$$

If we divide this equation through by 2 and substitute in $a = -GM/(2E/m)$ (from equation N12.8) and rearrange things, we find that

$$\varepsilon^2 = 1 + \frac{1}{(GM)^2}\left(\frac{L}{m}\right)^2\left(\frac{2E}{m}\right) \qquad \text{(N12.10)}$$

The link between energy, angular momentum, and eccentricity

This equation gives us the relationship between the orbit's eccentricity and the orbiting object's conserved energy and angular momentum. Note that E is negative (see equation N12.8), so $\varepsilon < 1$, as it should be for an elliptical orbit.

Exercise N12X.1

Verify equation N12.10.

Finally, note that if we solve $a = b/|1 - \varepsilon^2|$ (equation N12.2) for b, use $a = -GMm/2E$ (equation N12.8), and use $1 - \varepsilon^2 = -(L/m)^2(2E/m)/(GM)^2$ (from equation N12.10), we find that

$$b = a(1 - \varepsilon^2) = -\left(\frac{GMm}{2E}\right)\left[-\frac{1}{(GM)^2}\left(\frac{L}{m}\right)^2\left(\frac{2E}{m}\right)\right] = \frac{1}{GM}\left(\frac{L}{m}\right)^2 \qquad \text{(N12.11)}$$

This interesting result tells us that the constant b (for orbits around a given primary) is determined completely by L/m.

The point is that knowing r_c and $|\vec{v}_c|$ (or r_f and $|\vec{v}_f|$) allows us to compute both $2E/m$ and L/m, which in turn allows us to calculate b, a, and ε.

N12.3   Conservation Laws and Hyperbolic Orbits

The analogous results for hyperbolic orbits

We can do the same kind of analysis for hyperbolic orbits as well. This is somewhat trickier because a hyperbolic orbit does not have an easily analyzed far point. However, we can use the limiting behavior as the object goes to very large r as the equivalent of the far point in the elliptical orbit calculation. Even so, finding the equation for L/m in this limit is challenging, and the algebra after that is not much different than what we saw earlier, and so does not yield much illumination for the effort. (If you are interested, problem N12A.1 guides you through the derivation.) The final results, however, are very interesting and illuminating. For hyperbolic orbits, we have

$$E = +\frac{GMm}{2a} \qquad \text{(N12.12)}$$

(note that $E = -GMm/2a$ in the elliptical case),

$$\varepsilon^2 = 1 + \frac{1}{(GM)^2}\left(\frac{L}{m}\right)^2\left(\frac{2E}{m}\right) \qquad \text{(N12.13)}$$

just as we found in the elliptical case (but now E is positive so $\varepsilon > 1$), and

$$b = \frac{1}{GM}\left(\frac{L}{m}\right)^2 \qquad \text{(N12.14)}$$

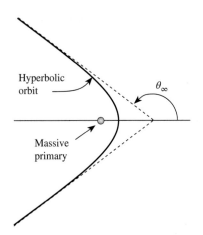

Figure N12.2
The definition of the asymptotic angle θ_∞ for a hyperbolic orbit.

also as we found in the elliptical case. Another interesting quantity for hyperbolic orbits is the angle θ_∞ that the orbiting object's nearly straight trajectory far from the primary makes with the line connecting the primary with the orbits closest point (see figure N12.2). In section N11.7, we saw that $\cos\theta_\infty = -1/\varepsilon$. Therefore

$$\tan\theta_\infty = \frac{\sin\theta_\infty}{\cos\theta_\infty} = \frac{\sqrt{1 - (1/\varepsilon)^2}}{-1/\varepsilon} = -\sqrt{\varepsilon^2 - 1} = -\frac{1}{GM}\left(\frac{L}{m}\right)\sqrt{\frac{2E}{m}} \qquad \text{(N12.15)}$$

where the last step follows from equation N12.13.

Note that equations N12.8, N12.10, N12.12, and N12.13 are consistent in implying that a negative system energy implies an elliptical orbit and a positive system energy implies a hyperbolic orbit. We can summarize this by the following statements:

An important result linking an orbit's energy to its shape

If $E < 0$ then $\varepsilon < 1$ (the orbit is *elliptical*)
If $E > 0$ then $\varepsilon > 1$ (the orbit is *hyperbolic*)

The case where $E = 0$ corresponds to the crossover between the two cases. This is of mathematical interest only (no real orbit will have its total energy *exactly* equal to zero), but the orbit in this case would be a parabola.

Table N12.1 Table of useful equations for solving orbit problems

Item	Equation(s)	Symbols
Formula for a conic section, Formula for a	$r = \dfrac{b}{1 + \varepsilon \cos\theta}, \qquad a = \dfrac{b}{\lvert 1 - \varepsilon^2 \rvert}$	r = distance from the primary b = scale constant ε = eccentricity θ = angle from the axis between the primary and the orbit's closest point
Definitions of E and L	$\dfrac{2E}{m} = \lvert \vec{v} \rvert^2 - \dfrac{2GM}{r} \quad \text{and} \quad \dfrac{L}{m} = r \lvert \vec{v} \rvert \sin\phi$	E = total system energy L = system's angular momentum
Formulas for ε and b	$\varepsilon^2 = 1 + \dfrac{1}{(GM)^2}\left(\dfrac{L}{m}\right)^2 \left(\dfrac{2E}{m}\right) \quad \text{and} \quad b = \dfrac{1}{GM}\left(\dfrac{L}{m}\right)^2$	G = gravitational constant m = satellite's mass M = primary's mass

Item	Elliptical Case	Hyperbolic Case	
Connection between a and E	$\dfrac{2E}{m} = -\dfrac{GM}{a}$	$\dfrac{2E}{m} = +\dfrac{GM}{a}$	$\lvert \vec{v} \rvert$ = satellite's speed at an instant ϕ = angle between $\vec{r}$ and $\vec{v}$ a = semimajor axis (for ellipse, half the widest diameter)
Location of extremes	$r_c = a(1 - \varepsilon) \quad a = \tfrac{1}{2}(r_f + r_c)$ $r_f = a(1 + \varepsilon) \quad \varepsilon = \dfrac{r_f - r_c}{r_f + r_c}$	$r_c = a(\varepsilon - 1)$ $\theta_\infty = \cos^{-1}(-1/\varepsilon)$	r_c = distance from primary to the closest point on the orbit r_f = distance from primary to the farthest point on the orbit
Other useful relations	$T^2 = \dfrac{4\pi^2}{GM}a^3$ (Kepler's third law)	$\tan\theta_\infty = -\sqrt{\varepsilon^2 - 1}$ $= -\dfrac{1}{GM}\left(\dfrac{L}{m}\right)\sqrt{\dfrac{2E}{m}}$	T = orbital period θ_∞ = value that θ approaches as $r \to \infty$ in a hyperbola

The preceding result completely makes sense, if you think about it. The basic distinction between an ellipse and a hyperbola is that a hyperbola goes to infinite r at certain angles. Since the system's gravitational potential energy goes to zero at infinity, if an orbiting object is to make it to infinity, it must have nonzero kinetic energy there, meaning that $E > 0$. If $E < 0$, the object cannot make it to infinity, and the orbit must therefore be elliptical.

N12.4 Solving Orbit Problems

The point of the last two sections is that if we know the values of $2E/m$ and L/m for a system involving a satellite orbiting a massive primary, then we can calculate a and ε (or b and ε), which in turn completely specify the shape of the satellite's orbit. Table N12.1 summarizes the crucial equations regarding orbits. Note that since E and L are conserved quantities for the orbit, we can calculate these quantities using position and velocity information at *any* point on the orbit.

So what is the minimum information we need to completely determine an orbit's shape and size? In general, we need to at least know the values of a suitable *pair* of quantities. The most basic pair are the constants appearing in the conic section equation (b and ε), but Table N12.1 shows that we can calculate these from other value pairs, such as: $2E/m$ and L/m; r_c and r_f; r_c and $\lvert \vec{v}_f \rvert$; r_f and $\lvert \vec{v}_f \rvert$; a and ε; a and θ_∞; and so on. We can also calculate the orbit from the values of $r, \lvert \vec{v} \rvert$, and ϕ at any given instant.

These equations allow us to solve an astonishing variety of interesting orbit problems. Examples N12.1 through N12.4 are just a sample.

We can use the equations in table N12.1 to solve a variety of orbit problems

Example N12.1

Problem: A strange object is discovered 22 AU (3.3 Tm) from the sun. Measurements show its velocity to be 11.2 km/s in a direction 169.7° from the line between the object and the sun (meaning that the velocity vector points mostly toward the sun). Is this a previously unknown member of the solar system, or is it an interloper from outside? How close will it get to the sun?

Solution The drawing below shows the situation.

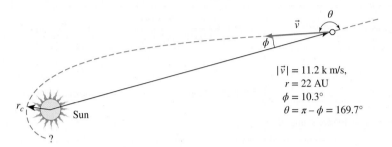

$$|\vec{v}| = 11.2 \text{ k m/s,}$$
$$r = 22 \text{ AU}$$
$$\phi = 10.3°$$
$$\theta = \pi - \phi = 169.7°$$

Since we know the object's initial distance r, its initial speed $|\vec{v}|$, and the angle ϕ between the $\vec{r}$ and $\vec{v}$ vectors, we should be able to determine the orbit. Calculating $2E/m$ will help us determine whether the orbit is elliptical or hyperbolic, which will answer the question about whether the object is a denizen of the solar system or an interloper from outside.

The value of GM for the sun is

$$GM = \left(6.67 \times 10^{-11} \frac{N \cdot m^2}{kg^2}\right)(1.99 \times 10^{30} \text{ kg})\left(\frac{1 \text{ kg} \cdot m/s^2}{1 \text{ N}}\right)$$

$$= 1.33 \times 10^{20} \text{ m}^3/\text{s}^2 \qquad \text{(N12.16)}$$

The value of $2E/m = |\vec{v}|^2 - 2GM/r$ for this object is therefore

$$\frac{2E}{m} = \left(11{,}200 \frac{m}{s}\right)^2 - \frac{2(1.33 \times 10^{20} \text{ m}^3/\text{s}^2)}{3.3 \times 10^{12} \text{ m}} = +4.5 \times 10^7 \text{ m}^2/\text{s}^2 \qquad \text{(N12.17)}$$

Since this is positive, the orbit is hyperbolic, so the object must be from *outside* the solar system (assuming it is not powered). Ominous! The system's angular momentum per unit mass will be

$$\frac{L}{m} = r|\vec{v}|\sin\theta = (3.3 \times 10^{12} \text{ m})(11{,}200 \text{ m/s})(\sin 169.7°)$$
$$= 6.6 \times 10^{15} \text{ m}^2/\text{s} \qquad \text{(N12.18)}$$

Therefore,

$$a = \frac{+GM}{(2E/m)} = \frac{1.33 \times 10^{20} \text{ m}^3/\text{s}^2}{4.5 \times 10^7 \text{ m}^2/\text{s}^2} = 3.0 \times 10^{12} \text{ m} \qquad \text{(N12.19}a\text{)}$$

$$\varepsilon = \left[1 + \frac{(4.5 \times 10^7 \text{ m}^2/\text{s}^2)(6.6 \times 10^{15} \text{ m}^2/\text{s})^2}{(1.33 \times 10^{20} \text{ m}^3/\text{s}^2)^2}\right]^{1/2} = 1.054 \qquad \text{(N12.19}b\text{)}$$

and the object's distance r_c from the sun at the orbit's vertex will be

$$r_c = a(\varepsilon - 1) = (3.0 \times 10^{12} \text{ m})(0.054) = 1.62 \times 10^{11} \text{ m} \qquad \text{(N12.19}c\text{)}$$

Note that r_c is nearly the same as the earth's orbital radius of 1.5×10^{11} m.

Exercise N12X.2

You can see that it is quite handy to know the value of GM. Show that GM for the earth is 3.99×10^{14} m³/s². (Similarly, GM for the moon is about 4.9×10^{12} m³/s², and for Jupiter it is about 1.3×10^{17} m³/s².)

Problem: A certain asteroid in an elliptical orbit is a distance $r_c = 3.5$ AU from the sun at the closest point in its orbit and a distance $r_f = 4.5$ AU at the farthest point. What is the period of its orbit in years?

Example N12.2

Solution The semimajor axis of the orbit is $a = \frac{1}{2}(r_c + r_f) = 4.0$ AU. The semi-major axis of the earth's orbit $a_e \approx 1$ AU (by definition). Since $T \propto a^{3/2}$,

$$\frac{T}{T_e} = \left(\frac{a}{a_e}\right)^{3/2} = \left(\frac{4.0 \text{ AU}}{1.0 \text{ AU}}\right)^{3/2} = 8.0 \qquad (N12.20)$$

Since the period of the earth's orbit $T_e = 1$ y by definition, the period of the asteroid's orbit must be 8.0 y.

Problem: At the point of its orbit closest to the earth, a satellite 200 km above the earth's surface is traveling at a speed of 10.0 km/s. Find the farthest distance this satellite gets from the earth, and determine its orbital period.

Example N12.3

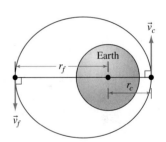

Solution See figure N12.3 for a sketch of the orbit. We are given that at the closest point in its orbit, the satellite is 200 km from the earth's surface, and thus 200 km + 6380 km = 6580 km = r_c from the earth's center, and that its speed $|\vec{v}_c| = 10$ km/s at this point. We want to find r_f.

The earth is extremely massive compared to any human-made satellite, so the massive-primary model we have developed in chapter N11 should work. We can calculate $2E/m$ and L/m using the values of r_c and $|\vec{v}_c|$ given for the close point, and we calculate anything we want from that.

With the given data and a result from exercise N12X.2, we find that

Figure N12.3
The situation described in example N12.3.

$$\frac{2E}{m} = |\vec{v}_c|^2 - \frac{2GM}{r_c} = \left(10{,}000 \, \frac{\text{m}}{\text{s}}\right)^2 - \frac{2(3.99 \times 10^{14} \text{ m}^3/\text{s}^2)}{6580 \times 10^3 \text{ m}}$$

$$= -2.1277 \times 10^7 \text{ m}^2/\text{s}^2 \qquad (N12.21)$$

$$a = \frac{GM}{(2E/m)} = \frac{3.99 \times 10^{14} \text{ m}^3/\text{s}^2}{-2.1277 \times 10^7 \text{ m}^2/\text{s}^2} = 1.8753 \times 10^7 \text{ m} \qquad (N12.22)$$

which (rounded) is 18,750 km. Since $r_f + r_c = 2a$,

$$r_f = 2a - r_c = 37{,}500 \text{ km} - 6580 \text{ km} = 30{,}900 \text{ km} \qquad (N12.23)$$

and according to the generalized version of Kepler's third law,

$$T = \sqrt{\frac{4\pi^2 (1.88 \times 10^7 \text{ m})^3}{3.99 \times 10^{14} \text{ m}^3/\text{s}^2}} = 25{,}600 \text{ s} \left(\frac{1 \text{ h}}{3600 \text{ s}}\right) = 7.1 \text{ h} \qquad (N12.24)$$

The units all work out here, and the answers seem plausible.

Example N12.4

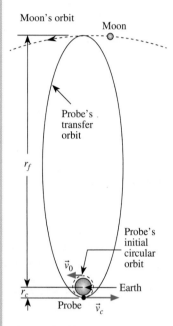

Figure N12.4
The situation described in example N12.4.

Problem: Suppose we have a space probe in a circular orbit 200 km above the earth and we want to put it into an elliptical orbit whose farthest point coincides with the moon. How much do the probe's rocket engines have to add to its speed to put it into this orbit? Assume the rocket engines fire for so brief a time that the object's distance from the earth does not change much during the time the engines are firing (that is, during the *burn* in astronautical language).

Solution Figure N12.4 shows a sketch of the situation. The earth is extremely massive compared to any probe, so the massive-primary model should be accurate. If the rocket engines fire only very briefly, the probe will be essentially at the same radius $r_c = 6580$ km both before and just after the engines fire. We also know $r_f = 384,000$ km from the inside front cover. We can use this information to calculate a, use that to calculate $2E/m$, and use that to calculate the speed v_c the probe must have just after the burn. We can calculate the probe's original speed $|\vec{v}_0|$ in its circular orbit by using equation N11.12. Since $2a = r_f + r_c$, we have

$$a = \frac{1}{2}(r_f + r_c) \quad \Rightarrow \quad \frac{2E}{m} = |\vec{v}_c|^2 - \frac{2GM}{r_c} = -\frac{GM}{a} = -\frac{2GM}{r_f + r_c} \qquad \text{(N12.25)}$$

$$\Rightarrow \quad |\vec{v}_c| = \sqrt{-\frac{2GM}{r_f + r_c} + \frac{2GM}{r_c}}$$

$$= \sqrt{-\frac{2(3.99 \times 10^{14}\ \text{m}^3/\text{s}^2)}{6,580,000\ \text{m} + 384,000,000\ \text{m}} + \frac{2(3.99 \times 10^{14}\ \text{m}^3/\text{s}^2)}{6,580,000\ \text{m}}}$$

$$= 10,920\ \text{m/s} \qquad \text{(N12.26)}$$

Since the probe's original circular orbit speed was

$$|\vec{v}_0| = \sqrt{\frac{GM}{r_c}} = \sqrt{\frac{3.99 \times 10^{14}\ \text{m}^3/\text{s}^2}{6,580,000\ \text{m}}} = 7790\ \text{m/s} \qquad \text{(N12.27)}$$

the engines need to boost the probe's speed by 3130 m/s.

To match the moon's speed at the far point would require another short burn to boost the probe's speed. Such a maneuver, which involves two short burns and an elliptical transfer orbit between one circular orbit and another is called a *Hohmann transfer*: this happens to be the way to move between the two circular orbits that minimizes the probe's total change in velocity (and thus minimizes the fuel required for the two burns).

Exercise N12X.3

Suppose you are in a spaceship that is orbiting a planet of mass M in a circular orbit of radius R. By what minimum factor would you have to increase your speed to go into a hyperbolic orbit?

TWO-MINUTE PROBLEMS

N12T.1 According to the text, an orbiting object having energy $E < 0$ follows an elliptical orbit. When we specify the numerical value for an object's energy, we are implicitly comparing it to the energy the object has in some reference situation. So, when we say that $E < 0$ in this case, we are comparing the energy of the orbiting object to the energy of an object with the same mass that is
A. At rest at the primary's center.
B. At rest on the primary's surface.
C. In a circular orbit just above the primary's surface.
D. At rest at $r = \infty$.
E. Some other situation (specify).

N12T.2 An object is discovered near the earth with values of $2E/m = 2.12 \times 10^7 \text{ m}^2/\text{s}^2$ and $L/m = 7.8 \times 10^{10} \text{ m}^2/\text{s}$. This object is in what kind of orbit around the earth?
A. Elliptical
B. Parabolic
C. Hyperbolic

N12T.3 The object described in problem N12T.2 has an orbital eccentricity of:
A. 1.08
B. 1.81
C. 0.81
D. 0.90
E. 1.35
F. Other (specify)

N12T.4 A comet is discovered in an elliptical orbit around the sun. Its closest distance from the sun is 1.0 AU, and measurements of its speed at this distance imply that its greatest distance from the sun is 7 AU. About how many years will pass between the comet's approaches to the sun?

A. 8 y
B. 16 y
C. 22.6 y
D. 64 y
E. 226 y
F. Other (specify)

N12T.5 The relationship $r_c|\vec{v}_c| = r_f|\vec{v}_f|$ for an elliptical orbit follows from the fact that the satellite's velocity is perpendicular to its position vector at the two extreme points and also from
A. Conservation of angular momentum.
B. Conservation of energy.
C. Conservation of momentum.
D. Newton's second law.
E. Some other properties of ellipses.
F. (The given relationship is false!)

N12T.6 Suppose a satellite orbits the earth so closely that it experiences some drag due to the earth's upper atmosphere. This will drain away some of the orbital energy of this system, converting it to thermal energy. If this happens fairly slowly, the satellite's orbit will remain nearly circular. What happens to the radius of this satellite's orbit as time passes in this case?
A. It slowly decreases.
B. It remains the same: just the satellite's speed decreases.
C. It slowly increases.

N12T.7 In the situation described in problem N12T.6, what do you think will happen to the satellite's speed as time passes?
A. It slowly decreases.
B. It remains roughly the same.
C. It slowly increases.

HOMEWORK PROBLEMS

Basic Skills

N12B.1 An object is discovered near the earth with values of E, L, and m such that $2E/m = -8.2 \times 10^6 \text{ m}^2/\text{s}^2$ and $L/m = 7.8 \times 10^{10} \text{ m}^2/\text{s}$. Find the eccentricity of this orbit and the radius of the closest point of the orbit to the earth, and classify the orbit as being elliptical, parabolic, or hyperbolic.

N12B.2 A space probe near the earth has values of E, L, and m such that $L/m = 7.8 \times 10^{10} \text{ m}^2/\text{s}$ and $2E/m = 0$. Find the eccentricity of this orbit and the radius of the closest point of the orbit to the earth, and classify the orbit as elliptical, parabolic, or hyperbolic.

N12B.3 Imagine you are in an orbit around the earth whose most distant point from the earth is 5 times farther from the earth's center than the closest point. If your speed is 10.0 km/s at the closest point, what is your speed at the farthest point?

N12B.4 A satellite orbits the earth in such a way that its speed at its point of closest approach is roughly 3 times its speed at the most distant point. How many times more distant from the earth's center is the farthest point than the closest point?

Modeling

N12M.1 A small asteroid is discovered 14,000 km from the earth's center moving at a speed of 9.2 km/s. Can you tell from the information provided whether this asteroid is in an elliptical or hyperbolic orbit around the earth? Is the *direction* of its velocity vector important in determining this? Please explain.

N12M.2 Astronomers spot a mysterious object moving at a speed of 21 km/s at a distance of 3.8 AU from the sun (where 1 AU = earth's orbital radius = 1.5×10^{11} m). Can you tell from the information provided whether this object is in an elliptical or hyperbolic orbit around the sun? Is the *direction* of its velocity vector important in determining this? Please explain.

N12M.3 Imagine you are in a circular orbit of radius $R = 7500$ km around the earth. You'd like to get to a geostationary space station whose circular orbit has a radius of $3R$, so you'd like to put yourself in an elliptical orbit whose closest point to the earth has radius R and whose most distant point has radius $3R$ (that is, a Hohmann transfer orbit). By what percentage would you have to increase your speed while at radius R to put yourself in this transfer orbit? (Assume your rocket engines fire for such a brief time that your distance from the earth remains essentially R throughout the time the engines are on.)

N12M.4 Imagine you are an astronaut in a circular orbit of $R = 6500$ km around the earth.
(a) What is your orbital speed?
(b) Say that you fire a rocket pack so that in a very short time you increase your speed in the direction of your motion by 20%. What are the characteristics of your new orbit? Calculate a, ε, T, r_c, and r_f for the new orbit.

N12M.5 A certain satellite is in an orbit around the earth whose nearest and farthest points from the earth's center are 7000 km and 42,000 km, respectively. Find the satellite's orbital speed at its point of closest approach.

N12M.6 An asteroid is in an orbit around the sun whose closest point to the sun has a radius R and whose most distant point has a radius $9R$, where R is equal to the radius of the earth's orbit = 1 AU = 1.5×10^{11} m.
(a) What is the asteroid's speed when it is closest to the sun?
(b) How does this compare with the earth's orbital speed?
(c) What is the period of this orbit, in years?

N12M.7 A new comet is discovered 6.6 AU from the sun (where 1 AU = earth's orbital radius = 1.5×10^{11} m) moving with a speed of 17 km/s. At that time, its velocity vector $\vec{v}$ makes an angle of 174.3° with respect to its position vector $\vec{r}$ (meaning that the comet's velocity points almost toward the sun).
(a) Is this comet in a hyperbolic orbit?
(b) What will be its distance from the sun at the point of closest approach?

N12M.8 Some recent space probes have made several hyperbolic passes past the earth to help give them the right direction and speed to go to their final destination. Imagine that one such probe has a speed of 12 km/s at a distance of 650,000 km from the earth. If the angle that the probe's velocity vector $\vec{v}$ makes with its position vector $\vec{r}$ at that time is 177°, how near will it pass by the earth at its point of closest approach?

N12M.9 A certain comet follows an elliptical orbit around the sun such that $r_c = r_e = 1$ AU and $r_f = 17$ AU, where $r_e = 1$ AU is the radius of the earth's orbit around the sun.

(a) What is the comet's period in years?
(b) Show that the value of $2E/m$ for this comet is $-\frac{1}{9}|\vec{v}_0|^2$, where $|\vec{v}_0| = \sqrt{GM/r_e}$ is the earth's orbital speed.
(c) Find the comet's speed at r_c as a multiple of $|\vec{v}_0|$.

Derivation

N12D.1 Argue that for a given value of angular momentum L, the smallest (most negative) energy that an orbiting object can have is

$$E = \frac{G^2 M^2 m^3}{2L^2} \qquad \text{(N12.28)}$$

and that the orbit in this case will be circular.

Rich-Context

N12R.1 You are the commander of the starship *Execrable*, which is currently in a standard orbit around a class-M planet whose mass is 4.4×10^{24} kg and whose radius is 6100 km. Your current circular orbit around the planet has a radius of $R = 50{,}000$ km. Your exobiologist wants to get in closer (say to an orbital radius of 10,000 km, or $R/5$) to look for signs of life. Your planetary geologist wants to stay at the current radius so that the entire face of the planet can be scanned with the sensors at once. Because you are tired of the bickering, you decide to put the *Execrable* into an elliptical orbit whose minimum distance from the planetary center is $R/5$ and whose farthest point is R from the same. Your navigational computers are down again (of course), so you have to compute by hand how to insert yourself in this new orbit. Your impulse engines are capable of causing the ship to accelerate at a rate of $1|\vec{g}| = 9.8$ m/s². In what direction should you fire your engines, relative to your current direction of motion (forward or backward)? For how many seconds?

N12R.2 You'd like to put your spaceship into a Hohmann transfer orbit between earth and Mars (that is, an elliptical orbit whose closest point to the sun is the same as the earth's orbital radius and whose farthest point is the same as Mars's orbital radius).
(a) If your spaceship is initially traveling around the sun with the same orbital speed as the earth and at the same distance from the sun as the earth is, by what factor will you need to increase your speed to put you into the transfer orbit? Assume you fire your rocket engines for a short enough time that your distance from the sun is still essentially 1.0 AU when the engines cease firing.
(b) How long will it take you to get to Mars following this orbit?
(c) Check your work using the Newton app. Set up the probe as having an initial position of 1.0 AU along the x axis, and give it an initial velocity parallel to the y axis whose magnitude is the value of $|\vec{v}_c|$ you calculated in part (b). I recommend a time step of about 0.01 y. Show that running some time steps does indeed produce an elliptical orbit having the desired value of $r_f = R$, and by counting time steps, verify the flight time you predicted in part (a).

N12R.3 Imagine you are the orbital engineer for the first NASA space shot to Ceres, the largest known asteroid. Ceres's nearly circular orbit around the sun has a radius of $R = 2.77$ AU. After being launched from the earth, the probe will initially be in a circular orbit around the sun with the same radius ($r_c = 1.0$ AU) and the same orbital speed ($|\vec{v}_c| = 6.28$ AU/y) as the earth's orbit. The probe's rocket engines will then fire briefly to increase the probe's speed to that speed $|\vec{v}_c|$ needed to put the probe into an elliptical orbit whose initial (and minimum) distance from the sun is $r_c = 1.0$ AU and whose final (and largest) distance from the sun is $R = 2.77$ AU (this is a Hohmann transfer orbit).
(a) How long will it take the probe to get from the earth to Ceres in such an orbit?
(b) What is the speed $|\vec{v}_c|$ that the probe has to have just after firing its engines to be inserted into this orbit? (Assume the duration of the boost is short enough that its distance from the sun is still nearly r_c just after the engines have been fired.)
(c) Check your work using the Newton computer program. Set up the probe as having an initial position of 1.0 AU along the x axis, and give it an initial velocity parallel to the y axis whose magnitude is the value of $|\vec{v}_c|$ that you calculated in part (b). I recommend a time step of about 0.01 y. Show that running some time steps does indeed produce an elliptical orbit having the desired value of $r_f = R$, and by counting time steps, verify the flight time you predicted in part (a).

N12R.4 Imagine that an object's velocity $\vec{v}_0$ far from the earth is such that it would miss the earth's *center* by a distance D if the earth had no gravity. What is the minimum value D can have if the object is to miss hitting the earth's *surface*? Use conservation of energy and angular momentum to determine D in terms of the earth's radius R, $|\vec{v}_0|$, and GM. (*Hint:* First explain why the constant value of L/m is equal to $D|\vec{v}_0|$. Then ensure that $r_c > R$.)

Advanced

N12A.1 Consider an object in a hyperbolic orbit around a massive primary.
(a) Argue that the object's speed $|\vec{v}|$ and angular momentum per unit mass at a point a very long distance from the primary are

$$|\vec{v}| \approx \sqrt{\frac{2E}{m}} \quad \text{and} \quad \frac{L}{m} = |\vec{v}|\, a\sqrt{\varepsilon^2 - 1} \quad \text{(N12.29)}$$

[*Hints:* The angular momentum formula is the tricky one. First argue that the angle ϕ between the velocity and the radius vector is given by $\tan\phi = r\,d\theta/dr$, where $d\theta$ is a small increment in the angle as the object moves outward and dr is the increment in the distance for that angle increment. Prove then that

$$\sin\phi = \frac{r(d\theta/dr)}{\sqrt{r^2(d\theta/dr)^2 + 1}} \quad \text{(N12.30)}$$

Now evaluate $d\theta/dr$ using the basic formula for a conic section (it is easiest to do this by taking differentials of $1 + \varepsilon\cos\theta = R/r$). Finally, evaluate $r\sin\phi$ in the limit that $\cos\theta$ goes to $-1/\varepsilon$ and then use the formula linking R to a to eliminate the R in favor of a. You should find that $r\sin\phi = a\sqrt{\varepsilon^2 - 1}$.]
(b) Use equations N12.29 and the formulas for $2E/m$ and L/m at $r = r_c$ to derive equations N12.12 through N12.14 for a hyperbolic orbit. [*Hint:* Use the calculation of the corresponding equations for the ellipse as a template, though there are some differences.]

N12X.1 Dividing equation N12.9 by two and substituting in $a = -GM/(2E/m)$, we get

$$\left(\frac{L}{m}\right)^2 - 2GM\left(\frac{-GM}{2E/m}\right) = \left(\frac{2E}{m}\right)\left(\frac{GM}{2E/m}\right)^2 (1 + \varepsilon^2)$$

$$\Rightarrow \left(\frac{L}{m}\right)^2 + \frac{2(GM)^2}{2E/m} = \frac{(GM)^2}{2E/m}(1 + \varepsilon^2)$$

$$\Rightarrow \frac{1}{(GM)^2}\left(\frac{2E}{m}\right)\left(\frac{L}{m}\right)^2 + 2 = 1 + \varepsilon^2 \quad \text{(N12.31)}$$

Subtracting 1 from both sides yields the desired result.

N12X.2 This is just a matter of substituting in numbers for G and M. If you disagree with the answer given, check that you have gotten the right numbers and that you have typed them correctly into your calculator.

N12X.3 To get into a hyperbolic orbit, the spaceship needs to have a speed $|\vec{v}_c|$ at radius R such that

$$0 < \frac{2E}{m} = |\vec{v}_c|^2 - \frac{2GM}{R} \Rightarrow |\vec{v}_c|^2 > \frac{2GM}{R} \quad \text{(N12.32)}$$

The circular orbit speed is (see equation N11.11)

$$|\vec{v}_0| = \sqrt{\frac{GM}{R}} \quad \text{(N12.33)}$$

So the factor by which we must increase the speed is

$$\frac{|\vec{v}_c|}{|\vec{v}_0|} = \frac{\sqrt{2GM/R}}{\sqrt{GM/R}} = \sqrt{2} \quad \text{(N12.34)}$$

Differential Calculus

NA.1   Derivatives

The definition of the derivative

The purpose of this appendix is to review (in the relatively informal language used by physicists) some basic principles of differential calculus, principles that we will use repeatedly throughout this unit and in subsequent units.

Consider a function $f(t)$, where t is some *arbitrary* variable (not *necessarily* time, though functions of time will be our main concern in this particular unit). We define the derivative of $f(t)$ with respect to t as follows:

$$\frac{df}{dt} \equiv \lim_{\Delta t \to 0} \frac{f(t + \Delta t) - f(t)}{\Delta t} \qquad (NA.1)$$

That is, the derivative is the limit as Δt goes to zero of the change in f as the variable t changes from t to $t + \Delta t$ divided by change Δt in that variable.

First example of calculating a derivative from the definition

The best way to show what we mean by "the limit as Δt goes to zero" here is to do an example. Suppose that $f(t) = t^2$. The derivative of $f(t)$ in this particular case is

$$\frac{df}{dt} \equiv \frac{d}{dt}[t^2] \equiv \lim_{\Delta t \to 0} \frac{(t + \Delta t)^2 - t^2}{\Delta t} \qquad (NA.2)$$

Multiplying out the binomial in the numerator yields

$$\frac{(t + \Delta t)^2 - t^2}{\Delta t} = \frac{t^2 - 2t\Delta t + \Delta t^2 - t^2}{\Delta t} = \frac{2t\Delta t + \Delta t^2}{\Delta t} = 2t + \Delta t \qquad (NA.3)$$

As Δt becomes smaller, this clearly gets closer and closer to $2t$. So

$$\text{if} \quad f(t) = t^2 \qquad \text{then} \qquad \frac{df}{dt} \equiv \lim_{\Delta t \to 0}(2t + \Delta t) = 2t \qquad (NA.4)$$

Second example of calculating a derivative from the definition

Here is a second example. Suppose that $f(t) = 1/t$. Its derivative is

$$\frac{df}{dt} = \frac{d}{dt}\left(\frac{1}{t}\right) \equiv \lim_{\Delta t \to 0}\left[\frac{1}{\Delta t}\left(\frac{1}{t + \Delta t} - \frac{1}{t}\right)\right] \qquad (NA.5)$$

If we put the expression in the brackets over a common denominator, we get

$$\frac{1}{\Delta t}\left(\frac{1}{t + \Delta t} - \frac{1}{t}\right) = \frac{1}{\Delta t}\left(\frac{t - \{t + \Delta t\}}{\{t + \Delta t\}t}\right) = \frac{1}{\cancel{\Delta t}}\frac{-\cancel{\Delta t}}{\{t + \Delta t\}t} = \frac{-1}{t^2 + t\Delta t} \qquad (NA.6)$$

In the limit that Δt becomes very small, $t^2 + t\Delta t \to t^2$, so

$$\text{If} \quad f(t) = \frac{1}{t} \qquad \text{then} \qquad \frac{df}{dt} \equiv \lim_{\Delta t \to 0}\left[\frac{-1}{t^2 + t\Delta t}\right] = \frac{-1}{t^2} \qquad (NA.7)$$

This illustrates the general technique: one typically finds that the difference $f(t + \Delta t) - f(t)$ can be written as something proportional to Δt, which cancels with the Δt in the denominator. One then sets $\Delta t = 0$ in what is left over.

NA.2   Some Useful Rules

In general, one can use the same general approach to show that

$$\text{If} \quad f(t) = t^n \quad \text{then} \quad \frac{df}{dt} = nt^{n-1} \qquad \text{(NA.8)} \qquad \text{Rule for simple powers}$$

for any integer $n \neq 0$. (Although the proof is more complicated, one can also show that the same applies to non-integer powers n.) Note also that

$$\text{If} \quad f(t) = c \quad \text{then} \quad \frac{df}{dt} = \lim_{\Delta t \to 0}\left[\frac{c - c}{\Delta t}\right] = \lim_{\Delta t \to 0}\left[\frac{0}{\Delta t}\right] = 0 \qquad \text{(NA.9)} \qquad \text{The derivative of a constant}$$

This case makes it clear that the *limit* of $0/\Delta t$ as Δt approaches zero is not the same as the *value* of $0/\Delta t$ at $\Delta t = 0$. Because $0/\Delta t$ is 0 for all nonzero Δt, the *limit* is zero (this is what $0/\Delta t$ *approaches*), even though $0/0$ is undefined.

The following useful theorems follow from the definition of the derivative. Let $f(t)$, $g(t)$, and $h(t)$ be arbitrary functions of t. We then have

$$\text{If} \quad f(t) = g(t) + h(t) \quad \text{then} \quad \frac{df}{dt} = \frac{dg}{dt} + \frac{dh}{dt} \qquad \text{(NA.10)} \qquad \text{The sum rule}$$

$$\text{If} \quad f(t) = g(t)h(t) \quad \text{then} \quad \frac{df}{dt} = \frac{dg}{dt}h + g\frac{dh}{dt} \qquad \text{(NA.11)} \qquad \text{The product rule}$$

$$\text{If} \quad f(t) = \frac{1}{g(t)} \quad \text{then} \quad \frac{df}{dt} = \frac{-1}{g^2}\frac{dg}{dt} \qquad \text{(NA.12)} \qquad \text{The inverse rule}$$

One can pretty easily prove these rules, as the example below illustrates:

Problem: Prove the product rule.   **Example NA.1**

Solution If $f(t) = g(t)h(t)$, then

$$\frac{df}{dt} \equiv \lim_{\Delta t \to 0}\left[\frac{g(t + \Delta t)h(t + \Delta t) - g(t)h(t)}{\Delta t}\right] \qquad \text{(NA.13)}$$

Adding zero to the numerator in the form $-g(t + \Delta t)h(t) + g(t + \Delta t)h(t)$ yields

$$\frac{df}{dt} \equiv \lim_{\Delta t \to 0}\left[\frac{g(t + \Delta t)h(t + \Delta t) - g(t + \Delta t)h(t) + g(t + \Delta t)h(t) - g(t)h(t)}{\Delta t}\right]$$

$$= \lim_{\Delta t \to 0}\left[g(t + \Delta t)\frac{h(t + \Delta t) - h(t)}{\Delta t} + \frac{g(t + \Delta t) - g(t)}{\Delta t}h(t)\right] \qquad \text{(NA.14)}$$

But the limit of a sum is the sum of the limits, and the limit of a product is the product of the limits, so

$$\frac{df}{dt} \equiv \lim_{\Delta t \to 0}[g(t + \Delta t)]\lim_{\Delta t \to 0}\left[\frac{h(t + \Delta t) - h(t)}{\Delta t}\right] + h(t)\lim_{\Delta t \to 0}\left[\frac{g(t + \Delta t) - g(t)}{\Delta t}\right]$$

$$= g\frac{dh}{dt} + h\frac{dg}{dt} \qquad \text{(NA.15)}$$

as claimed.

The constant rule

The constant and product rules together directly imply that when c is a constant,

$$\text{if} \quad f(t) = ch(t) \qquad \text{then} \quad \frac{df}{dt} = \frac{dc}{dt}h + c\frac{dh}{dt} = c\frac{dh}{dt} \qquad \text{(NA.16)}$$

This, in conjunction with the sum and product rules and equation NA.8, makes it pretty easy to evaluate the derivative of any polynomial in t.

Example NA.2

Problem: What is the derivative of $f(t) = at^3 + bt + c$, where a, b, and c are constants?

Solution By the sum rule, we have

$$\frac{df}{dt} = \frac{d}{dt}[at^3] + \frac{d}{dt}[bt] + \frac{dc}{dt} = a\frac{d}{dt}[t^3] + b\frac{dt}{dt} + 0 \qquad \text{(NA.17a)}$$

where in the second step, I applied the constant rule. Using equation NA.8, we find that

$$\frac{df}{dt} = a[3t^2] + b[1] = 3at^2 + b \qquad \text{(NA.17b)}$$

Exercise NAX.1

Use the rules we have stated so far to calculate the derivative of the function $f(t) = (at + b)^{-1}$, where a and b are constants.

NA.3   Derivatives and Slopes

The derivative is the slope of a line tangent to the graph of the function $f(t)$

We can better understand the *meaning* of the derivative of a function $f(t)$ by considering a graph of the function $f(t)$. This section reviews the graphical interpretation of the derivative.

Recall the definition of the derivative (see equation NA.1):

$$\frac{df}{dt} \equiv \lim_{\Delta t \to 0} \frac{\Delta f}{\Delta t} \equiv \lim_{\Delta t \to 0}\frac{f(t + \Delta t) - f(t)}{\Delta t} \qquad \text{(NA.18)}$$

Figure NA.1 illustrates that on a graph of $f(t)$ plotted versus t, the ratio $\Delta f/\Delta t$ for a nonzero Δt is equal to the slope of a straight line drawn between the points on the curve at t and $t + \Delta t$. As Δt approaches zero, the value of this slope approaches a specific value that equals the slope of a line drawn tangent to the graph at t. We call this the **slope** of $f(t)$ at that point. Therefore, we can interpret df/dt evaluated at t as giving the slope of the line tangent to the curve of $f(t)$ at that point t.

NA.4   The Chain Rule

The **chain rule** is one of the most important and useful theorems of differential calculus. Assume that $f(u)$ is a function of some variable u, which in turn is a

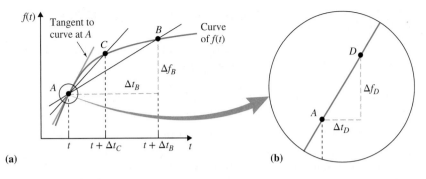

Figure NA.1
(a) The ratio $\Delta f_B/\Delta t_B$ specifies the slope of the line going through points A and B. The ratio $\Delta f_C/\Delta t_C$ specifies the slope of the line going through points A and C. The latter is closer to the slope of the line tangent to the curve of $f(t)$ at point A (light-colored straight line). (b) As Δt gets very small, the curve and the tangent line become indistinguishable, and $\Delta f/\Delta t$ specifies the slope of both in the neighborhood of A.

function $u(t)$ of t. This means that f is also a function of t, and therefore has a meaningful derivative with respect to t. The chain rule claims that

$$\frac{df}{dt} = \frac{df}{du}\frac{du}{dt} \qquad \text{(NA.19)}$$

The statement of the chain rule

This theorem looks relatively pedestrian, but in fact is very useful, as the following example illustrates.

Problem: Calculate the derivative of $f(t) = (at^2 + bt + c)^5$.

Example NA.3

Solution We could evaluate this derivative by multiplying out the polynomial and then taking the derivative of each term, using the constant rule and the rule for the derivatives for simple powers. But this would be *very* tedious. It is much simpler to define $u(t) \equiv at^2 + bt + c$, which means that $f(u) = u^5$. The chain rule then tells us that

$$\frac{df}{dt} = \frac{df}{du}\frac{du}{dt} = \left(\frac{d}{du}u^5\right)\left(\frac{d}{dt}[at^2 + bt + c]\right)$$

$$= (5u^4)(2at + b) = 5(at^2 + bt + c)^4(2at + b) \qquad \text{(NA.20)}$$

The physicist's "proof" of this theorem, by the way, is as follows.

A "physicist's proof"

$$\frac{df}{dt} = \frac{df}{dt}\frac{du}{du} = \frac{df}{du}\frac{du}{dt} \qquad \text{(NA.21)}$$

Physicists tend to think of df, du, and dt as simply being "sufficiently small" numbers, so multiplying and dividing by du seems perfectly natural. Mathematicians worry more about the definitions of the derivatives, the limiting process, and all the things that might go wrong with this cross-multiplication, but the result is the same. At the very least, you can think of "multiplying top and bottom by du" as a useful *mnemonic* for the chain rule formula.

Exercise NAX.2

Use the chain rule $df/dt = (df/du)(du/dt)$ to calculate the derivative of the function $f(t) = 1/(t^2 - b^2)^3$.

NA.5   Derivatives of Other Functions

Derivatives of some other useful functions

We will sometimes make use of the following derivatives:

$$\frac{d}{d\theta} \sin\theta = \cos\theta \quad \text{and} \quad \frac{d}{d\theta} \cos\theta = -\sin\theta \tag{NA.22}$$

$$\frac{d}{du} e^u = e^u \quad \text{and} \quad \frac{d}{du} \ln u = \frac{1}{u} \tag{NA.23}$$

We will often see such functions in contexts where the angle θ or the quantity u depends on t: for example, $\theta(t) = \omega t$ or $u(t) = -bt$, where ω and b are constants. In such cases, we can use the chain rule to evaluate the derivatives of such functions with respect to time. For example, if $\theta(t) = \omega t$, then

$$\frac{d}{dt} \sin\omega t = \frac{d}{dt} \sin[\theta(t)] = \frac{d\sin\theta}{d\theta}\frac{d\theta}{dt} = \cos\theta\frac{d}{dt}\omega t = \omega\cos\omega t \tag{NA.24}$$

Similarly, if $u(t) = -bt$, then

$$\frac{d}{dt} e^{-bt} = \left(\frac{d}{du} e^u\right)\left(\frac{du}{dt}\right) = e^u\frac{d}{dt}(-bt) = e^u(-b) = -be^{-bt} \tag{NA.25}$$

Exercise NAX.3

Use this technique to find the time derivative of $f(t) = A\cos(\omega t + \theta_0)$, where A, ω, and θ_0 are constants.

I strongly recommend you memorize the rules in this appendix: they will be used often enough in the reading that you will go crazy if you have to look everything up. If your calculus skills are rusty, I recommend practicing with some of the homework problems for this appendix (answers are given at the end of the book).

Using WolframAlpha

However, for derivatives of really nasty functions, you can always use WolframAlpha. For example, one *can* evaluate the derivative of a function such as $f(t) = t^2(e^{b/t} + 1)^{-3/2}$ using only the rules described in this section, but if you go to www.wolframalpha.com and type in "take the derivative of t^2/(exp(b/t)+1)^(3/2) with respect to t" it will say:

$$\frac{df}{dt} = \frac{3be^{b/t}}{2(e^{b/t} + 1)^{5/2}} + \frac{2t}{(e^{b/t} + 1)^{3/2}} \tag{NA.26}$$

which would take some considerable time to find by hand (with a low probability of being right). But DO NOT use WolframAlpha for routine calculations of simple derivatives: you will greatly enhance your ability to read physics texts if you practice doing simple integrals so that eventually you can do them in your head. You will only learn to do this with practice.

HOMEWORK PROBLEMS

Basic Skills

NAB.1 Calculate the derivatives of the following functions (where a, b, and c are constants).
(a) $at^2 + b$
(b) $1/ct^3$
(c) $b/(1 - at^2)$

NAB.2 Use the chain rule to calculate the derivatives of the following functions (where a, b, and c are constants).
(a) $3(t^3 + b^3)^4$
(b) $\sqrt{t^2 - b^2}$
(c) $\cos(bt^2)$

NAB.3 Use the chain rule to calculate the derivatives of the following functions (where a, b, and c are constants).
(a) $5b(bt - 1)^{-4}$
(b) $(bt + 1)^{3/2}$
(c) $\sin(bt + \pi)$

Derivations

NAD.1 Using the basic approach illustrated in section NA.1, prove that if $f(t) = at$ (where a is a constant), then $df/dt = a$. (Do not use equation NA.8.)

NAD.2 Using the basic approach illustrated in section NA.1, prove that if $f(t) = t^{-2}$, then $df/dt = -2t^{-3}$. (Do not use equation NA.8.)

NAD.3 Prove that the *sum rule* (equation NA.10) follows from the definition of the derivative. (You may assume without proof that the limit as $\Delta t \to 0$ of the sum of two quantities is the same as the sum of the limits of each quantity separately.)

NAD.4 Prove that the *inverse rule* (equation NA.12) follows from the definition of the derivative. (You may assume without proof that the limit as $\Delta t \to 0$ of the product of two quantities is the same as the products of the limits of each quantity separately.)

NAD.5 Use the chain rule to prove the inverse rule.

NAD.6 Use the chain rule to prove this very useful result:

$$\frac{d}{dt}(t + c)^n = n(t + c)^{n-1} \qquad \text{(NA.27)}$$

ANSWERS TO EXERCISES

NAX.1 Using the inverse rule, we have

$$\frac{df}{dt} = -\frac{1}{(at + b)^2}\frac{d}{dt}[at + b] = \frac{-a}{(at + b)^2} \qquad \text{(NA.28)}$$

NAX.2 Define $u(t) \equiv t^2 - b^2$. Then

$$\frac{d}{dt}\frac{1}{(t^2 - b^2)^3} = \frac{d}{du}u^{-3}\frac{du}{dt} = -3u^{-4}\frac{d}{dt}(t^2 - b^2)$$

$$= \frac{3}{(t^2 - b^2)^4}(2t + 0) = -\frac{6t}{(t^2 - b^2)^4} \qquad \text{(NA.29)}$$

NAX.3 If we define $\theta(t) = \omega t + \theta_0$, then the derivative is

$$\frac{df}{dt} = A\frac{d}{dt}\cos[\theta(t)] = A\left(\frac{d}{d\theta}\cos\theta\right)\frac{d\theta}{dt}$$

$$= A(-\sin\theta)\frac{d}{dt}(\omega t + \theta_0) = -A\omega\sin(\omega t + \theta_0) \qquad \text{(NA.30)}$$

Integral Calculus

NB.1 Antiderivatives

The definition of the antiderivative

The antiderivative $F(t)$ of a function $f(t)$ is defined to be any function whose derivative is $f(t)$. For example, one possible antiderivative for $f(t) = at$ (where a is a constant) is $F(t) = \frac{1}{2}at^2$ because

$$\frac{dF}{dt} = \frac{d}{dt}\left(\frac{1}{2}at^2\right) = \frac{1}{2}a\frac{d}{dt}(t^2) = \frac{1}{2}a(2t) = at \qquad \text{(NB.1)}$$

Note that there is not a unique antiderivative for a given function. If $F(t)$ is an antiderivative of $f(t)$, then so is $F(t) + C$ (where C is a constant), since

$$\frac{d}{dt}[F(t) + C] = \frac{dF}{dt} + \frac{dC}{dt} = \frac{dF}{dt} = f(t) \qquad \text{(NB.2)}$$

So the antiderivative $f(t)$ is really a whole *family* of functions $F(t)$ that differ by an additive constant. We therefore typically write the antiderivative of $f(t)$ as $F(t) + C$, where C is an unspecified constant we call a **constant of integration**.

NB.2 Definite Integrals

The definition of a definite integral

Suppose we graph $f(t)$ as a function of t. We define the **definite integral** of $f(t)$ from a point t_A to a point t_B on the horizontal axis as follows:

$$\int_{t_A}^{t_B} f(t)\, dt \equiv \text{the total area under the curve of } f(t) \text{ between } t_A \text{ and } t_B \qquad \text{(NB.3}a)$$

(considering the area above the t axis to be positive and the area below the axis to be negative). Figure NB.1 shows that we can approximate this area by a set of thin bars of width Δt, where the ith bar's height is $f(t_i)$, where i is an integer index, and $t_i = t_A + \left(i - \frac{1}{2}\right)\Delta t$, the value of t halfway across the bar's width. Perhaps you can convince yourself that as the bar width Δt goes to zero, we can write the area under the function's curve as

Figure NB.1
(a) The definite integral of a function $f(t)$ is the area under a graph of $f(t)$ between t_A and t_B (see the shaded regions). The area above the axis is positive, and the area below is negative. (b) The area is approximately equal to the total area of the bars shown, where each bar is Δt wide and as tall as the function's value across the bar.

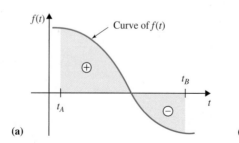

(a)

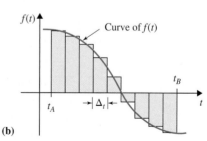

(b)

$$\int_{t_A}^{t_B} f(t)\,dt \equiv \lim_{\Delta t \to 0} \sum_{i=1}^{N} f(t_i)\,\Delta t \qquad (NB.3b)$$

The definite integral as a sum

where $N = (t_B - t_A)/\Delta t$ is the number of bars in the set. Indeed, the standard symbol $\int f\,dt$ is meant to be a stylized version of what is on the right, with the S-like integral symbol standing for "sum." Physicists (more than mathematicians) often think of an integral as being an actual sum of small quantities (as we will see, especially in unit E).

The following properties of the integral follow from this definition:

$$\int_{t_A}^{t_B} [f(t) + g(t)]\,dt = \int_{t_A}^{t_B} f(t)\,dt + \int_{t_A}^{t_B} g(t)\,dt \qquad (NB.4)$$

The sum rule

$$\int_{t_A}^{t_B} c\,f(t)\,dt = c \int_{t_A}^{t_B} f(t)\,dt \qquad (NB.5)$$

The constant rule

$$\int_{t_A}^{t_C} f(t)\,dt = \int_{t_A}^{t_B} f(t)\,dt + \int_{t_B}^{t_C} f(t)\,dt \quad \text{(when } t_C > t_B > t_A\text{)} \qquad (NB.6)$$

The limit rule

The last equation says that the total area under the curve between t_A and t_C (where $t_C > t_B > t_A$) is the sum of the areas between t_A and t_B and between t_B and t_C (which is not really very surprising). Equations NB.4 and NB.5 are pretty easy to prove from equation NB.3b, if you think about it.

NB.3   The Fundamental Theorem

At first glance, it may not seem that the antiderivative of a function $f(t)$ would have anything to do with the *integral* of $f(t)$. However, the **fundamental theorem of calculus** asserts that if $F(t)$ is any antiderivative of $f(t)$, then

$$F(t) - F(t_0) = \int_{t_0}^{t} f(t)\,dt \qquad (NB.7)$$

The fundamental theorem of calculus

where t_0 is some arbitrary fixed value of the variable t. Note that when we use the variable t as the upper limit of integration, then the value of the integral (the area from t_0 to t) is essentially a function of t. (Some people don't like to use the same variable symbol for the integral's upper limit as for the integration variable, but physicists are not usually so finicky.)

We can prove this theorem as follows. The definition of the derivative implies that the derivative of the right side of equation NB.7 is

$$\frac{d}{dt}\left[\int_{t_0}^{t} f(t)\,dt\right] = \lim_{\Delta t \to 0} \frac{1}{\Delta t}\left[\int_{t_0}^{t+\Delta t} f(t)\,dt - \int_{t_0}^{t} f(t)\,dt\right] \qquad (NB.8)$$

However, as figure NB.2 shows, the difference between the area under the curve from t_0 to t and that from t_0 to $t + \Delta t$ (as Δt becomes very small) is essentially the area of a bar having width Δt and height $f(t)$. Therefore

$$\frac{d}{dt}\left[\int_{t_0}^{t} f(t)\,dt\right] = \lim_{\Delta t \to 0}\left[\frac{1}{\Delta t} f(t)\Delta t\right] = f(t) \qquad (NB.9)$$

The integral is thus indeed an antiderivative of $f(t)$: $F(t) = \int_{t_0}^{t} f(t)\,dt + C$. So

$$F(t) - F(t_0) = \int_{t_0}^{t} f(t)\,dt + C - \left[\int_{t_0}^{t_0} f(t)\,dt + C\right] = \int_{t_0}^{t} f(t)\,dt \qquad (NB.10)$$

since the area under the curve from t_0 to t_0 is clearly zero. Q. E. D.

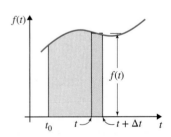

Figure NB.2
The difference between the area from t_0 to t and that from t_0 to $t + \Delta t$ is approximately equal to the area of the colored bar, which has width Δt and height $f(t)$. This approximation becomes exact in the limit that $\Delta t \to 0$.

NB.4   Indefinite Integrals

The indefinite integral $\int f(t)\,dt$ is *any* one of the possible antiderivatives of the function $f(t)$ (usually the one with the simplest algebraic form). You will find it useful to memorize the following indefinite integrals, where b, c, and ω are all constants:

The integral is the opposite of the derivative

$$\int \frac{df}{dt}\,dt = f(t) \quad \text{for } \textit{any} \text{ function } f(t) \tag{NB.11}$$

The constant rule

$$\int c\,dt = ct \tag{NB.12}$$

Simple powers

$$\int t^n\,dt = \frac{t^{n+1}}{n+1} \quad \text{for } n \neq 1 \tag{NB.13}$$

Simple powers of binomials

$$\int (bt+c)^n\,dt = \frac{(bt+c)^{n+1}}{b(n+1)} \quad \text{for } n \neq 1 \tag{NB.14}$$

Trigonometric functions

$$\int \sin(\omega t + b)\,dt = -\frac{1}{\omega}\cos(\omega t + b) \tag{NB.15}$$

$$\int \cos(\omega t + b)\,dt = \frac{1}{\omega}\sin(\omega t + b) \tag{NB.16}$$

Exponentials

$$\int e^{bt}\,dt = \frac{1}{b}e^{bt} \tag{NB.17}$$

Logarithms

$$\int \frac{dt}{t} = \ln t \quad (\text{assuming } t > 0) \tag{NB.18}$$

One can pretty easily prove any of these indefinite integrals by taking the derivative of both sides.

Definite integrals from indefinite integrals

Once you know a function's indefinite integral, you can find its definite integral from t_0 to t by simply evaluating the indefinite integral at these two points and subtracting. For example

$$\int_{t_0}^{t} c\,dt = \left[ct\right]_{t_0}^{t} \equiv ct - ct_0 \tag{NB.19}$$

This works no matter which member of the family of possible antiderivatives has been selected as the indefinite integral. This is because equation NB.10 in the proof of the fundamental theorem applies to *any* antiderivative of a given function $f(t)$: whatever arbitrary constant went into the specification of the indefinite integral cancels out in the definite integral.

NB.5   Substitution of Variables

An example of substitution of variables

Beyond the integrals listed above (which every well-educated person should know), calculating integrals can become really hairy. An important method for evaluating more complicated integrals is the technique of **substitution of variables**. To illustrate the process, consider the following integral (where b is a constant):

$$\int \frac{t\,dt}{\sqrt{b^2 t^2 + 1}} \tag{NB.20}$$

This looks like an appallingly difficult function to integrate. However, imagine that we define a new variable $u \equiv b^2 t^2 + 1$. Note that this means that

$$\frac{du}{dt} = b^2 (2t) + 0 = 2b^2 t \quad \Rightarrow \quad \frac{du}{2b^2} = t\, dt \tag{NB.21}$$

If we substitute these results into equation NB.20, we find that

$$\int \frac{t\, dt}{\sqrt{b^2 t^2 + 1}} = \int \frac{du}{2b^2 u^{1/2}} = \frac{1}{2b^2} \int u^{-1/2}\, du \tag{NB.22}$$

where I have used the constant rule in the last step. But the last integral involves a simple power that we can evaluate using equation NB.13:

$$\frac{1}{2b^2} \int u^{-1/2}\, du = \frac{1}{2b^2} \frac{u^{+1/2}}{(+1/2)} = \frac{u^{1/2}}{b^2} = \frac{\sqrt{b^2 t^2 + 1}}{b^2} \tag{NB.23}$$

Exercise NBX.1

Use substitution of variables to evaluate the integral

$$\int \sqrt{at + b}\, dt \tag{NB.24}$$

Check your work by taking the derivative of the result and showing that you recover the integrated function.

Please note that if you substitute variables in a *definite* integral, you must also change the limits of the integration to reflect the new variable. For example, consider a definite integral involving the case just discussed

$$\int_{t_A}^{t_B} \frac{t\, dt}{\sqrt{b^2 t^2 + 1}} = \frac{1}{2b^2} \int_{u_A}^{u_B} u^{1/2}\, du = \frac{1}{b^2} \left[u^{1/2} \right]_{u_A}^{u_B} = \frac{\sqrt{u_B} - \sqrt{u_A}}{b^2} \tag{NB.25}$$

where $u_A \equiv u(t_A) = b^2 t_A^2 + 1$ and $u_B \equiv u(t_B) = b^2 t_B^2 + 1$. For example, if we are evaluating the integral on the left from $t = 0$ to $t = 2$ s, and $b = 0.5$ s^{-1}, then we would evaluate the integral on the right from $u(0) = b^2 0^2 + 1$ to $u(2\text{ s}) = (0.5\text{ s}^{-1})^2 (2.0\text{ s})^2 + 1 = 1 + 1 = 2$. Equivalently, one can substitute the definition of u in terms of t back into the final expression for the evaluated integral and then use the original limits:

$$\int_{t_A}^{t_B} \frac{t\, dt}{\sqrt{b^2 t^2 + 1}} = \frac{1}{b^2} \left[u^{1/2} \right]_{u_A}^{u_B} = \frac{1}{b^2} \left[\sqrt{b^2 t^2 + 1} \right]_{t_A}^{t_B} = \frac{\sqrt{b^2 t_B^2 + 1} - \sqrt{b^2 t_A^2 + 1}}{b^2} \tag{NB.26}$$

Both approaches are correct. The point is that you should *think carefully* about what happens to the limits in definite integrals when you change variables.

Another issue that might arise when you change variables is that you might find yourself integrating from a large value of u to a smaller value. In such a case, it is helpful to note that

$$\int_{u_A}^{u_B} f(u)\, du = -\int_{u_B}^{u_A} f(u)\, du \tag{NB.27}$$

When we substitute variables in a definite integral, we must adjust the limits of integration

Swapping the limits changes the integral's sign

Exercise NBX.2

Use substitution of variables to integrate

$$\int_b^\infty \frac{b}{t^2} e^{b/t}\, dt \tag{NB.28}$$

Again, check by taking the derivative of the result and showing that you recover the original function. (Note that if t has units of seconds, so does b.)

NB.6   Looking Up Integrals

No mechanical method exists for determining integrals

While one can calculate the derivative of essentially any function by a diligent but mechanical application of the chain rule and other rules, no foolproof mechanical process exists for evaluating integrals. Some simple-looking integrals cannot be evaluated at all in symbolic form, and others require advanced techniques or tricks.

Tables of integrals

A good **table of integrals** provides a list of indefinite integrals for a long but well-organized list of functions. I like the table at

http://en.wikipedia.org/wiki/Lists_of_integrals

because (unlike many other online tables) it has some useful definite integrals that don't have simple indefinite integrals. But all of the integral tables I can find online are short compared to those one can find in printed books, and I also worry about the accuracy.

Using WolframAlpha

Because of this, WolframAlpha (www.wolframalpha.com) is often a better resource. To use WolframAlpha, type in something like "integrate t/sqrt(b^2*t^2+1) with respect to t" to evaluate the integral we discussed in the last section.

However, using substitution of variables to simplify an integral as far as possible will make it much easier to find an integral in a table or to get something useful out of WolframAlpha (which is easy to confuse, poor thing). Also note that WolframAlpha does not always deliver the result in the nicest form: you may have to do some post-processing on the result.

Feel free to look up integrals! (But cite your source!)

Feel free to use either a table or WolframAlpha for integrals that are more complex than the simple integrals in equations NB.12 through NB.18 (which you really should memorize), but indicate your appreciation of the poor selfless mathematicians who worked out all those integrals by *citing your source*.

HOMEWORK PROBLEMS

Basic Skills

NBB.1 Integrate the following functions from 0 to t (b is a constant). **(a)** $f(t) = b^2 t^2$ **(b)** $f(t) = 1/\sqrt{bt+1}$

NBB.2 Integrate the following functions from 0 to t (b and T are constants). **(a)** $f(t) = e^{-bt} + 1$ **(b)** $f(t) = b(t+T)^{3/2}$

NBB.3 Use WolframAlpha to evaluate the integral of

$$\int \frac{te^{at}\, dt}{(at+1)^2} \tag{NB.29}$$

Derivations

NBD.1 Show that equations NB.12 through NB.18 are correct by taking the derivatives of the results.

NBD.2 Use substitution of variables to evaluate the integral of $f(t) = (1 - bt)^3$.

NBD.3 Use substitution of variables to evaluate

$$\int_0^{2s} \frac{\sin \omega t\, dt}{(1 + \cos \omega t)^2} \quad \text{where} \quad \omega = \frac{\pi}{4s} \tag{NB.30}$$

ANSWERS TO EXERCISES

NBX.1 Define the function $u \equiv at + b$ and note that $du/dt = a + 0 = a$, meaning that $du/a = dt$. Substituting this into the integral yields

$$\int \sqrt{at+b}\, dt = \int \sqrt{u}\, \frac{du}{a} = \frac{1}{a} \int u^{1/2}\, du$$

$$= \frac{1}{a} \frac{u^{3/2}}{3/2} = \frac{2u^{3/2}}{3a} = \frac{2}{3a}(at+b)^{3/2} \tag{NB.31}$$

NBX.2 Define $u \equiv b/t$ and note that $du/dt = -b/t^2$, meaning that $dt = (-t^2/b)\, du$. Substituting this into the integral yields

$$\int_b^\infty \frac{b}{t^2} e^{b/t}\, dt = \int_{u(b)}^{u(\infty)} \frac{b}{t^2} e^u \left(-\frac{t^2}{b}\right) du = -\int_1^0 e^u\, du$$

$$= +\int_0^1 e^u\, du = [e^u]_0^1 = e^1 - 1 = 1.718\ldots \tag{NB.32}$$

Note how I had to change the limits of integration.

Index

NOTE: Page numbers followed by *f* and *t* indicate figures and tables, respectively.

Periodic Table of the Elements

Legend: Atomic number / Symbol / Atomic mass

1		
H		
1.008		

Group																	
1 1A	2 2A	3 3B	4 4B	5 5B	6 6B	7 7B	8 8B	9 8B	10	11 1B	12 2B	13 3A	14 4A	15 5A	16 6A	17 7A	18 8A
1 **H** 1.008																	2 **He** 4.003
3 **Li** 6.941	4 **Be** 9.012											5 **B** 10.81	6 **C** 12.01	7 **N** 14.01	8 **O** 16.00	9 **F** 19.00	10 **Ne** 20.18
11 **Na** 22.99	12 **Mg** 24.31											13 **Al** 26.98	14 **Si** 28.09	15 **P** 30.97	16 **S** 32.07	17 **Cl** 35.45	18 **Ar** 39.95
19 **K** 39.10	20 **Ca** 40.08	21 **Sc** 44.96	22 **Ti** 47.88	23 **V** 50.94	24 **Cr** 52.00	25 **Mn** 54.94	26 **Fe** 55.85	27 **Co** 58.93	28 **Ni** 58.69	29 **Cu** 63.55	30 **Zn** 65.39	31 **Ga** 69.72	32 **Ge** 72.59	33 **As** 74.92	34 **Se** 78.96	35 **Br** 79.90	36 **Kr** 83.80
37 **Rb** 85.47	38 **Sr** 87.62	39 **Y** 88.91	40 **Zr** 91.22	41 **Nb** 92.91	42 **Mo** 95.94	43 **Tc** (98)	44 **Ru** 101.1	45 **Rh** 102.9	46 **Pd** 106.4	47 **Ag** 107.9	48 **Cd** 112.4	49 **In** 114.8	50 **Sn** 118.7	51 **Sb** 121.8	52 **Te** 127.6	53 **I** 126.9	54 **Xe** 131.3
55 **Cs** 132.9	56 **Ba** 137.3	57 **La** 138.9	72 **Hf** 178.5	73 **Ta** 180.9	74 **W** 183.9	75 **Re** 186.2	76 **Os** 190.2	77 **Ir** 192.2	78 **Pt** 195.1	79 **Au** 197.0	80 **Hg** 200.6	81 **Tl** 204.4	82 **Pb** 207.2	83 **Bi** 209.0	84 **Po** (210)	85 **At** (210)	86 **Rn** (222)
87 **Fr** (223)	88 **Ra** (226)	89 **Ac** (227)	104 **Rf** (267)	105 **Db** (268)	106 **Sg** (271)	107 **Bh** (272)	108 **Hs** (270)	109 **Mt** (276)	110 **Ds** (281)	111 **Rg** (280)	112 **Cn** (285)	113 **Nh** (284)	114 **Fl** (289)	115 **Mc** (288)	116 **Lv** (293)	117 **Ts** (294)	118 **Og** (294)

Lanthanides:

58 **Ce** 140.1	59 **Pr** 140.9	60 **Nd** 144.2	61 **Pm** (147)	62 **Sm** 150.4	63 **Eu** 152.0	64 **Gd** 157.3	65 **Tb** 158.9	66 **Dy** 162.5	67 **Ho** 164.9	68 **Er** 167.3	69 **Tm** 168.9	70 **Yb** 173.0	71 **Lu** 175.0

Actinides:

90 **Th** 232.0	91 **Pa** (231)	92 **U** 238.0	93 **Np** (237)	94 **Pu** (244)	95 **Am** (243)	96 **Cm** (247)	97 **Bk** (247)	98 **Cf** (251)	99 **Es** (252)	100 **Fm** (257)	101 **Md** (258)	102 **No** (259)	103 **Lr** (262)

Short Answers to Selected Problems

[Note that most of the derivation (D) problems as well as a number of other problems have answers given in the problem statement. These problems are also useful for practice, but their answers are not reiterated here.]

Chapter N1

B1 (a) The cannon is much more massive than the ball, and so responds less. (b) The earth is even more massive. **B3** 2600 N backward. **B5** (a) a (b) $2at - 2b^{-2}t^{-3}$ (c) $2a^2t + 2ab$. **B7** [0.5 m/s, 4 m/s, 2 m/s] **B9** Zero. **B11** 5.6 m/s^2 upward. **B13** 3 m/s^2. **M2** 54 kg. **M3** 17,000 m/s^2. **M5** (a) m and s^{-1}, (b) $t = 3\pi/2\omega$, (c) $mA\omega^2$ (d) $m(A\omega^2 + |\vec{g}|)$. **M7** (a) $0.067m|\vec{g}|$, (b) same. **R3** Braking is better.

Chapter N2

B7 (a) Buoyant force, (b) lift force. **B9** To the right. **B11** 5.2 m/s^2 down and to the right. **M1** Banking provides an inward force component. **M3b** Net force must be inward, not zero. **M5** (a) 735 N, (b) 30 hp. **M7a** $M|\vec{a}|/(|\vec{g}| + |\vec{a}|)$. **M9c** 6000 N. **R1** 53 mi/h (so post 40 mi/h). **R3** Greater than 19.6 m/s.

Chapter N3

B3 30.6 m. **B5** 10 cm, 2.45 cm. **M3** The first car is faster (25 m/s vs. 18 m/s) but the other is further ahead (90 m vs. 62.5 m). **M5** (a) −120 m/min, (b) $2bt$, (c) 55 min, −4400 m. **M7** (a) s^{-1} (b) 3.1 m/s, (c) 2.23 m. **M9** 0.2 s (note that initial velocity arrow should be 7 cm long; acceleration arrow 1.1 cm long). **D1** (a) $\frac{1}{2}bt^2$ (b) $\frac{1}{2}bt(t - 2T)$. **R1** I catch the car at 3.6 mi.

Chapter N4

B1 $m|\vec{g}|/\cos\theta$. **B3** 70 g. **B5** (a) $\frac{1}{5}L|\vec{F}_{CR}|$ toward the viewer, (b) $\frac{4}{5}L|\vec{F}_{CL}|$ away from the viewer (c) $\frac{3}{10}LM|\vec{g}|$ toward the viewer. **M1** Roughly 180 lbs, if the person can pull with 100 lbs of force. **M3** 270 N. **M5** No. **M7** 57 N. **M9** $\frac{1}{2}m|\vec{g}|\tan\theta$. **R1b** About 700 N.

Chapter N5

B1 0.68. **B3** 0.41. **B5** 45 N. **B7** 74 m. **M1** 2.9 m/s^2. **M3** 8.3 N up the incline. **M5** 3.6 N. **M7** Yes, $b = 10$ μg/s. **M9** 1.3 m/s^2 upward. **M11** ≥ 4.6 s. **M13** ≥ 0.21. **R1** ≈ 0.91. **R3** Yes, if you have antilock brakes. Drag is not significant.

Chapter N6

B1 (a) yes, (b) no, (c) no. **B3** 1.5 N. **M1** One team can exert a greater static friction force on the ground. **M3** 86,400 N, 84,000 N. **M6** (a) 240 N, (b) 0.43. **M7** ≤ 1400 N. **M9** 0.30 m/s^2 down the incline. **R1** The mule *can* exert a bigger static friction force on the ground than the plow. **R3** *Maybe* Pat can without the backpack, but definitely with the backpack. **A1** $m_n = (0.9)^{n-1}M$, total $= 10M[1 - 0.90^N]$

Chapter N7

B1 510 m. **B3** 10.7 km. **B5** [0.49, 0.81, −0.32]. **B7** D, E, F. **M1** 49 m/s. **M3** $\mu_s \geq 0.22$. **M5** (a) ≥ 0.74, (b) Not reasonable, (c) 36°. **M7** $2\pi[L\cos\theta/|\vec{g}|]^{1/2}$. **M9** 0.21. **M11** 2.5R. **D4c** $r = |\vec{p}|/|q||\vec{B}|$. **R1** Reduce speed to 22 m/s or use a concrete road.

Chapter N8

B1 24 m/s. **B3** 146 km/h. **B5** (a) Zero, (b) 4 m/s^2. **B7** 9.8 m/s^2. **M1** 1.94 m/s. **M3** (a) 2.2 m/s^2, (b) 140 N, (c) zero, (d) 140 N. **M5** 152 lbs. **M7** 0.8 s compared to 1.1 s. **M9** 2.9 s. **D2b** 1.4 fm. **D3** 15 μm/s^2. **R1** 26th or 27th floor. **R3** Yes.

Chapter N9

B1 No, 29° shift in direction. **B3** 2.24 s. **B5** 5.1 s. **B7** 8.5 m/s. **M1** 7.5 m/s. **M3** 15.3 m/s. **M5** ≥ 9.5 m/s. **M7** 880 m. **M9** (a) 130 m/s, (b) 23 s, (c) drag is significant, (d) 161 m/s, 25.6 s. **M11** (a) Range 245 m, height 46 m, time 6.1 s, (b) 133 m, 32 m, 5.1 s, (c) 140 m, 37.5 m, 5.8 s, yes. **R1** 990 m/s. **R3** $\frac{2}{3}D$.

Chapter N10

B1 (a) 3.3 s, (b) 0.30 Hz, (c) 1.9 rad/s. **B3** 20.5 N/m. **B5** π. **B7** 3.6 s. **M1** 0.45 s. **M3** Longer. **M5** (a) 6.0 cm, (b) 98 N/m, (c) 0.77 m/s. **M7** 1.2 Hz. **M9** Longer. **M11** (a) N, m^{-1}, (b) *Hint:* $\sin\theta \approx \theta$ for small θ in radians, (c) 5.0 Hz. **R1** 56.58 kg. **A1** (a) $r_0 = c^3/b^2$, (b) $k_s = ab^8/c^9$, (c) $b^4[a/mc]^{1/2}/2\pi c^4$.

Chapter N11

B1 44,300 km. **B3** 42,300 km. **B5** 29.7 km/s. **B7** 164 y. **B9** 83 ms. **B11** 1.4 y. **M1** Yes (CM is 4670 km from the earth's center). **M3** The ratio is 0.0056. **M5** (a) Yes, (b) yes. **M8** (a) 2.9×10^{11} m, (b) 1.017 times larger. **D1** $T = 2\pi[m/k_s]^{1/2}$. **D3** $T^2 = (4\pi^2m/b)R$. **R1** Yes. **R2** (b) $\Delta E = -\frac{1}{4}GM^2/R_0$. (d) 3.5×10^{29} W. **R3** 46,870 km or 38,660 km.

Chapter N12

B1 0.83, 8300 km, elliptical. **B3** 2 km/s. **M1** Hyperbolic, no. **M3** Increase speed by 22.5%. **M5** 9.90 km/s. **M7** (a) Yes, (b) 0.070 AU. **M9c** $\sqrt{17/9}|\vec{v}_0|$. **R1** Fire engines in the direction of the ship's motion (to slow it down) for 104 s. **R3** (a) 1.29 y, (b) boost speed by 21%.

Appendix NA

B1 (a) $2at$, (b) $-3c/t^4$, (c) $2abt(1 - at^2)^{-2}$. **B3** (a) $-20b^2(bt - 1)^{-5}$, (b) $\frac{3}{2}b(bt + 1)^{1/2}$, (c) $b\cos(bt + \pi)$.

Appendix NB

B1 (a) $\frac{1}{3}b^2t^3$, (b) $2b^{-1}(\sqrt{bt + 1} - 1)$ **B3** $e^{at}/[a^2(at + 1)]$. **D3** $(2/\pi)$ s.